Process Intensification

Process Intensification: Faster, Better, Cheaper presents basic concepts and applications of process intensification (PI) and links their common effects across processes. It defines two fundamental parameters, PI factor, and Cost Impact (CI) factor, and uses these to analyze various applications where Process Intensification has been carried out.

Process Intensification principles have, in the past, been applied to diverse fields, ranging from biodiesel production to offshore processing, and this book unifies these aspects to identify the common factors that drive process enhancements. Each chapter investigates a specific application, discusses the key PI principles, and includes problem sets and examples. The book also provides case studies and real-world examples throughout the chapters.

Features:

- Explores Cost Impact of Process Intensification, and their relative magnitudes, as a universal metric.
- Covers a range of industrial applications, including heat and mass transfer, atomization and comminution, and enhanced oil recovery.
- Discusses the application of Process Intensification for clean coal technology and environmental remediation.
- Includes end-of-chapter problems, examples, and case studies.

The book suits senior undergraduate chemical and mechanical engineering students taking courses in Process Design, Process Optimization, Process Synthesis, and Process Intensification. Graduate students undertaking research in Process Intensification will also find the material in the book illustrative and instructive.

Prof. Ramamurthy Nagarajan is Alumni Community Chair Professor in the Department of Chemical Engineering at the Indian Institute of Technology Madras, having previously served as the Department Head from October 2018 to November 2021, and prior to that for two terms as the Institute's first-ever Dean of Alumni & International Relations. He obtained his B.Tech. in Chemical Engineering in 1981 from IIT Madras, and a Ph.D. in the same field from Yale University (New Haven, CT, USA) in 1986.

Process Intensification
Faster, Better, Cheaper

Ramamurthy Nagarajan

CRC Press
Taylor & Francis Group
Boca Raton London New York

CRC Press is an imprint of the
Taylor & Francis Group, an **informa** business

Cover image: www.shutterstock.com

First edition published 2023
by CRC Press
6000 Broken Sound Parkway NW, Suite 300, Boca Raton, FL 33487-2742

and by CRC Press
4 Park Square, Milton Park, Abingdon, Oxon, OX14 4RN

CRC Press is an imprint of Taylor & Francis Group, LLC

© 2023 Ramamurthy Nagarajan

ISBN: 978-1-032-25477-7 (hbk)
ISBN: 978-1-032-25481-4 (pbk)
ISBN: 978-1-003-28342-3 (ebk)
ISBN: 978-1-032-25486-9 (eBook+)

DOI: 10.1201/9781003283423

Typeset in Times
by KnowledgeWorks Global Ltd.

Access the Support Material: https://routledgetextbooks.com/textbooks/instructor_downloads/

Dedication

I owe my career in Chemical Engineering to my bachelor's alma mater, and current employer—Indian Institute of Technology Madras. The Institute made ChE a "cool" subject, and it remains so even now. Prof. M.S. Ananth—my instructor then for Thermodynamics and Fluid Mechanics, later Director of IITM when I was recruited as a faculty— was and is a role model.

My research career was launched at Yale University, which I had originally selected as my graduate school as it had more Film Societies than the others I had applied to! My Thesis Advisor, Prof. Daniel Rosner, imbibed in me an appreciation for intellectual rigor and a systematic approach to problem analysis. Given his work habits and dedication, it is no surprise that while many of his students have retired, he remains active in research.

During my 15-year industrial career with IBM's Storage Systems Division in San Jose, I was fortunate to work with some wonderful colleagues, and under some exceptional managers—in particular, Ron Weaver, who oversaw my initial foray into the industry. Since my return to academics 18 years ago, I have been surrounded by a galaxy of brilliant colleagues, Staff, and students, who have kept my creative juices flowing. I wish to thank my research scholars and project Staff who contributed so much to my research endeavors. In particular, the knowledge generated by Dr. K.R. Gopi, Dr. R. Vetrimurugan, Dr. T.K. Jagannathan, Dr. B. Ambedkar, Dr. N.P. Dhanalakshmi, Dr. Srivalli Hariharan, Dr. S. Balakrishnan, Dr. S. Ponmani, and Ms. Sreenita Bhattacharya has contributed greatly to this book.

Part of this book was written during my sabbatical with Purdue University as Robert T. Henson Distinguished Visiting Scholar in the Davidson School of Chemical Engineering. I am grateful to Prof. Rakesh Agrawal for hosting my stay, and to the Department Head, Prof. Sangtae Kim, for enabling and enriching my stay.

On the family front, Usha, my wife of 36 years, has been a friend and companion, nonpareil. She has been there for me every step of the way—from Morgantown, WV to Chennai, India. She knows me better than myself, as I do her. Our first daughter, Swati, left us too early at 3 years of age, and yet, left us with a lifetime of memories. Our son, Amrit, and second daughter, Vani, are a constant source of joy and wisdom, leaving us to wonder at our good fortune.

Finally, my parents, as well Usha's, made us what we are today. My sister, Sudha, as well as my wife's, Ranjani, have been a constant source of support and encouragement. This book is dedicated to our extended families.

Contents

Preface...xi

Chapter 1   What is Process Intensification?..1

    1.1   Process..1
    1.2   Intensification 2
    1.3   Process Intensification 3
    Exercise Questions 5
    References 5

Chapter 2   PIF/CIF Case Studies 7

    2.1   The Cookie Factory 7
    2.2   Tolerancing of Critical Dimensions in a Hard Disk Drive 9
    2.3   Microcontamination Control...12
    2.4   Call Centers: The Business Case for Offshoring....................15
    2.5   Clean Coal Technology 16
    Exercise Questions 20
    References 22

Chapter 3   Common Process Intensification Parameters....................................25

    3.1   Temperature..25
    3.2   Pressure 26
    3.3   Time...28
    Exercise Questions 31
    References 32

Chapter 4   Process Intensification Driven by External Fields 33

    Exercise 35
    References 35

Chapter 5   Acoustic Intensification of Processes—Mechanisms Involved.........37

    5.1   Surface Cleaning 40
    5.2   Heat Transfer Enhancement in Furnace Tubes.......................42
    5.3   Degradation of Methyl Violet...42
    5.4   Sono-Fragmentation of Coal 42
    5.5   Destratification 43
    5.6   Effervescent Atomization...43
    5.7   Coal Beneficiation 44
    Exercise Questions 47
    References 48

Chapter 6   Process Intensification Fields: Case Studies 51

    6.1   Microwave 51
        6.1.1   Heating and Drying..51
        6.1.2   Applications in Food Industry.....................................52
        6.1.3   Waste Treatment under Microwave Irradiation...........53
        6.1.4   Pyrolysis of Biomass Waste.......................................54
    6.2   Laser 55
        6.2.1   Surface Hardening..55
        6.2.2   Surface Cleaning..56
        6.2.3   Ablation 56
    6.3   Thermophoresis 58
        6.3.1   Bulb Blackening 58
        6.3.2   Hot Corrosion 60
        6.3.3   Particle Thermophoresis...61
    6.4   Electrophoresis 64
        6.4.1   Biomolecular Separations...64
        6.4.2   Transport in Microchannels 65
        6.4.3   Electrostatic Precipitator 66
    6.5   Magnetophoresis..67
        6.5.1   Separation...68
        6.5.2   Mixing 69
        6.5.3   Trapping 69
    6.6   Acoustophoresis...70
        6.6.1   Continuous Flow Concentration of Cells
             and Particles 71
        6.6.2   Filtration/Removal of Suspended Particles.................71
        6.6.3   Particle Separation, Mixing, and Focusing 72
    Exercise Questions 74
    References 75

Chapter 7   Acoustic Fields Coupled to Liquids: Basics and Some Applications 79

    7.1   Destratification/Mixing...84
    7.2   Effervescent Atomization 94
    7.3   Nanoparticle Synthesis 103
        7.3.1   Bottom-Up Methods...104
        7.3.2   Top-Down Methods..107
    Exercise Questions 114
    References 116

Chapter 8   Intensification of Transport Phenomena and Chemical Reactions.........119

    8.1   Heat Transfer 119
        8.1.1   Surface Roughness 119
        8.1.2   Flow Disruptions 119
        8.1.3   Channel Curvature 121

8.1.4 Re-Entrant Obstructions 121
8.1.5 Secondary Flows 122
8.1.6 Out-of-Plane Mixing 124
8.1.7 Fluid Additives 124
8.1.8 Vibration 125
8.1.9 Electrostatic Fields 125
8.1.10 Flow Pulsation 126
8.1.11 Variable Roughness Structures 127
8.1.12 Acoustic Enhancement 127
8.2 Mass Transfer (Without Chemical Reaction) 134
8.3 Chemical Reactions 136
8.4 Mass Transfer (With Chemical Reaction) 138
Exercise Questions 141
References 141

Chapter 9 Intensification of Heat Transfer: Case Studies 147

9.1 Heat Transfer in Furnace Tubes 147
9.2 Heat Transfer in Tank with Side Heaters 154
9.2.1 Quiescent Conditions 156
9.2.2 Flow Conditions 159
9.3 Nanofluids 163
Exercise 169
References 170

Chapter 10 Mass Transfer Rate Enhancement: Case Study 173

10.1 Acoustic Intensification of Dye Uptake by Leather 173
10.2 Acoustic Intensification of Gas-Liquid Mass Transfer 181
Exercise 188
References 189

Chapter 11 Pollution Abatement and Microcontamination Control:
Case Study 191

11.1 Methyl Violet Degradation 191
11.2 Component Surface Cleaning in Microelectronics
Manufacturing 197
11.2.1 Spray Cleaning 199
11.2.2 Spin-Rinse-Dryer Cleaning 200
11.2.3 Vapor Degreasing 201
11.2.4 Chemical Cleaning 201
11.2.5 Solvent Cleaning 202
11.2.6 Mechanical Agitation Cleaning (Undulation
and Sparging) 203
11.2.7 Manual Cleaning (Swabbing and Wiping) 203

11.2.8   Specialty Cleaning (Plasma, UV, and
UV/Ozone, CO_2 Snow, Supercritical Fluid, etc.) ...203
11.2.9   Supercritical Fluid Cleaning 204
11.2.10 CO_2 Snow Cleaning 205
11.2.11 Ultrasonic Cleaning 205
11.2.12 Component Cleaning in the HDD Industry 209
Exercise Questions 210
References 210

Chapter 12 Formulation of Nanoemulsion: Case Study 213

12.1   Formulation Process 215
12.2   Ultrasound Application 216
12.3   Experimental Setup and Procedure 220
12.4   Results and Discussion 223
12.4.1   Emulsion Droplet Size Distribution 223
12.4.2   Stability Analysis 229
References 235

Chapter 13 Enhanced Oil Recovery: Case Study 237

13.1   Experimental Setup and Methodology 239
13.2   Results and Discussion 244
Exercise 249
References 249

Chapter 14 In Conclusion.. 251

References 253

Index...255

Preface

This book was motivated by two wishes of mine—to write a technical book that captures the learnings of my body of research and to write it in a style that sets it apart from the traditional technical book. I believe I have at least partially succeeded.

"Process Intensification" is a topic that, to me, captures the essence of chemical engineering. First, identify the process, then own it! I recall my encounter with a Chemical Engineer who was managing a software group in IBM's Santa Teresa Labs. When I asked him why IBM thought him a good fit for the position, he pointed out that only ChE dealt with processes in such excruciating detail as part of the curriculum. We can design, scale, model, simulate, manage, control, and optimize processes in our sleep. "Intensification" is a different challenge, though. Even chemical engineers need to get out of their comfort zone to do it well. "Better, cheaper, faster" is the PI mantra, and it resonates with me. Reading through the many papers and books on the subject, it became clear to me that there is a lack of a unified vision of the Process Intensification Factor across applications. An even more glaring omission is the simultaneous consideration of process intensification and associated cost-impact factors. This book is an attempt to bridge this gap.

There is an obvious bias in the book toward one particular technique for process intensification—deployment of high-intensity, high-frequency acoustic fields as an intensifier. My years of working with ultrasonic and megasonic cleaners while at IBM made me realize how effective these fields were in removing surface contaminants down to the molecular and nanometer dimensions. When I later switched sides and became an academic, I built up a state-of-the-art ultrasonic lab at IIT Madras and started exploring these fields as an intensifier for virtually every process—transport, thermodynamic, kinetic, unit operations. The results were surprisingly consistent and exceedingly positive. Hence, the emphasis you would doubtless observe. Most of the Case Studies presented to demonstrate the dual concepts of Process intensification and Cost impact are based on results obtained and analyses performed in our own laboratories at IIT Madras.

I am not a big reader of technical books, or even non-fiction as a genre. I'm a fiction addict, especially pop fiction that keeps the reader on tenterhooks even as the serial-killer protagonist does likewise to his victims. I love short, crisp words, sentences, paras, chapters, books, etc. The un-put-downable nature of the reading material is, to me, de rigueur. I have tried to imbue this book with that sense and sensibility, though realizing frequently that the subject matter sometimes simply did not lend itself to this mode of writing. It is very easy to slip into the dry, didactic style when writing on scientific matters. It is a trap I've tried to avoid, not always with success. If I start sounding like your worst memories of boring lecturers, just skip ahead to the next chapter. Or, take it in small dozes (pun intended!), a spoonful at a time.

Go forth and intensify!

Prof. Ramamurthy Nagarajan

1 What is Process Intensification?

1.1 PROCESS

Merriam-Webster defines a process as "a series of actions or operations conducing to an end, a continuous operation or treatment especially in manufacture", whereas the Cambridge Dictionary defines it as "a series of actions that you take in order to achieve a result". These definitions reappear in most standard references, with minor variations.

The word "process" suggests an orderly progression, from start to end, with outcomes achieved by deliberate actions. Life is a process that begins with birth, and ends with.... Well, you know what. Every act that we indulge in affects what happens subsequently, although the cause-and-effect relationship may not always be obvious.

Learning is a process, some would say lifelong. There is structured learning, and then there is the learning by observing, by doing, by failing then succeeding, etc. The process in many cases is a journey to a destination, a means to an end. There are many phases to running a process:

1. Process Definition—to produce a desired end product or service, given a set of inputs and operational conditions
2. Process Analysis—an in-depth review of what a suitable process would consist of, and what it would be capable of producing
3. Process Modeling—an idealized picture of the process that is mathematically tractable, yet deviates least from reality
4. Process Simulation—a predictive methodology that attempts to take the process model and extract from it future prospects for the business
5. Process Design—the first import of the process from the realm of imagination into physical space
6. Process Scale-Up—an initial ramp from concept to pre-production scale
7. Process Implementation—the first on-the-ground running of the process to produce the first lot of honest-to-goodness products
8. Process Debugging—fixing the blemishes in the first production lot so that subsequent lots come out in better shape
9. Process Monitoring—data collection and interpretation to keep track of key metrics in the process and product
10. Process Control—data-based action and reaction to ensure that the process stays in a "sweet spot"
11. Process Optimization—minimizing deviations from the target state of the process, achieving the "FBC" (Faster, Better, Cheaper) trinity
12. Process Stabilization—with the start-up fun and games over, holding 'er steady as she goes

DOI: 10.1201/9781003283423-1

13. Process Scale-Down—as demand recedes, and clamor grows for the next-gen product, taking the foot off the accelerator
14. Process End-of-Life—putting the brakes on, shutting down production, saying the teary-eyed farewells

Do all processes follow this sequence? Surely not, but this is fairly representative of a typical industrial process. If allowed to run on its own, it will produce output, and make some money. But human need/greed being what it is, would we be content to let that happen?

No.... So, what do we do next? INTENSIFY!

1.2 INTENSIFICATION

"Intensification" is defined variously as: "the fact of becoming greater, more serious, or more extreme" (Cambridge Dictionary), "an increase in strength or magnitude" (vocabulary.com), etc. "Intensity" conveys a sense of focus and concentration, and is generally considered a desirable characteristic for athletes and actors, perhaps not so for engineers and scientists. When an emotion is intensified, it can become an obsession. When adulation is intensified, it can turn into fanaticism. Conflicts can intensify into wars. Weather systems intensify into storms. When an economic slowdown intensifies, depression is just around the corner. Thus, in the way we live our lives, "intense" can have a negative connotation as well.

In physics, intensity of radiant energy is the power transferred per unit area, where the area is measured on the plane perpendicular to the direction of propagation of the energy; In photometry and radiometry, *intensity* has a different meaning: it is the luminous or radiant power per unit solid angle. In optics, *intensity* can mean any of radiant intensity, luminous intensity, or irradiance, depending on the background of the person using the term. Radiance is also sometimes called *intensity*, especially by astronomers and astrophysicists, and in heat transfer (Wikipedia).

Agricultural intensification can be defined as an "increase in agricultural production per unit of inputs, which may be labor, land, time, fertilizer, seed, feed or cash" (fao.org). Giddens defines globalization as the "intensification of worldwide social relations linking distant localities in such a way that local happenings are shaped by events occurring many thousands of miles away and vice versa" (http://www.glopp.ch/A3/en/multimedia/giddens.pdf). In photography, intensification refers to "increase in the density and contrast of (a photographic image) by chemical treatment" (Merriam-Webster dictionary).

Clearly, intensification is a term that conveys a sense of significant change, for better or worse. It requires an active intervention in some cases or happens on its own in others. Controlled intensification, however, does necessitate controlled intervention with a predictable outcome. A student's interest in a subject can be intensified over time by a good teacher. A researcher's investigative abilities can be intensified over the duration of a dissertation by an experienced Guide. A husband's skill in performing household chores can be intensified over the course of a marriage by a patient wife. "Fanning the flame", on the other hand, is an example of uncontrolled intensification. The endpoint is not always where you expect it to be. If you blow too hard, you can even extinguish the flame.

To intensify is indeed to provide more entropy to a system through an application of an external force, whether it be thought, words, or deed. There is an art to it since the relationship between input and outcome is not always quantifiable nor reproducible. However, there is one context in which intensification has a very specific connotation, and has evolved into a precise science.

1.3 PROCESS INTENSIFICATION

Taken together, "Process Intensification" is a phrase common to many disciplines, Chemical Engineering being a prominent example. That is not surprising given that, of all engineering disciplines, Chemical Engineering is the one that devotes maximum attention to the study of processes. Process Identification, Analysis, Modeling, Simulation, Control, Optimization—all are courses in any undergraduate curriculum in Chemical Engineering. Process Intensification is also increasingly being viewed as an arrow in the quiver of chemical engineers.

Process Intensification (PI) is a revolutionary approach to process and plant design, development, and implementation. It presents significant scientific challenges to chemists, biologists, and chemical engineers while developing innovative systems and practices which can offer a drastic reduction in chemical and energy consumption, improvements in process safety, decreased equipment volume and waste formation, and increased conversions and selectivity toward the desired product(s). In addition, they can offer a relatively cheaper and more sustainable process option. The study of process intensification can be broadly divided into two areas: Process Intensifying Equipment, and Process Intensifying Methods (Unit Operations). The former includes spinning disk reactor, static mixer reactor, static mixing catalysts, monolithic reactors, heat-exchange reactors, supersonic gas/liquid reactors, jet impingement reactor, and rotating packed bed reactor for carrying out chemical reactions. For purposes other than chemical reactions, static mixers, compact heat exchangers, microchannel heat exchangers, rotor/stator mixers, rotating packed beds, and centrifugal adsorbers are used. The latter includes multifunctional reactors, hybrid separators, alternative sources of energy, and other methods. The alternative source of energy can be by the application of electric fields, ultrasound, solar energy, microwaves, and plasma technology.

Collins Dictionary defines process intensification in Chemical Engineering as "a change made to a process to make it work in a smaller volume for the same performance; aimed at reducing the size of equipment by orders of magnitude; involving a move away from large-scale batch reactions, towards flow reactors in which the actual volume in which the reaction occurs is very small"—in other words, implementation of technologies that replace large, expensive, energy-intensive equipment, or processes with ones that are smaller, less costly, and more efficient.

In the European Roadmap of process intensification, PI is defined as "a set of innovative principles applied in process and equipment design, which can bring significant benefits in terms of process and chain efficiency, lower capital and operating expenses, higher quality of products, less wastes and improved process safety".

Process Intensification is defined by Tsouris and Porcelli (2003) as a revolutionary approach to process and plant design, development, and implementation. PI is about providing a chemical process with the precise environment it needs to flourish, which results in better products, and processes that are safer, cleaner, smaller, and cheaper. The term "process intensification", thus, refers to technologies that replace large, expensive, energy-intensive equipment, or processes with ones that are smaller, less costly, and more efficient.

Jachuck et al. (1997) have defined PI as "a term used to describe the strategy of making dramatic reductions in the size of a chemical plant in order to achieve a given production objective".

Stankiewicz and Moulijn (2000) have defined PI as "any chemical engineering development that leads to a substantially smaller, cleaner, safer, and more energy-efficient technology".

A general definition of PI is simply any intervention that makes the process run better, faster, and cheaper. While all three may not be achievable at the same time, the process owner can decide which of these attributes is of higher priority and intensify the process accordingly. If quality is the major differentiator for the product, "better" would appear to deserve the highest priority; if time to market is key to profits, "faster" would be a good metric to focus on. If the market considers price to be the primary selling point, "cheaper" is the way to go. But in that case, the cost impact associated with the process intensification needs to be quantified with the same precision as the intensification itself.

While process intensification certainly sounds like a good thing to attempt, suitable quantification of the resultant outcome is frequently lacking. This requires that key process or product parameters first be identified so that their values prior to and post-intensification can be measured. If this were possible, a Process Intensification Factor (PIF) can clearly be defined as:

$$PIF = \left(\text{Parameter Value Post} - \text{Intensification} \right) / \left(\text{Parameter Value Pre} - \text{Intensification} \right) \quad (1.1)$$

The denominator may also be denoted as "baseline value".

It is not uncommon for process intensification efforts to involve an element of cost increase, and this needs to be accounted for in assessing the desirability of the intensification from a business economics point of view. The Cost Impact Factor (CIF) needs to be therefore quantified in a manner analogous to PIF as:

$$CIF = \left(\text{Cost Associated with Intensified Process} \right) / \left(\text{Baseline Cost} \right) \quad (1.2)$$

The ratio (PIF/CIF) is thus a metric that enables a fair assessment of the impact of process intensification on the bottom line as far as the business is concerned, and can subsequently enable optimization exercises.

In the next chapter, we'll look at some practical examples that illustrate this concept.

EXERCISE QUESTIONS

1. From your everyday life, list 10 processes and their purposes.
2. For these processes, propose ways to intensify the processes identified, and identify the desired outcomes.
3. For the proposed intensification techniques, estimate PIF and CIF values. What do the estimates tell you?

REFERENCES

Jachuck, R.J., J. Lee, D. Kolokotsa, C. Ramshaw, P. Valachis, and S. Yanniotis, "Process Intensification for Energy Saving", *Appl. Therm. Eng.*, vol. 17, no. E-10, 1997, pp. 861–867.

Stankiewicz, A.I. and J.A. Moulijn, "Process Intensification: Transforming Chemical Engineering", *Chem. Eng. Prog.*, January 2000, pp. 22–34.

Tsouris, C. and J.V. Porcelli, "Process Intensification: Has Its Time Finally Come?", *Chem. Eng. Prog.*, vol. 99, 2003, pp. 50–53.

2 PIF/CIF Case Studies

2.1 THE COOKIE FACTORY

We all devour cookies like monsters, and we like them baked just right. The hotter you set the oven, the more chewy and moist the cookies turn out, unless you make it too hot. In that case, they can become "sticky". But if you set the temperature too low, they come out crumbly. Let us say that 380°F is the ideal baking temperature to get the best cookies and that they crumble if baked at or below 350°F, and stick if baked above 400°F. How can the cookie maker decide on the optimum baking temperature? If cost were no constraint, it would be 380°F—the most "intensified" process condition. If quality were not a consideration, it would be 350°F. Clearly, the correct setting lies between the two, but how do you identify it in an objective manner?

The relevant cost of manufacturing—let us call it C_M—is, in this case, quantifiable, assuming that it is mostly related to the baking cost. We can use a formula along the lines of:

$$C_M = C_{M,350F} + a_M \left[(T_b - 350)/350 \right] \qquad (2.1)$$

where T_b is the baking temperature setting in the oven, $C_{M,350F}$ is the cost of baking at 350°F, and the model assumes a linear relationship between heating cost and the temperature differential from the baseline of 350°F. The critical parameter in this equation is a_M, which represents the magnitude of heating cost rise with temperature.

The cost of quality—call it C_Q—is a little more difficult to define. Genuchi Taguchi-san had defined the Cost of Quality as the loss to society when any desired characteristic deviates from its ideal value. In this case, it may be taken as a measure of revenue loss associated with "crumbly" or "sticky" cookies, and stated in the following form:

$$C_Q = 0 + a_Q \left[(T_b - 380)/380 \right]^2 \qquad (2.2)$$

Since 380°F is the "ideal" temperature, C_Q will be clearly zero at that temperature. Analogous to a_M, the critical parameter in this equation is a_Q, which represents the magnitude of revenue loss with a deviation from 380°F.

The key for a process manager is to realize that the total cost to business, C_T, is the sum of costs of manufacturing and cost of quality, i.e.:

$$C_T = C_M + C_Q \qquad (2.3)$$

This total cost is to be minimized if the business were to maximize its profitability. The ratio (a_M/a_Q) is an indicator of the relative weightage given by the cookie

DOI: 10.1201/9781003283423-2

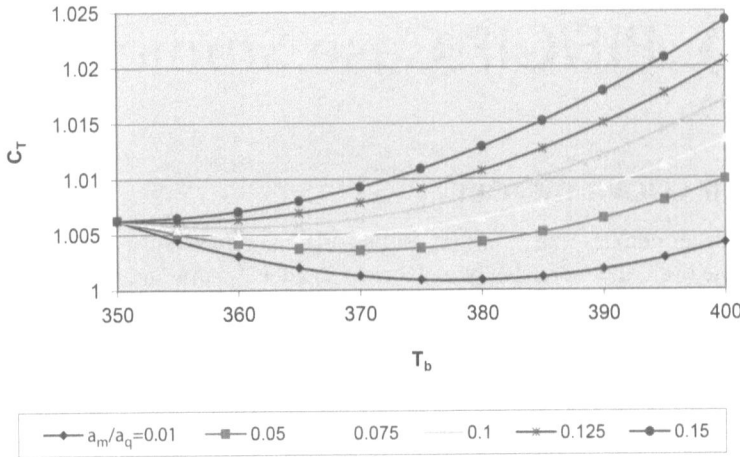

FIGURE 2.1 Dependence of total cost of cookie baking on baking temperature.

company to the cost of making the cookies to the cost of business loss associated with less-than-perfect cookies. The higher this ratio, the less the maker is committed to customer satisfaction from a quality viewpoint, but more attuned to the customer's price sensitivity. Figure 2.1 shows the dependence of total cost, C_T, on the baking temperature, T_b, for values of (a_M/a_Q) varying in the range of 0.01 to 0.15.

The minima in these lines clearly represent the lowest-total-cost conditions and the optimum baking temperature, $T_{b,opt}$. This parameter is plotted in Figure 2.2 against (a_M/a_Q) for values ranging from 0 to 0.16.

This is an illustration of the cost-constrained quality intensification approach, where the ideal operating value of the intensification parameter—cookie baking

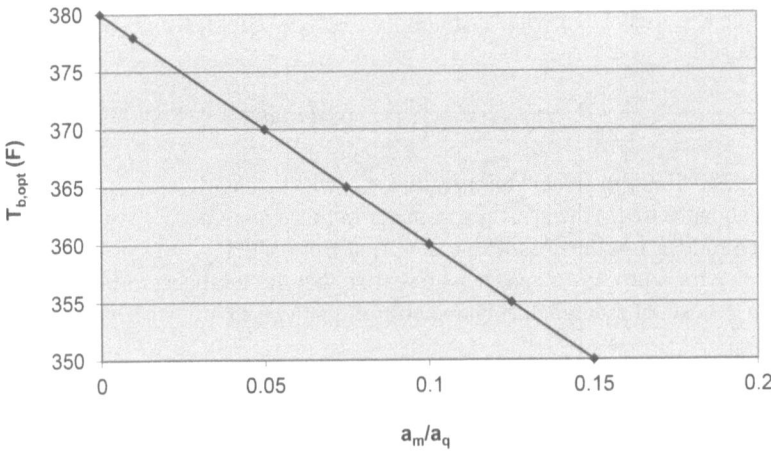

FIGURE 2.2 Dependence of $T_{b,opt}$ on (a_M/a_Q).

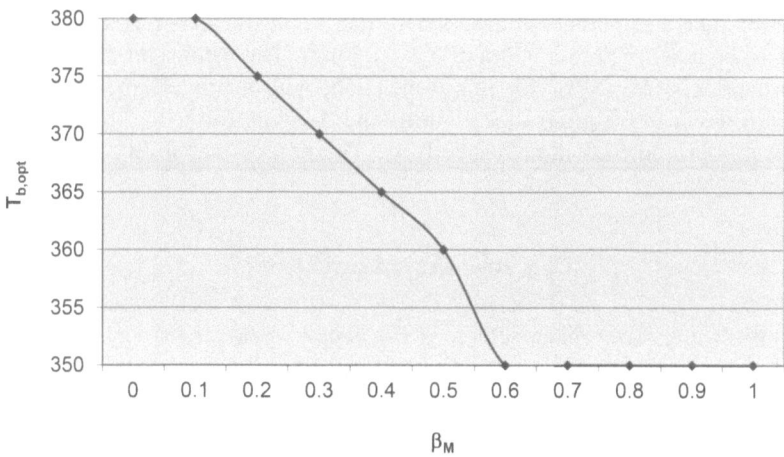

FIGURE 2.3 Dependence of $T_{b,opt}$ on weighting factor, β_M.

temperature—is selected on the basis of the estimated costs of manufacturing and quality (or lack thereof).

A more refined approach is to assign weighting factors to the costs of manufacturing and quality. If β_M is the weighting factor assigned to manufacturing cost, $(1-\beta_M)$ would be the weighting factor assigned to quality cost. The form of a total cost function to be minimized then becomes:

$$C_T = \beta_M C_M + (1 - \beta_M)C_Q \qquad (2.4)$$

If price is the primary market differentiator, β_M approaches 1; and if quality is the primary market leverage, β_M approaches 0. In general, $0 \le \beta_M \le 1$.

For an (a_M/a_Q) value of 0.1, $T_{b,opt}$ may then be plotted as a function of β_M as shown in Figure 2.3.

You now have a rational basis for setting the optimum process conditions (though it must be said that human decisions are not always based on rational thinking!).

2.2 TOLERANCING OF CRITICAL DIMENSIONS IN A HARD DISK DRIVE

Magnetic data storage devices, also known as hard disk drives (HDDs), are electromechanical assemblies designed and manufactured with the highest precision. Given that the read/write head flies at 20,000 rpm above the disk at a distance of a few nanometers, precision in mechanical dimensions is clearly a critical necessity. Appropriate tolerancing in design is required in order to achieve acceptable device reliability while minimizing the cost of manufacturing. A classic illustration of Process Intensification Factor (PIF) versus Cost Impact Factor (CIF), obviously!

Let us take a hypothetical case where there is a slot in the HDD cover that needs to be machined to a width of 5 ± 0.05 mm. The vendor promises statistical process control (SPC) for this dimension, with a process mean at 5 mm, standard deviation (sigma) of 0.01 mm, and 3 sigma lower and upper control limits at 4.97 and 5.03 mm, respectively. The process capability index, C_p, is then given by:

$$C_p = \text{Tolerance} / (6 \text{ sigma}) = 1.67 \qquad (2.5)$$

However, the HDD manufacturer is not happy whenever the slot width is not exactly 5 mm. They want to see a minimal drift from the ideal or target value. On the other hand, the vendor is not happy whenever the process needs to be tightened toward the "ideal target" of 5 mm. It requires manual intervention, additional inspection, rework, etc.—all cost adders.

Another factor is that the HDD manufacturer makes and markets the drive for a spectrum of end-users, ranging from high-end enterprise systems (who value reliability) to home PCs (where price is the major differentiator). The same slotted cover is used for low-end as well as high-end products.

Given this scenario, what would be the optimum cost-quality strategy for the vendor and the customer to follow? The solution, in this case, is given by a rigorous cost-benefit analysis, in which PIF and CIF play a central role.

A good first step is to quantify the pain and the gain by following various machining strategies. Table 2.1 summarizes a set of illustrative data.

Here, the Quality Loss Cost is estimated based on monetization of the extent of the unhappiness of the HDD manufacturer and the associated loss of future business.

Figure 2.4 presents the cost data graphically. It may be observed that the optimum setting for slot width that minimizes total cost is 5.01 mm.

In Figure 2.4, the implicit assumption is that the costs of quality and manufacturing are weighted equally in the decision-making process. However, depending on the market segment at which the finished HDD is targeted, the weighting factor for each

TABLE 2.1

Costs Associated with Various Machining Settings

Slot Width, mm	Quality Loss Cost, C_Q (Rs)	Mfg Cost, C_M (Rs)	Total Cost, C_T (Rs)
5.03	15	0	15
5.02	11	2	13
5.01	3	6	9
5	0	14	14
4.99	4	8	12
4.98	13	4	17
4.97	20	0	20

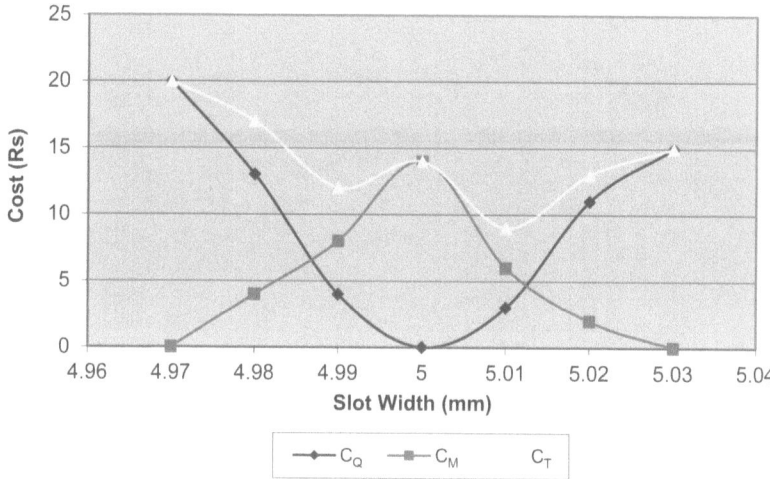

FIGURE 2.4 Costs of quality and manufacturing, and total cost (by simple addition) as a function of slot width.

may be quite different. This can be captured in the analysis by defining a weighting parameter, α_Q, such that:

$$C_T = \alpha_Q C_Q + (1 - \alpha_Q) C_M \qquad (2.6)$$

α_Q will approach zero for drives being sold to the low-end consumer market and tend to 1 for selling to high-end customers with high-reliability requirements. The data in Figure 2.4 have been replotted in Figure 2.5 for values of α_Q ranging from 0.2 to 0.8.

It is apparent that the optimum setting for slot dimension now depends on the relative importance accorded to quality and cost of manufacturing, as expressed via α_Q. This explicit dependence is plotted in Figure 2.6.

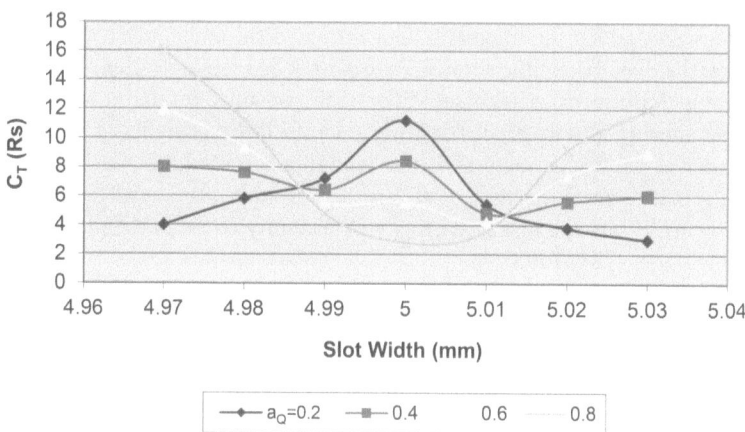

FIGURE 2.5 Total cost variation as a function of the weighting parameter.

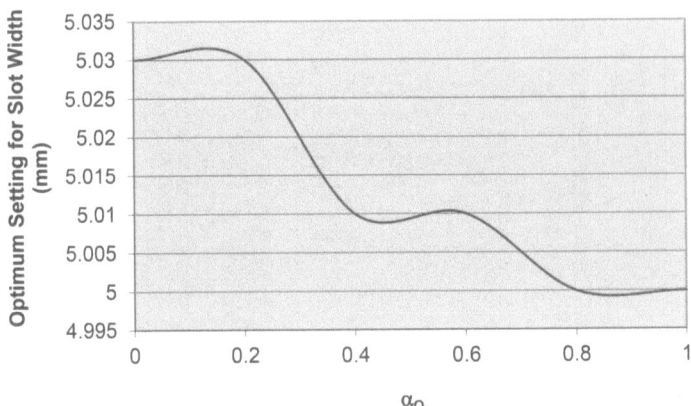

FIGURE 2.6 Variation in optimum slot width with weighting factor, α_Q.

This example is intended to serve as an illustration of the methodology of giving weightage to both quality and cost parameters, and arriving at a rational techno-economic decision based on business consideration. However, in a notoriously competitive business such as hard drive manufacturing, reason, again, may not always prevail. In fact, many HDD manufacturers lose money with every drive sold, but they gladly accept the trade-off of gaining or keeping market share.

2.3 MICROCONTAMINATION CONTROL

In the manufacture of electronic components and devices, such as semiconductor chips and (again!) hard disk drives, particulate and chemical contaminants have a significant effect on the production line yield and field reliability. For example, a single particle of size 20 nm, if it happens to be in the wrong place at the wrong time, can cause a head crash in a hard drive. It can get caught between the read/write head and get dragged along the magnetic data storage disk, and cause a circumferential scratch on the disk. Along the scratch, data cannot be read or written, and this would constitute a "hard error". Chemical vapors can condense on the head, and alter its trajectory, causing it to fly higher or lower than the design height (which is typically of the order of a few nanometers). If the head flies higher than intended, read/write sensitivity is lost, resulting in a "soft error". If it makes the head fly lower, it will eventually crash into the disk, which would certainly result in data loss and a hard error (Nagarajan et al., 1998).

Similarly, in silicon fabs, contaminants can cause semiconductors to become conductors by "bridging", thereby compromising the functionality of the chip. In lithography, since typical film thicknesses are much smaller than pattern feature sizes, defects that are as small as one-hundredth of the lithographic dimension must be controlled. Scaling device dimensions by a factor of 1/3 to 1/2 will require almost a factor of 10 reduction in particulate levels in order to maintain the pre-scaling yield. To achieve a 78% yield (0.25 defects/cm^2) in a typical submicron process containing

250 process steps, each step must contribute no more than 0.001 killer defects/cm^2 on average (Osburn et al., 1988).

Contaminants are contributed by several sources (Welker et al., 2006)—the cleanroom facility, equipment and tooling, components, processes, people, garments and consumables, chemicals, etc. Of these, the cleanroom represents the highest capital investment and contributes the most to the selling price of the manufactured or assembled product. Hence, any savings in cleanroom cost while maintaining the same quality level of the product would constitute a "process intensification" in the "cheaper" context (out of the trio of better, cheaper, faster).

Cleanrooms are classified on the basis of contaminant levels per unit volume of air. As per the old Fed-STD-209E, cleanroom "levels" were defined on the basis of the number of particles per cubic foot larger than 0.5 μm. For example, Class 100 defines a facility having 100 particles of size > 0.5 μm per cu.ft. of air, whereas a Class 10 facility would have 10 particles of size > 0.5 μm per cu.ft. of air (Figure 2.7). The newer ISO Standard 14644-1 redefines these as Level 4 and Level 5, respectively, and also lowers the smallest controlled size to 0.1 μm. An ISO 5 cleanroom will have 100,000 particles per cubic meter of size > 0.1 μm, and an ISO 4 cleanroom will have 10,000 particles in the same size range (Table 2.2).

Thus, going from ISO Class 4 to Class 5 can potentially introduce into the process 10X the number of particles, with the associated degradation of yield and reliability. However, the cost savings can be significant since facility construction cost scales roughly as the square of the reduction in contaminant loading. This is certainly tempting from a cost-of-manufacturing viewpoint. But what about the cost of quality?

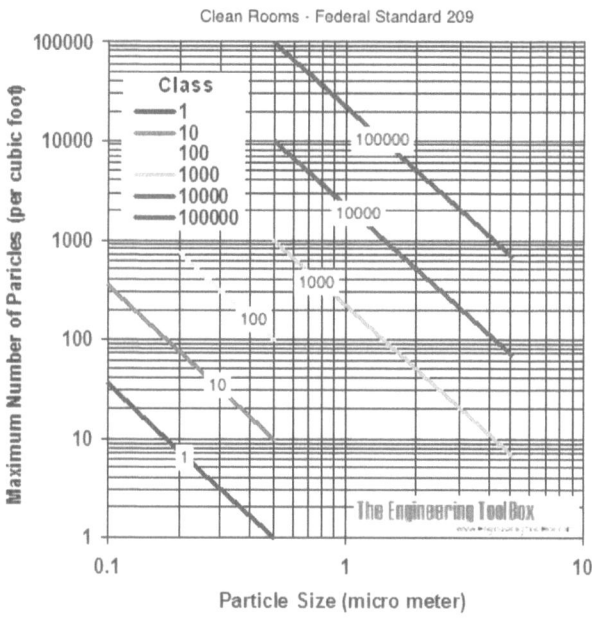

FIGURE 2.7 FED-STD-209E cleanroom cleanliness levels.

TABLE 2.2
ISO Cleanroom Classes

	Number of Particles per Cubic Meter by Micrometer Size					
CLASS	0.1 μm	0.2 μm	0.3 μm	0.5 μm	1 μm	5 μm
ISO 1	10	2				
ISO 2	100	24	10	4		
ISO 3	1,000	237	102	35	8	
ISO 4	10,000	2,370	1,020	352	83	
ISO 5	100,000	23,700	10,200	3,520	832	29
ISO 6	1,000,000	237,000	102,000	35,200	8,320	293
ISO 7				352,000	83,200	2,930
ISO 8				3,520,000	832,000	29,300
ISO 9				35,200,000	8,320,000	293,000

Line yield is measured traditionally as the ratio of good (shippable) products delivered to the end-of-line to the number of products started, while field reliability is measured commonly as the mean time between failures (MTBF). MTBF is taken to be the predicted elapsed time between inherent failures of a mechanical or electronic system, during normal system operation, and is calculated as the arithmetic mean (average) time between failures of a system. Another measure is Annualized Failure Rate (AFR)—the probable percent of failures per year, based on the manufacturer's total number of installed units of similar type. AFR is an estimate of the percentage of products that will fail in the field due to a supplier cause in one year.

If, say, an HDD manufacturer were to decide that manufacturing in an ISO Class 5 cleanroom will reduce their facility cost by 100 times compared to a Class 4 cleanroom operation, the risk element in the decision would be the degradation in quality metrics such as yield and MTBF/AFR values, and the consequent customer dissatisfaction that can lead to loss of market share. Rational decision-making in this scenario would require that pilot-scale trials be run in both facility types to assess the effect of facility cleanliness on the critical quality indices.

A leading manufacturer of hard drive XY ran such a trial, and obtained the data (see Table 2.3).

As Run # 3 data show, sometimes unexpected results are obtained even in carefully designed experiments. This is typically because not all relevant variables have been accounted for and controlled. In the case of the HDD yield, it is possible that one or more of the other sources of contamination might have gained ascendancy during that particular run and virtually negated the effect of the cleanroom environment.

To obtain MTBF estimates, a previously formulated relationship with line yield was used:

$$MTBF = (MTBF)_{ideal} \times (Yield_{actual} / Yield_{ideal})^2 \qquad (2.7)$$

TABLE 2.3
Process Yields in ISO Class 4 and
Class 5 Facilities during Pilot Study

Run #	Yield in ISO Class 4 (%)	Yield in ISO Class 5 (%)
1	71	62
2	73	68
3	70	72
4	66	64
5	74	70
6	74	66
7	73	60
8	69	68
9	72	65
10	71	71
Average	71.3	66.6

While manufacturing line yield is essentially an internal metric, it does indicate the robustness of product and process design. Therefore, it is not surprising that many manufacturers have found a strong correlation between process yield and product reliability in the field. The measured MTBF does influence buying decisions, especially in high-end applications where quality and reliability are key. In unique cases such as hard drives, which are very complex electromechanical assemblies that are yet regarded as mere commodities in the market, the need to balance quality and cost is paramount.

In this instance, the PIF is the associated cost reduction in manufacturing—10, whereas the Cost Impact Factor (CIF) is a little more nebulous to ascertain. The 14% degradation in MTBF may be sufficient for several large-enterprise customers to prefer a competitor who can offer a higher MTBF. The revenue loss could amount to several hundreds of millions of dollars. A clever strategy for the manufacturer, in this case, would be to manufacture the hard drives separately for low-end (e.g., domestic) consumers and high-end customers (e.g., banks, airlines). For the former, manufacturing the same drive in an ISO Class 5 cleanroom facility may be acceptable, given the price sensitivity; for the latter, the facility may be maintained at a higher level of cleanliness in order to ensure short-term yield and long-term reliability.

2.4 CALL CENTERS: THE BUSINESS CASE FOR OFFSHORING

Many consumer industries in richer economies have cultivated outsourcing of labor-intensive functions such as Customer Service to regions where labor costs are relatively low. India, in particular, combines low labor costs with a high degree of comfort with English as a spoken language, as well as familiarity and expertise with IT systems. India has naturally emerged as a hub for outsourcing operations (as immortalized in the hit TV serial, Outsourced!). However, the net cost savings

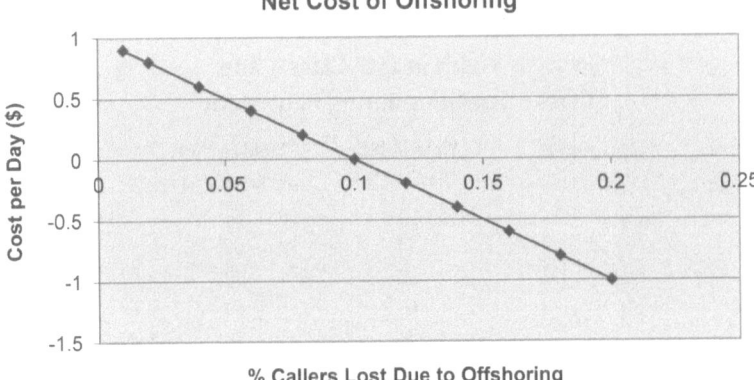

Net Cost of Offshoring

FIGURE 2.8   Case of savings of $1 per call due to outsourcing.

associated with outsourcing is not easy to calculate, given customer proclivities and preferences. For example, foreign accents bother many. Some resent outsourcing as a business philosophy, given the resulting loss of local jobs.

Call Centers in low-cost areas are cheaper to operate, but customer unhappiness is an ever-present risk. The issues can be cultural or technical (though the latter is less likely). Thus, there is a need to balance operational cost savings versus quality cost incurred, reflected in the loss of customers unhappy with the performance on offer. Several U.S. and European companies have indeed stopped outsourcing Call Centers, preferring to live with the higher labor cost in their own geographies. So, how can we formulate a business case for outsourcing? Let us take an illustrative example.

The Customer Services division of an American bank receives "M" inquiries per day. They outsource "X" % of the calls to India and handle (100-X)% in the U.S. The savings per call for an Indian operator is estimated to be $y. Thus, the total savings per day = $(MX/100) y.

Of the X% callers routed to India, "q" % take their business elsewhere due to various causes of dissatisfaction; the average loss per caller is estimated to be $"r", implying that the average loss per day = $(MX/100) (q/100) r.

For the case of savings per call of $1 for an outsourced operator, and average loss of business per lost customer of $1,000. Figure 2.8 displays the net cost of outsourcing. Figure 2.9 presents the case when savings per call is $10.

For net savings to be positive, it is evident that [y − (qr/100)] > 0. Only the magnitude of net savings depends on M and X values. If q = 2, and r = 1,000, y must be at least 20 for break-even! If q = 2, and r = 10,000, y must be at least 200 for break-even!! Now, with these metrics in front of you, make the right call!

2.5   CLEAN COAL TECHNOLOGY

Like it or not, coal is not going away. Projections are that coal will fuel power plants for the foreseeable future. Global coal consumption has increased by 25 million tonnes of oil equivalent (Mtoe) in 2017, with this consumption being

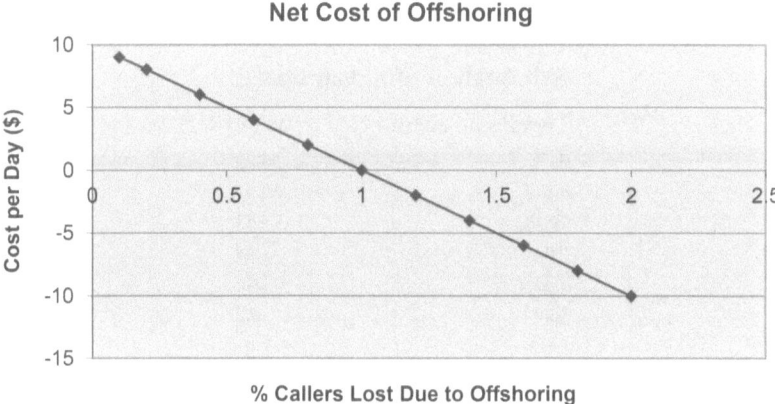

FIGURE 2.9   Case of savings of $10 per call due to outsourcing.

mainly driven by India and China. Coal production has also been growing at a rate of 3.2%, the fastest since 2011 (Srivalli, 2019). As carbon emissions increase and energy efficiency decreases, the thrust to move toward renewable sources of energy has motivated governments across the world to look at alternative sources of fuel. However, the fact that energy consumption has also increased in parallel, makes it quite challenging to make the shift immediately. Asia Pacific accounts for 41% of the coal reserves, mainly split between Australia, India, and China. This, combined with huge coal reserves, which are sufficient to meet the demands for more than a century, implies that coal is here to stay.

Indian bituminous coal has an ash content of about 35–38%, much higher than coal mined in other countries. Over a period of time, the quality of coal available has also degraded as the higher quality reserves are nearly fully utilized, leaving only low-grade, high-ash, low-calorific-value coal for use. Sulfur levels in Indian coal are lower compared to coal imported from Australia or Indonesia, where the majority of the import happens. However, importing coal poses many challenges, the primary one being cost, followed by transportation. Currently, power plants tackle these issues by using blended coal (a mix of Indian and foreign coal) so that such issues are addressed partially.

The main constituents of coal are fixed carbon, moisture, volatile matter (includes a varying percentage of hydrocarbons), and ash. Each of these constituents can be quantified via proximate analysis, performed in accordance with ASTM D3172-13. Moisture can be estimated by measuring the mass loss that occurs during the combustion of coal under specified conditions of temperature, pressure, and time, while volatile matter can be estimated by measuring mass loss on combustion of coal under rigidly controlled conditions. The mass of residue left after burning coal gives the ash quantity in the sample. A sum of percentages of moisture, volatile matter, and ash, when subtracted from a hundred, yields the percentage of fixed carbon present in the sample.

TABLE 2.4
Ash Analysis of Indian Coal

Constituents of Ash	Wt (%)
SiO_2	56.52
Al_2O_3	30.41
Fe_2O_3	5.63
TiO_2	1.44
P_2O_5	1.81
MnO	0.2
CaO	1.81
MgO	1.04
SO_3	0.52
Na_2O	0.3
K_2O	0.32

Ash contains oxides of silicon, aluminum, iron, titanium, manganese, sodium, potassium, calcium, magnesium, etc. A typical ash analysis of Indian coal is included in Table 2.4.

Ash, on combustion in the boiler, gets converted into fly ash and bottom ash. Bottom ash, being heavier, is collected out of the system, while fly ash particles, being finer in size, are carried away by flue gas. These tend to stick to heat transfer surfaces that they come in contact with. The ash particles that stick to the surface of the boiler subsequently grow in the form of layered deposits, as more particles arrive and add layers to the deposit. These deposits alter the heat transfer surfaces on which flue gas flows, obstruct the natural flow of gas, cause erosion of boiler surfaces, and eventually lead to frequent power shutdowns and reduced efficiencies if left unattended for long periods of time. Ash deposits on heat transfer surfaces can be in solid or molten form depending on where they are found and on the temperatures of the flue gas as well as the surface temperature. Based on these factors, they are classified as fouling and slagging, respectively.

Sodium and potassium, on combustion, leave the ash since they are more volatile, and react with gaseous species to form NaOH and KOH, respectively. The NaOH and KOH react with calcium, magnesium, and alumino-silicates in the temperature range of 1300–1850 K and form a sticky layer that acts like a "glue". This contributes to the adhesion of ash particles to the surface of boilers. At temperatures of 1300–1400 K, sodium reacts with sulfur oxides to form sodium sulfate which begins to condense from 1200 to 1300 K (Walsh et al., 1992). The melting temperature of sodium sulfate is 1157 K and its dew point is 1250 ± 50 K. If all other conditions are favorable, deposition of fly ash on heat transfer surfaces occurs between the melting point and dew point of sodium sulfate (Ross et al., 1988). It is thus evident that the presence of alkali metals plays a major role in fouling deposit formation.

Several approaches have been employed to restrict particle deposition on heat transfer surfaces. The method selected must be technically and economically viable, in addition to providing a suitable remedy to prevent fouling and slagging. Some of the mitigation strategies being considered for the prevention of fouling and slagging are the use of soot blowers and active cooling of surfaces with air, the use of additives to scavenge condensable vapors that assist in fouling, and coal wash to remove impurities including ash. A coal-wash method that has evinced interest in recent times is the use of sonication for intensification of the process of impurity removal from coal.

Ultrasound is defined as sound whose frequency is higher than the upper limit of human hearing. A piezo-electric crystal undergoes alternating expansion and contraction cycles when an AC is applied to it. These occur at the same frequency at which the AC changes polarity, leading to the generation of an oscillating acoustic field. Two major phenomena occur when a system is exposed to ultrasound: cavitation and acoustic streaming (Suslick et al., 1986). When sonication is applied to a liquid medium, a negative pressure is created. When the negative pressure caused is large enough, the distance between the molecules of the liquid becomes greater than the intermolecular distance of the particles that is required to hold the liquid in place. This leads to breakdown of the liquid and causes what is known as "cavitation". Cavitation can be of two types: stable cavitation and transient cavitation. Stable cavitation occurs at low intensities, and the vacuum bubbles which are formed oscillate about a mean equilibrium size for many cycles. Transient cavitation occurs at high intensities, and the bubbles keep growing till they collapse violently in the compression cycle as shown by Ginsburg et al. (1998). During transient bubble collapse, very high temperatures, of nearly thousands of degree Celsius, and high pressures (several hundreds of atmospheres), are created, leading to the enhancing effects of ultrasound. Cavitation mainly occurs at lower frequencies. As the frequency increases, there is less time for the bubble to grow before collapsing. Since cavitation energy is related to bubble volume before collapse, this results in a reduction in cavitation intensity.

Acoustic streaming occurs mainly due to viscous attenuation and wave interaction with solid boundaries. When it occurs outside the boundary layer, it is called "Eckart streaming", and when it occurs inside the boundary layer, it is termed "Schlichting streaming" (Gale and Busnaina, 1999). The boundary layer thickness is inversely proportional to the square root of frequency and therefore decreases with increasing frequency. The acoustic streaming velocity increases with increasing frequency and power. This leads to high viscous stress, and, in turn, to accelerated flow.

The combined action of cavitation and acoustic streaming, where cavitation breaks open the coal matrix and the streaming flow leaches out sub-surface impurities, has been demonstrated to be very effective in cleaning of pulverized coal prior to combustion (Ambedkar et al., 2011b; Balakrishnan et al., 2015; Srivalli and Nagarajan, 2018). In this context, the PIF is fairly straightforward to estimate, as a ratio of the percentage of impurities in coal post-wash under the baseline process versus (sono-)intensified process. The CIF can certainly be a significant factor here given that new process steps need to be added. It is imperative, therefore, that an optimal process be defined that best balances cost and benefit. Table 2.5 presents a set of data obtained under process conditions optimized for each impurity. It is interesting, but entirely coincidental, that in the case illustrated, all (PIF/CIF) ratios turn

TABLE 2.5
PIF and CIF under Optimized Coal Wash Conditions

Process	PIF = [Post-Wash Level under Intensified Process]/ [Post-Baseline-Wash Level]	CIF = [Intensified Process Cost]/ [Baseline cost]	Ratio
Ash removal	48/8 = 6	1.2	5
Total sulfur removal	80/10 = 8	1.6	5
Alkali removal	60/6 = 10	2.0	5

out to be identical. Whether the ratio is acceptable from a profitability viewpoint is entirely up to the manufacturer to decide based on market forces. The more important point here is that there is a methodology involved that demands full appreciation from the decision-makers.

The case studies presented in this chapter are intended to drive home the point that process intensification typically comes with a cost impact, and the latter must be taken into account when deciding whether/when/how to intensify. Process intensification has been a human pursuit since the dawn of civilization, fire being the first and perhaps foremost example. It is drilled into our collective mindset that progress depends on innovation. The first weapon devised out of sharpened sticks greatly enhanced early man's survival chances, as did the rock and the catapult. If the human skin protected bones and internal organs, clothing "intensified' the protection by protecting the body from external elements. If hunting together improved chances of bagging the game and returning alive, organizing into communities "intensified" this social contract. If mating ensured procreation, marriage "intensified" this contract between man and woman.

From a process intensification viewpoint, the parameters leveraged most widely are not too numerous. We will list a few in the next chapter.

EXERCISE QUESTIONS

1. Chicago pizzerias pride themselves on the quality of their deep-dish pizza. Typically, slow cooking is recommended to make sure that the dough has enough time to rise, and the toppings cook well and evenly. Although this can be a good option at home, commercial pizzerias have to handle the throughput as well. Time is money, so to speak. Joe's Pizzeria prices their full 14" pie at $20, and finds that they can process any number from 60 to 80 pizzas an hour given their oven capacity. They do have some data that suggest that customer satisfaction falls off from 90% to about 75% in a linear manner as the oven times decrease between these limits. Assist Joe in making the necessary trade-offs here. Make any assumptions required and develop a full business case. You can model two extremes—a virtual monopoly for Joe, and a perfect marketplace where quality (and cost) are king.

2. A long-haul commercial flight from Abu Dhabi to Chicago ORD takes about 15 hours. The flight always takes off 30–45 minutes late from Abu Dhabi as U.S. Customs and Immigration checks are carried out at Abu Dhabi, and there are always associated delays. The airlines have a choice to make: They can make up part or all of the delays by increasing flight speed or can compromise their on-time record by flying at a lower speed that conserves expensive jet fuel. The latter may appear to be better to reduce operating costs, but customer happiness is at least partly linked to on-time arrival, and there is a potential loss of revenue as a result. Set up a PIF/CIF analysis treating jet fuel cost and "quality cost" as variables. What advice would you have for the airline executive?

3. Deriving weighting factors for cost and quality can sometimes be a highly subjective exercise, especially if based on customer polls. People are not always sticklers for the truth when responding to such polls, primarily as they do not want to come across as "cheapskates". A careful review of buying patterns can be a more reliable indicator. A classic test case is generic versus brand-name products. Take one example—say, breakfast cereals—and compare sales figures. Do this across market segments— children, adults, seniors, pets—and you may find how much value society places on each!

4. A testing of HDDs made by leading manufacturers was carried out by Backblaze, and reported in https://platinumdatarecovery.com/blog/most-reliable-brand. Relevant data are excerpted in the table below. The column AFR represents the annualized failure rate. What data would you require from the manufacturers and from customers to carry out a PIF/CIF analysis for this case, and how would you go about it?

Backblaze Hard Drives Quarterly Failure Rates for Q1 2021							
Reporting Period: 1/1/2021 through 3/31/2021 Inclusive							
MFG	Model	Drive Size	Drive Count	Avg. Age (months)	Drive Days	Drive Failures	AFR
HGST	HMS5C4040ALE640	4TB	3,163	59.4	281,692	5	0.65%
HGST	HMS5C4040BLE640	4TB	12,738	53.4	1,146,496	10	0.32%
HGST	HUH728080ALE600	8TB	1,077	37.7	97,027	4	1.50%
HGST	HUH721212ALE600	12TB	2,599	18.0	233,948	5	0.78%
HGST	HUH721212ALE604	12TB	5,685	3.7	308,793	6	0.71%
HGST	HUH721212ALN604	12TB	10,825	24.0	974,310	9	0.34%
Seagate	ST4000DM000	4TB	18,882	65.3	1,701,967	59	1.27%
Seagate	ST6000DX000	8TB	886	71.8	79,740	-	0.00%
Seagate	ST8000DM002	8TB	9,744	54.1	878,106	26	1.08%
Seagate	ST8000NM0055	8TB	14,419	44.2	1,297,674	31	0.87%
Seagate	ST10000NM0086	10TB	1,200	41.6	108,057	6	2.03%
Seagate	ST12000NM0007	12TB	13,702	31.9	1,732,307	66	1.39%
Seagate	ST12000NM0008	12TB	20,085	12.3	1,764,318	41	0.85%
Seagate	ST12000NM001G	12TB	9,029	7.5	704,446	12	0.62%

(Continued)

Seagate	ST14000NM001G	14TB	5,977	5.9	538,401	13	0.88%
Seagate	ST14000NM0138	14TB	1,675	3.8	135,157	9	2.43%
Seagate	ST16000NM001G	16TB	2,459	1.7	54,177	1	0.67%
Toshiba	MD04ABA400V	4TB	99	70.3	8,910	-	0.00%
Toshiba	MG07ACA14TA	14TB	27,336	8.6	2,165,421	34	0.57%
Toshiba	MG07ACA14TEY	14TB	405	3.8	33,831	1	1.08%
Toshiba	MG08ACA16TEY	16TB	1,014	4.1	91,260	-	0.00%
WDC	WUH721414ALE6L4	14TB	8,400	3.8	640,767	10	0.57%
WDC	WLH721816ALE6L0	16TB	520	0.4	4,853	-	0.00%
	Totals		**171,919**		**14,981,658**	**348**	0.85%

BACKBLAZE

5. Many organizations outsource work to contract employees in order to minimize the overhead associated with own staff. The long-term wisdom of this is debatable, again given the loss in quality that can easily happen. There is an optimum size for such a contract workforce, as well as an ideal set of tasks for them to perform. Direct interfacing with customers is, for example, best left to home-grown talent. Analyze an employer from this viewpoint and recommend an appropriate strategy.

6. Energy and the environment are two sides of the same coin, and decisions regarding the security of one over the other may as well be made by tossing that coin. Growing economies demand energy, while mature economies take it as a matter of right. While carbon footprint is important to track and control, the human impetus for progress must always be taken seriously. A return to nature is, well, basically unnatural. Many cleaner sources of energy cost more per unit. While this cost is easy to estimate, the benefits of burning cleaner fuel cannot be as easily monetized. A classic example is hybrid versus electric cars. There are strong lobbies on both sides of the debate. Investigate this further and formulate your own argument in favor of one or the other.

REFERENCES

Ambedkar, B., T.N. Chintala, R. Nagarajan, and S. Jayanti, "Feasibility of Using Ultrasound-Assisted Process for Sulfur and Ash Removal from Coal", *Chem. Eng. Process. Process Intensif.*, vol. 50, no. 3, 2011, pp. 236–246.

Balakrishnan, S., V.M. Reddy, and R. Nagarajan, "Ultrasonic Coal Washing to Leach Alkali Elements from Coals", *Ultrason. Sonochem.*, vol. 27, 2015, pp. 235–240.

Gale, G.W. and A.A. Busnaina, "Roles of Cavitation and Acoustic Streaming in Megasonic Cleaning", *Part. Sci. Technol.*, vol. 17, 1999, pp. 229–238.

Ginsburg, E., M.D. Kinsley, and A. Quitral, "The Power of Ultrasound", *Adm. Radiol. J.*, vol. 17, no. 5, 1998, pp. 17–20.

Nagarajan, R., L. Nebenzahl, J. Wong, L. Volpe, and G. Whitney, "Chemical Integration and Contamination Control in Hard Disk Drive Manufacturing", *J. Institute Environ. Sci. Technol.*, vol. 41, no. 5, 1998, pp. 31–35.

Osburn, C.M., H. Berger, R.P. Donovan, and G.W. Jones, "Effects of Contamination on Semiconductor Manufacturing Yield", *J. Institute Environ. Sci. Technol.*, vol. 31, no. 2, 1988, pp. 45–57.

Ross, J., R. Anderson, and R. Nagarajan, "The Effect of Sodium on Deposition in a Simulated Combustion Gas Turbine Environment", *Energy Fuels*, vol. 2, 1988, pp. 282–289.

Srivalli, H. and R. Nagarajan, "Mechanistic Study of Ultrasound-Assisted Solvent Leaching of Sodium and Potassium from an Indian Coal Using Continuous and Pulsed Modes of Operation", *Chem. Eng. Communic.*, vol. 206, no. 2, 2018, pp. 207–226.

Srivalli, H., "Mechanistic Study of Leaching of Sodium and Potassium from an Indian Coal Using Ultrasonic and Additive Methods", Ph.D. Thesis, Indian Institute of Technology Madras, 2019.

Suslick, K.S., D.A. Hammerton, and R.E. Cline, "The Sonochemical Hotspot", *Am. Chem. Soc.*, vol. 108, no. 7, 1986, pp. 5641–5642.

Walsh, P.M., A.F. Sarofim, and J.M. Beer, "Fouling of Convection Heat Exchangers by Lignitic Coal Ash", *Energy Fuels*, vol. 6, 1992, pp. 709–715.

Welker, R.W., R. Nagarajan, and C.E. Newberg, *Contamination and ESD Control in High-Technology Manufacturing*, Wiley-Interscience, 2006.

3 Common Process Intensification Parameters

3.1 TEMPERATURE

The most obvious one is temperature. Higher temperatures make endothermic reactions run faster, improve flowability of liquids, enhance mixing of solids in liquids, increase diffusion rates, and raise melting and boiling rates. Lower temperatures increase freezing and solidification rates, etc.

An increase in temperature typically increases the rate of chemical reactions. It will raise the average kinetic energy of the reactant molecules. Therefore, a greater proportion of molecules will have the minimum energy necessary for an effective collision. The Arrhenius equation provides the temperature dependence of the rate of a chemical reaction:

$$k = A\,e^{(-Ea/RT)} \tag{3.1}$$

where

   k = rate constant of the reaction
   A = Arrhenius Constant
   E_a = Activation Energy for the reaction (in joules mol^{-1})
   R = Universal Gas Constant, and
   T = Temperature in absolute scale (in kelvins).

Thus, the reaction rate, which is typically defined as a multiple of the rate constant with reactant concentrations raised to appropriate powers, will rise exponentially with temperature. Clearly, temperature is the most sensitive "lever" to influence reaction rates.

Mass, heat, and momentum transport rates in fluids are less sensitive to temperature, as may be verified by examining the temperature coefficients of the respective constitutive properties. For low-density gases, dynamic viscosity, μ, is proportional to temperature raised to a power between 0.5 and 1.0 (Rosner, 1986). Liquid viscosity, on the other hand, decreases with increasing temperature as (Rosner, 1986):

$$\mu = \mu_\infty . \exp\left(E_\mu / RT\right) \tag{3.2}$$

where E_μ is the "activation energy for fluidity", and the pre-exponential factor, μ_∞, the (hypothetical) dynamic viscosity at infinite temperature.

In the temperature range of 200–1300 K, the effect of temperature on the thermal conductivity of most common substances is quite modest. For dilute solute diffusion in low-density gases, the effective diffusivity scales as T^n, where $n \geq 3/2$. The n values for binary mixtures typically vary between 1.6 and 2.0, with 1.8 being a representative value.

For dilute solutes in dense vapors or liquids, the well-known Stokes-Einstein equation yields the following temperature dependence of diffusivity:

$$D_i = k_B T / (3\pi\mu\sigma_{i,eff})$$ (3.3)

where k_B is the Boltzmann constant, μ is the Newtonian viscosity of the host fluid, and $\sigma_{i,eff}$ is the effective molecular diameter of solute i. The same relationship also applies to the Brownian diffusivity of particles in a gas, provided the particle diameter is large compared to the prevailing mean free path.

For solute diffusion in ordered solids, diffusivity has a more sensitive (exponential) dependence on temperature since there is a need to overcome an energy barrier in moving from one interstitial site to another.

If we were to estimate the Process Intensification Factors (PIF) in each of these cases as the enhanced value of the reaction/diffusion rate parameter divided by the baseline value, for the same change in temperature of, say, 10%, they can range in value from 1.1 to as high as 20–50. The Cost Impact Factor (CIF), of course, stays the same in all cases. Hence, it is important to first identify whether the process of interest is reaction-rate-controlled or transport-controlled. In the latter case, the temperature may not provide as significant a lever as in the former. This can also be interpreted in terms of sensitivity factors—i.e., change in dependent parameter for a unit change in the independent variable. The higher the sensitivity, the greater will be the "bang for the buck". Sensitivity analysis is, therefore, a useful tool in process design, optimization, and ruggedization.

After temperature among key process parameters comes…. Pressure.

3.2 PRESSURE

Pressure also has different effects depending on whether a chemical process is reaction-rate-controlled or transport-controlled. In the case of homogeneous gas-phase reactions, as the pressure of gaseous reactants is increased, there are more reactant particles in a given volume. This is clear from the ideal gas equation:

$$pV = nRT$$ (3.4)

which can be rewritten as:

$$p = (n / V)RT$$ (3.5)

In other words, gas molecule volumetric concentration is directly proportional to pressure when the temperature is constant, hence the higher reactivity. There will be more collisions, and hence the reaction rate is increased. The higher the pressure of reactants, the faster the rate of a reaction will be. When pressure is decreased, molecules do not collide as often and the rate of reaction decreases. Changing the pressure on a reaction that involves only solids or liquids has no effect on its rate. Thus, pressure variations are only effective in altering reaction rates, and that too in a linear fashion, in the case of gas-phase systems. Operating equipment in high-pressure settings is, on the other hand, quite complex, and requires substantial capital investment.

When a process is diffusion-limited, pressure can have a significant effect. Solute diffusivities in low-density gases are inversely proportional to pressure for the same reason that reaction rates are directly proportional—crowding of molecules, which in this context induces a reduction in the mean free path. Since liquids and solids are incompressible, pressure has a negligible effect on corresponding diffusivities.

Pressure has an intensifying effect on several other phenomena, such as:

- Surface adsorption, where increased pressure will have a linear effect on the number of molecules striking the surface and getting adsorbed; this will reach a plateau value once adsorption sites are saturated
- Crystallization, whose rate increases with increasing pressure until limited by any phase transformations
- Solubilization of gas molecules in a liquid, which follows a linear relationship with pressure (Henry's Law)
- Adhesion to surfaces, as well as agglomeration and aggregation in solution, which forces increase with pressure albeit to varying extents
- Outgassing from materials, which is enhanced at lower pressures
- Evaporative drying, where again lower pressures increase the rate of moisture removal
- Casting, where the percentage and average diameter of porosities decreases, and the density increases with the increase of the casting pressure
- Pelletization, where pellet density can increase dramatically as the pelleting pressure increases
- Spraying, in which process increasing pressure increases nozzle flow rate, reduces median droplet size, and typically increases spray fan angle
- Corrosion, whose rate can be heavily influenced by pressure both directly, by speeding the corrosion process, and indirectly, by affecting scale formation, fluid flow, and the fugacity of gases present in the environment
- Nucleation, which higher pressures inhibit or stimulate when the partial molecular volume of solute is, respectively, larger or smaller in the nucleating phase than in the solution (Kashchiev and van Rosmalen, 1995)
- Boiling point, which becomes higher when pressure increases, and lower when pressure decreases (for example, atmospheric pressure as elevation increases)

Clearly, pressure is as much of an "intensifier" as the temperature is, but its effects tend to be more linear. The Cost Intensification Factor associated with increasing or lowering pressure is considerably higher than comparable temperature variations. Hence, varying pressure for the purpose of process intensification should only be pursued after careful consideration of the techno-economic impacts involved.

Now, how about time? Can that be considered a process intensifier?

3.3 TIME

As the title of a Stephen King book says, everything's eventual. What can happen, will happen. All processes are unsteady, albeit to varying time scales. So, when can time be considered an "intensifier"? That would seem to depend on whether the effect of time is linear, as in a natural progression from one stage to the next, or nonlinear, as in a sudden change that occurs and sustains at some point in time. We have all experienced this in our lives—the day after the warranty expires, the appliance is sure to malfunction, right? Empirical observation, but universally subscribed to....

A more serious example is the time-dependent behavior of gel formation or curing. The viscosity value increases exponentially after reaching a threshold time when gel formation or curing begins. "Breakthrough" phenomena such as these are examples of highly nonlinear dynamics. An author whose contributions have been rejected over many years by a publisher may suddenly find himself or herself with a book on the bestseller list. This happened to JK Rowling! Conversely, a sports megastar who refuses to quit while at the top of his/her game may see a sudden drop-off in performance year-to-year, and leave in infamy. I wouldn't name names here, but we can all think of a few. Time, they say, heals all wounds. But some fester, and drive precipitate action of an extreme nature. Thus, terrorism thrives.

Global warming is a horrific illustration of a breakthrough phenomenon. Since the Industrial Revolution, the global annual temperature has increased in total by a little more than 1°C, or about 2°F. Between 1880—the year that accurate recordkeeping began—and 1980, it rose on average by 0.07°C (0.13°F) every 10 years. Since 1981, however, the rate of increase has more than doubled: For the last 40 years, we've seen the global annual temperature rise by 0.18°C, or 0.32°F, per decade. The result? A planet that has never been hotter.... Nine of the 10 warmest years since 1880 have occurred since 2005—and the 5 warmest years on record have all occurred since 2015 (NRDC, Global Warming 101). The slope change since the early '70s is clear from Figure 3.1. The earth's ocean temperatures are getting warmer, too (Figure 3.2).

Another example of time as an intensifier: While in 2009, only 0.7% of internet traffic worldwide was conducted on mobile phone devices, that number had jumped to 50.3% by 2020, and the rise was not gradual, as may be observed in Figure 3.3.

And one more: The COVID-19 pandemic caused a huge disruption in the 2020-21 timeframe, and previously stable metrics—e.g., air traffic (World Economic Forum Infographic, 2020) were suddenly impacted (Figure 3.4).

A linear change in a process metric with time does not typically encourage running the process longer, since the cost may also be rising linearly. If the delta change in cost with time is significantly lower than the associated delta change in a key

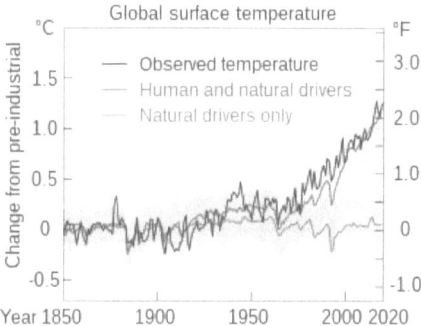

FIGURE 3.1 Global warming as a function of time (https://en.wikipedia.org/wiki/Attribution_of_recent_climate_change).

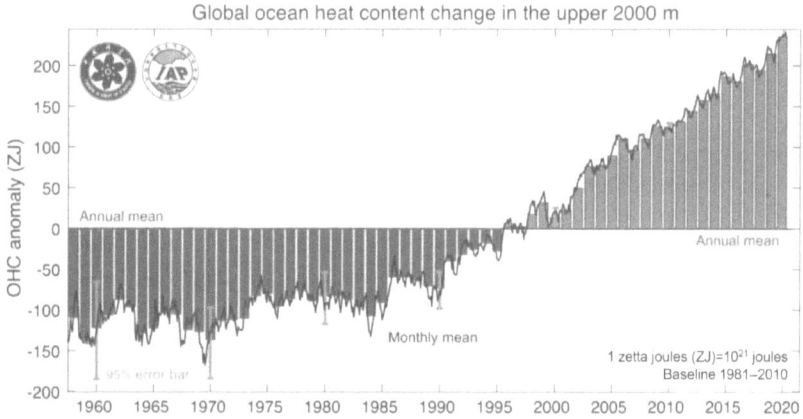

FIGURE 3.2 Ocean warming as a function of time (https://news.ucar.edu/132773/2020-was-record-breaking-year-ocean-heat).

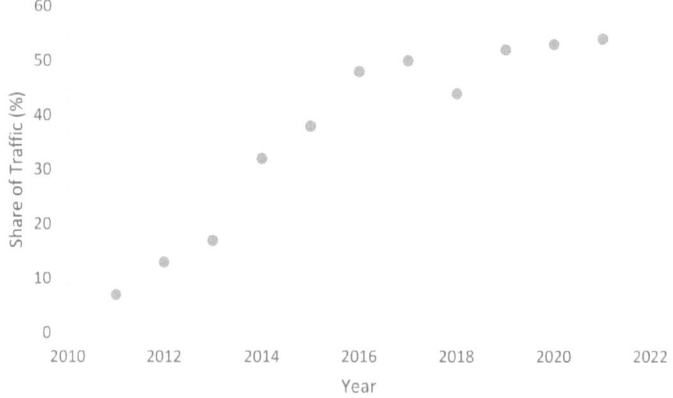

FIGURE 3.3 Smartphone usage for web access as a function of time (https://www.oberlo.com/statistics/mobile-internet-traffic).

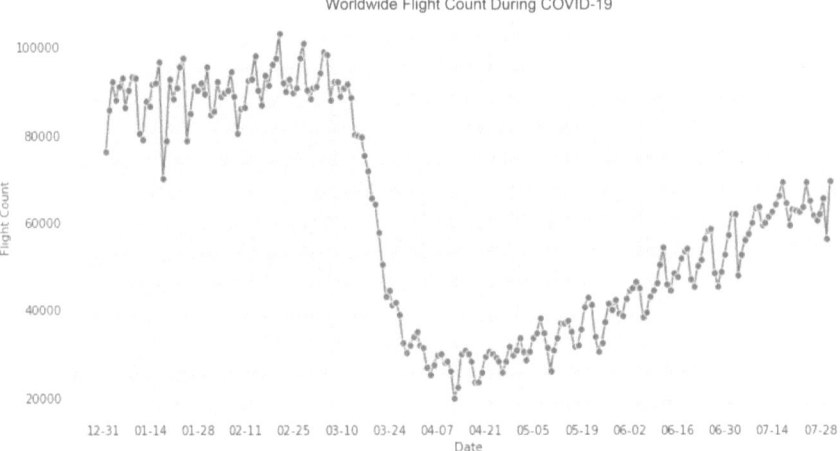

FIGURE 3.4   Air traffic volume during the COVID-19 pandemic.

process deliverable, lengthening of the process may be justified. But in the case of duration, process intensification is more commonly described as achieving the same end result in less time. This is frequently achieved by varying temperature and pressure, as described in previous sections, or by applying external fields that propel the process forward. So, what are some of these external fields? We will see this in the next few chapters.

PIF/CIF SUMMARY

In the case of temperature, pressure, and time variations for the purpose of process intensification, it is important to quantify the PIF and CIF objectively. First, the desired outcome of the process must be identified, along with its baseline (pre-intensification) value—e.g., reactant-to-product conversion in a given time. Then, its intensified value must be measured. The ratio will yield the PIF as a function of various process parameters whose effect is being assessed.

Similarly, costs corresponding to the baseline and intensified processes must be estimated, and their ratio taken to be CIF, again as a function of process variables under investigation. For example, if the temperature is being leveraged as the primary process intensifier, the differential cost to bring the process from the baseline temperature to the "intensified" temperature must be evaluated. In rare cases, process intensification may even be achieved with an associated cost reduction, but this requires some creative out-of-box thinking. It may not even be feasible in many instances.

The ratio (PIF/CIF) can then be taken as the objective function to be optimized, with suitable weighting factors and relevant constraints applied.

EXERCISE QUESTIONS

1. An argument advanced in favor of veganism is that cooking meat requires much higher energy than that required by plant-based products. In general, slow cooking at a lower temperature preserves flavor and nutrients, whereas higher temperatures do speed up the process. Closing the vessel while cooking is reported to reduce energy consumption by up to 30%. Please combine these considerations into the formulation of a strategy to cook meat with a minimum impact on the environment. Or, pledge to turn vegan!

2. Pressure is harder and more expensive to measure and control. Both low- and high-pressure operations involve considerable complexity and associated risk. However, there are applications where pressure is the most suitable "process intensifier". A good example is supercritical fluid processing, used for applications ranging from the decaffeination of coffee beans to the removal of winding oil from actuator coils. Extraction conditions for supercritical carbon dioxide are above the critical temperature of 31°C and critical pressure of 74 bar. The properties of the supercritical fluid can be altered by varying the pressure and temperature, allowing selective extraction. For example, volatile oils can be extracted from a plant with low pressures (100 bar). Pressure vessels are a process requirement, given the need to "dial" the pressure as needed.

   Extraction is a diffusion-based process, in which the solvent is required to diffuse into the matrix and the extracted material to diffuse out of the matrix into the solvent. Diffusivities are much faster in supercritical fluids than in liquids, and therefore extraction can occur faster. In addition, due to the lack of surface tension and negligible viscosities compared to liquids, the solvent can penetrate more into the matrix inaccessible to liquids. An extraction using an organic liquid may take several hours, whereas supercritical fluid extraction can be completed in 10–60 minutes. Given this scenario, how would you justify the use of supercritical fluid extraction if you were a Process Engineer in a decaffeination plant? When would the added cost and complexity make sense?

3. While all processes are ultimately time-dependent, quasi-steady conditions can certainly be achieved for long periods. The associated stability is a key requirement for successful and profitable manufacturing. Process equipment deterioration with time is a real concern for manufacturers, and is one phenomenon that can accelerate in a non-linear manner, especially as warranties expire! Tool wear is a contributor to product quality loss and cost as real-time maintenance and adjustments on the fly become increasingly necessary. Mating surfaces become rougher, which can lead to particle generation, material loss, and other issues which can affect productivity and throughput. Modeling tool wear and tool life to formulate effective maintenance and replacement strategies is a critical activity in the industry. The model must accurately capture the effect of time in accelerating the wear mechanism,

and also build in the cost for mitigating the effects of wear. For a typical manufacturing process, formulate such a model and examine its implications for quality and cost.

REFERENCES

Kashchiev, D. and G.M. van Rosmalen, "Effect of Pressure on Nucleation in Bulk Solutions and Solutions in Pores and Droplets", *J. Coll. Interf. Sci.*, vol. 169, no. 1, 1995, pp. 214–219.

Rosner, D.E., *Transport Processes in Chemically Reacting Flow Systems*, Butterworths, 1986.

World Economic Forum, 2020, https://www.weforum.org/agenda/2020/04/infographic-covid19-air-traffic-aviation-coronavirus-international-travel/.

4 Process Intensification Driven by External Fields

Intensification of a process requires an impetus, whether it be in the form of temperature, pressure, or time, or by means of an externally applied field. Examples of such fields are:

- Gravitational
- Thermal
- Electrical
- Magnetic
- Acoustic, etc.

This may be illustrated by taking the case of convective mass transfer of a solute in a carrier fluid. It could be entirely due to the motion of the fluid, or may also be associated with "drift" induced by an external field (Rosner, 1986).

The convective mass flux of species i, $(\mathring{\mathbf{m}}_i'')_{\mathrm{conv}}$ (where bold denotes a vector, ° denotes per unit time, and "denotes per unit area) may be related to the carrier fluid mass flux, $\mathring{\mathbf{m}}''$, via the local mass fraction ω_i as:

$$(\mathring{\mathbf{m}}_i'')_{\mathrm{conv}} = \omega_i \, \mathring{\mathbf{m}}'' \tag{4.1}$$

Here, $\mathring{\mathbf{m}}''$ is related to the prevailing fluid velocity, $\mathbf{v}$ and the fluid density, ρ, as:

$$\mathring{\mathbf{m}}'' = \rho \mathbf{v} \tag{4.2}$$

If an applied force then acts on the fluid, it can induce an additional solute velocity of $\mathbf{c}_i$, and a corresponding mass flux $(\mathring{\mathbf{m}}_i'')_{\mathrm{drift}}$, which may be defined as:

$$(\mathring{\mathbf{m}}_i'')_{\mathrm{drift}} = \rho_i \mathbf{c}_i \tag{4.3}$$

Thus, in the absence of a diffusive contribution, the total "intensified" mass flux of species i may be written as:

$$\mathring{\mathbf{m}}_i'' = (\mathring{\mathbf{m}}_i'')_{\mathrm{conv}} + (\mathring{\mathbf{m}}_i'')_{\mathrm{drift}} = \rho_i \mathbf{v} + \rho_i.\mathbf{c}_i = \rho_i \left(\mathbf{v} + \mathbf{c}_i \right) \tag{4.4}$$

Therefore, the ratio $(\mathbf{c}_i/\mathbf{v})$ yields an insight into the augmentation in mass flux attributable to the external field, and may be viewed as the Process Intensification Factor (PIF) in this context.

DOI: 10.1201/9781003283423-4

The drift velocity, c_i may be related to the acceleration acting on it, g_i as:

$$c_i = m_i g_i / f_i \qquad (4.5)$$

where m_i is the mass of molecule or particle i, and f_i represents a prevailing friction coefficient (or inverse of mobility) relating the drag force to the local drift velocity. c_i will clearly vary depending on the applied force field. For example, when the field is gravitational, it will become the "settling" or "sedimentational" velocity, denoted by $c_{i,s}$. When the field is electrostatic, "electrophoretic migration" takes place, with a velocity denoted as $c_{i,e}$. This is an important parameter in the case of charged particles in field-containing environments—e.g., fly-ash removal in an electrostatic precipitator.

Thermophoresis, associated with the drift of heavy molecules and particles in a temperature gradient, has a characteristic velocity denoted by $c_{i,T}$. Many aerosol systems—such as soot and ash particles depositing on cooled heat-exchanger surfaces—show a significant extent of thermophoretically driven mass transport.

Other drift mechanisms include diffusiophoresis, defined as the migration of a colloidal particle in a solution in response to the macroscopic concentration gradient of a molecular solute that interacts with the surface of the particle; osmophoresis, in which the particle is a vesicle composed of a body of fluid surrounded by a semipermeable membrane and the driving force for the migration is the osmotic pressure gradient along the particle surface caused by the solute concentration gradient prescribed in the ambient solution (Li, 2008); and photophoresis, caused by the transfer of photon momentum to a particle by refraction and reflection (Ashkin, 2000). In each case, the ratio of the corresponding drift velocity to the prevailing fluid velocity will serve as an indicator of the mass transfer intensification effect.

Another example of phoretic augmentation is "Stefan flow" (Rosner, 1986). In non-dilute systems, the mass transfer process itself can rise to convection normal to the surface. If chemical species A is being transferred into a carrier fluid B with a mainstream concentration of A of $\omega_{A,\infty}$, and if B does not penetrate the surface, a convective flow called "Stefan Flow" or "Stefan Blowing" is induced to balance the diffusional influx of species B. The blowing parameter, B_m, may then be estimated as (Rosner, 1986):

$$B_m = v_w \delta_m / D_A = \left(\omega_{A,w} - \omega_{A,\infty}\right)/\left(1 - \omega_{A,w}\right) \qquad (4.6)$$

Here, v_w is the fluid velocity at the interface, δ_m is the thickness of the mass transfer boundary layer under Stefan flow conditions, D_A is the Fick diffusion coefficient of the solute, and $\omega_{A,w}$ is the solute concentration at the interface with the surface.

If $B_m \ll 1$, as is the case in dilute systems where both $\omega_{A,w}$ and $\omega_{A,\infty}$ are small compared to unity, Stefan flow is negligible. However, if $\omega_{A,w} \neq \omega_{A,\infty}$, and $\omega_{A,w} \rightarrow 1$, this flow becomes significant (e.g., at surface temperatures approaching the boiling point of a liquid surface).

It is well known that in the case of diffusional transport from a sphere under completely quiescent conditions, the mass-transfer Nusselt number, $Nu_m = 2$. However, if B_m is not negligible, the Nusselt number then becomes:

$$Nu_m = 2\left[\ln\left(1 + B_m\right) / B_m\right] \tag{4.7}$$

where the "reduction factor" due to Stefan blowing is given by:

$$F_m = \ln\left(1 + B_m\right) / B_m \tag{4.8}$$

In the case of condensation, v_w, and hence B_m, are negative, resulting in an enhancement of Nu_m due to the associated "suction".

Heat transfer will be affected in the same way by Stefan flow, and the corresponding B_h may be expressed as:

$$B_h = v_w \delta_h / \alpha \tag{4.9}$$

where δ_h is the thickness of the heat transfer boundary layer under Stefan flow conditions and α is the thermal diffusivity ($k/\rho c_p$) with k being the thermal conductivity of the surrounding fluid, ρ its density, and c_p the fluid heat capacity at constant pressure.

Among external energy sources that have shown significant process intensification effects in practice, acoustic, microwave, and laser-based technologies are prominent. A more detailed treatment of these is presented in subsequent chapters. In particular, high-frequency acoustic fields coupled with liquids produce extraordinary PI consequences, and merit a relatively more detailed treatment in the remainder of this tome, starting with the next chapter which outlines the prevailing phenomena.

EXERCISE

The following papers illustrate applications of Stefan flow (Latiff et al., 2016), thermophoresis (Mensch and Cleary, 2017) and photophoresis (Krauss and Wurm, 2005), respectively. In each case, estimate the associated PIF. Is there sufficient information available in these papers to estimate the Cost Impact Factors as well? If so, estimate CIF as well.

REFERENCES

Ashkin, A., "History of Optical Trapping and Manipulation of Small-Neutral Particle, Atoms, and Molecules", *IEEE J. Select. Top. Quant. Electron.*, vol. 6, 2000, pp. 841–856.

Krauss, O. and G. Wurm, "Photophoresis and the Pile-up of Dust in Young Circumstellar Disks", *Astrophys. J.*, vol. 630, no. 2, 2005, pp. 1088–1092.

Latiff, N.A., M.J. Uddin, and A.I.M. Ismail, "Stefan Blowing Effect on Bioconvective Flow of Nanofluid Over a Solid Rotating Stretchable Disk", *Propuls. Power Res.*, vol. 5, no. 4, December 2016, pp. 267–278.

Li, D., *Encyclopedia of Microfluidics and Nanofluidics*, Springer, 2008.

Mensch, A. and T.G. Cleary, "Quantifying Thermophoretic Deposition of Soot on Surfaces", Suppression, Detection and Signaling Research and Applications Conference, College Park, MD, 2017 [online], https://tsapps.nist.gov/publication/get_pdf.cfm?pub_id= 923392.

Rosner, D.E., *Transport Processes in Chemically Reacting Flow Systems*, Butterworths, Stoneham, MA, 1986, pp. 311–314.

5 Acoustic Intensification of Processes— Mechanisms Involved

Application of ultrasound on liquids such as water leads to two types of physical mechanisms: acoustic cavitation and acoustic streaming. The former refers to the implosion of bubbles resulting in a shock wave that travels in all directions, while the latter is described by a flow that is unidirectional in nature along the normal to the transducer in the ultrasonic equipment. When a medium is exposed to sound waves of frequency above 18 kHz (Suslick, 1988), it results in the propagation of oscillating compression and rarefaction cycles. When a local pressure gradient develops due to a difference in ambient pressure and vapor pressure of the liquid, vacuum bubbles are formed and implode (Ginsburg et al., 1998). The associated local increases in temperature and pressure are reported to be up to 5000 K and 1000 atm, respectively.

When the ambient pressure of a liquid is reduced adequately, it can boil without heating. This principle is used in an ultrasonic tank where the reduction of local pressure results in formation and implosion of bubbles that grow in size till they can no longer sustain themselves, leading to a violent bubble collapse which causes the release of shock waves that traverse the medium. Cavitation can be stable (occurs at lower intensities where the bubbles oscillate about their mean position) or transient (occurs at higher intensities where the bubbles grow and collapse violently). As rarefaction and compression occur in alternating cycles, bubbles form and grow during the former and collapse in the latter. As the frequency increases, the number of bubbles formed increases correspondingly (Suslick et al., 1999). However, the size of the bubble decreases with increasing frequency and leads to reduced intensity of cavitation (which scales as the volume of the bubble prior to implosion) as the frequency increases. This explains the pronounced effect of cavitation intensity at lower frequencies and its rapid weakening at frequencies approaching the Megahertz range (Figure 5.1). The acoustic streaming velocity, on the other hand, scales as the square of frequency (Figure 5.2), and its effect intensifies in the higher range of ultrasonic frequency (400 kHz and above)—the so-called transonic and megasonic ranges.

The net effect of cavitational and streaming forces is plotted in Figure 5.3 as a function of frequency. It indicates that a minimum is reached in the 100–150 kHz range, where neither effect is substantial.

Cavitation intensity depends on a number of factors. While the force of cavitation is inversely proportional to the cube of frequency, other properties of the medium such as surface tension, vapor pressure, temperature, viscosity, and density play a key role as well. For example, altering the temperature results in a change in ambient pressure and the gradient developed becomes higher leading to higher cavitation

DOI: 10.1201/9781003283423-5

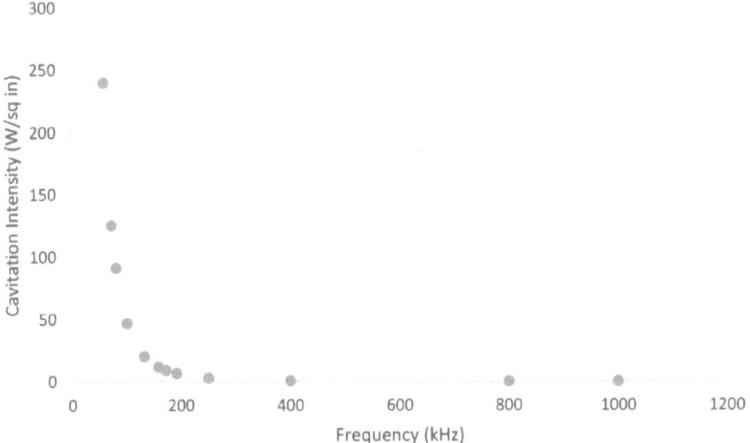

FIGURE 5.1   Effect of ultrasonic frequency on cavitation intensity in coupled fluid.

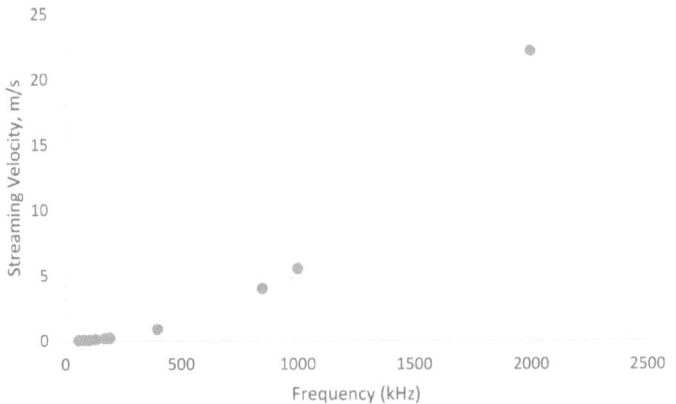

FIGURE 5.2   Effect of ultrasonic frequency on streaming velocity in coupled fluid.

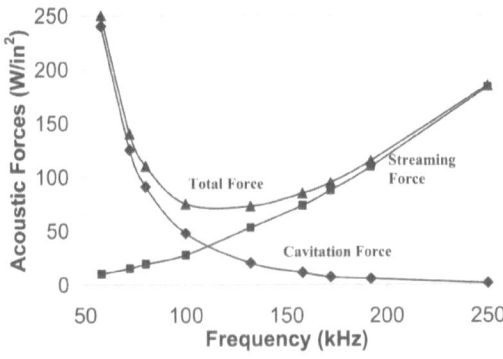

FIGURE 5.3   Acoustic force as a function of frequency.

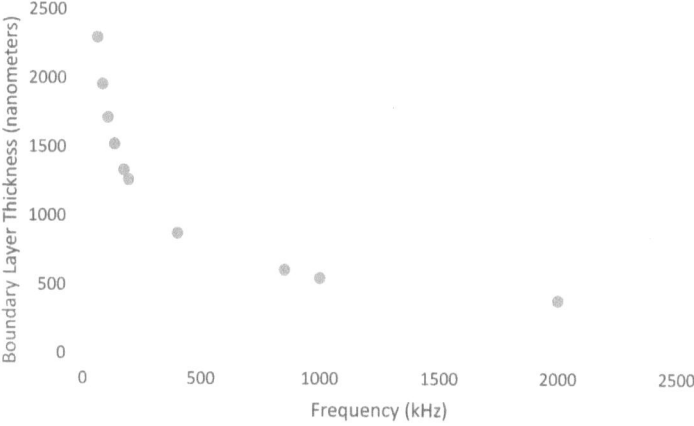

FIGURE 5.4 Effect of ultrasonic frequency on boundary layer thickness on immersed surface.

effects. Similarly, lowering the viscosity leads to a better response of the liquid to bubble growth and collapse, and increases cavitation effects in turn. Each of these properties can be altered depending on the medium used for bubble growth and collapse.

In the megasonic frequency range, acoustic streaming dominates, with a velocity that scales as f^2, even as cavitation decays as $1/f^3$. In addition, the thickness of the momentum-transfer boundary layer adjacent to the immersed surface scales as $1/f^{0.5}$ (Bakhtari et al., 2006). Megasonic frequencies reduce the boundary layer to less than 0.5 micrometers (μm) deep, compared to the boundary layer of 2.5 μm produced by lower ultrasonic frequencies (Figure 5.4). With a thinner boundary layer, more particles are exposed to the bulk fluid removal forces. This is especially important for detaching particles in the submicron range and for exposing smaller surface features to fresh chemistry and fluid flow forces.

Although ultrasound has been used for several decades in diverse fields to intensify processes (Laborde et al., 2000), the pace of applications has accelerated in recent times. Cavitation can have two types of effects on a chemical reaction: it can either accelerate the reaction or promote reactions that would not have occurred otherwise. The latter happens due to the dissociation of sonicated water into hydrogen ions and hydroxyl radicals which, in turn, leads to multiple back reactions and formation of new products.

Ultrasound behaves differently in liquid and liquid-solid media compared to gas medium. Ultrasound in an aqueous medium produces highly reactive species such as OH radicals, H_2O_2, and ozone which are strong oxidizing agents of high oxidation potential (2.8, 1.8, and 2.1 volts, respectively). These radicals are capable of initiating and enhancing oxidation and reduction reactions. Oxidation occurring due to ultrasound is called "Advanced Oxidation Process" (AOP). Lindstrom and Lamm (1951) first suggested the mechanism for this reaction, followed by many researchers who proved it in different ways through experiments.

Webster (1963) explained the effect of cavitation on liquid phase reactions as follows: Two classes of chemical effects are induced by ultrasonic cavitation. The first is the acceleration of reactions, and the second is the initiation of two reactions that would not otherwise occur; this takes place predominantly in an aqueous medium. Under the action of cavitation, water decomposes into hydrogen ions and OH radicals.

$$H_2O \rightarrow H^+ + OH^- \tag{5.1}$$

The predominant back reactions attendant to this process are:

$$OH + OH \rightarrow H_2O_2 \tag{5.2}$$

$$H + H \rightarrow H_2 \tag{5.3}$$

The products of these reactions are then responsible for secondary reactions involving dissolved substances. The reacting ions or molecules will be selectively subjected to reduction or oxidation according to their properties and structure. The oxidation of dissolved substances is detectable in the absence of dissolved oxygen. In its presence, the rate of formation of hydrogen peroxide is increased, with a consequent increase in the rate of oxidation; this effect has been attributed to the occurrence of the reaction

$$H + O_2 \rightarrow HO_2 \tag{5.4}$$

followed by

$$HO_2 + HO_2 \rightarrow H_2O_2 + O_2 \tag{5.5}$$

The mechanical effects associated with cavitation are also of significance in a number of systems. The formation and collapse of bubbles at a surface lead to a sudden inflow of liquid at the surface, and since this inflow occurs at a high local temperature, pressure, and velocity, it is an important mechanism in ultrasonic surface cleaning applications. Cavitation has also found abundant applications in wastewater treatment, rupture of cells, synthesis of nanoparticles, etc. While some processes may be highly sensitive to cavitation, others might need a combination of cavitation and acoustic streaming to demonstrate noteworthy intensification effects. Given these distinct and unique characteristics of high-frequency acoustic fields, the challenge is to deploy them in an effective and cost-effective manner. The next sections will briefly outline applications where acoustic intensification has been carried out with success. These are described in greater detail in subsequent chapters.

5.1 SURFACE CLEANING

Particle contamination has a significant effect on in-process yields, final outgoing quality, and reliability in customer applications of many microelectronic devices. In integrated circuit manufacture, particulate-related defects include circuit damage, pattern disruption, bridging of separated features, shorting of conductors, masking of areas that are supposed to be etched, and unmasking of areas that are not supposed to be etched (Cooper, 1986). According to a common rule of thumb, in order to cause physical and mechanical defects, a particle must be at least as large as one-tenth the minimum circuit feature size—0.1 μm in the case of 1 μm feature sizes (Fisher, 1990). However, even smaller particles can cause "killer" defects via chemical interactions or interactions with thin layers. Crystal-originated particles are known to cause a local reduction of the dielectric breakdown voltage of the insulating silicon dioxide (SiO_2) layer in metal-oxide-semiconductor (MOS) structures (Huth et al., 2001). Complementary metal-oxide-semiconductor (CMOS) on silicon-on-insulator (SOI) offers the potential for a significant reduction in chip power without sacrificing performance but is severely affected by many types of particle defects (metallic, oxide, organic) in the 0.2–2 μm size range (Mendicino et al., 1995). In thin-film deposition processes such as sputtering and plasma-enhanced chemical vapor deposition, particle contamination causes direct yield loss or scrap due to point defects, as well as reliability failures by film failure during product use (Selwyn et al., 1998).

In hard disk drives (HDDs), where head flying heights over the disk have dropped to the order of a few nanometers, particulate contamination has always played a central role in determining the quality and reliability of the drive (Nebenzahl et al., 1998). Particles trapped between the head and disk can cause irrecoverable data losses in several ways—head crash, circumferential scratching, embedding on the disk, thermal asperities, etc. Particles captured on the air-bearing surfaces of the slider can alter the trajectory of the flying head, causing the head to move closer to the disk than designed (potential for hard contact) or farther from the disk (potential for loss of read/write signal), depending on how the flying dynamics are affected. Larger particles entrained in the airflow within the disk enclosure (DE) can damage the disks, the head, or both by high-velocity impact. As particle concentration within the DE increases, the durability of the head/disk interface decreases (Yoon and Bhushan, 2001) with harder and larger particles causing earlier failures. Hence, the HDD industry has historically emphasized thorough cleaning of the technology components (such as disk and slider) and the myriad mechanical components comprising a drive. As a 1997 survey by Nagarajan (1997) shows, sonication is the predominant mode of cleaning in the HDD industry, although that mode is used far more aggressively for the mechanical components than for the head and disk. The ceramic used in the magneto-restrictive (MR) head, and the glass substrate that is increasingly used for the data disk, are both fragile materials that can easily be damaged by high-energy ultrasonics. However, functional cleanliness requirements still mandate that submicron particles be removed completely from these surfaces, thus necessitating the use of higher frequencies that exhibit minimal cavitation. Megasonics, in combination with strong chemistry, is also the method of choice in cleaning semiconductor wafers (Kuehn et al., 1996). Ultrasonic cleaners are used to remove large particles from wafers after sawing, and to remove lapping and polishing compounds after these operations (Menon, 1990); however, the use of ultrasonic cleaners continues to be limited by surface-damage considerations.

5.2   HEAT TRANSFER ENHANCEMENT IN FURNACE TUBES

Dhanalakshmi (2013) studied the effect of ultrasound-assisted heat transfer enhancement in furnace tubes. This work focused on the heat transfer characteristics of water
flowing in a tube exposed to heat in the presence and absence of ultrasound. It was
found that the critical parameter in the process is the ratio of characteristic ultrasonic
field velocity and prevailing flow velocity. Though heat transfer enhancement due to
ultrasonic exposure depended very much on the flow conditions, i.e., the Reynolds
number, heat transfer enhancement rate (HTER) at 20 kHz frequency increased
till a Reynolds number of 500 was reached; it decreased as the Reynolds number
increased from 500 to 1200 and showed no significant change thereafter. This study
was done with three different frequencies—20, 33 kHz, and a combination of 20
and 33 kHz. While 20 kHz showed much better HTER compared to 33, 20+33 kHz
was slightly better than 20 kHz alone. This illustrates that while 20 kHz alone is
a first-order effect, using a combination of frequencies introduces a second-order
effect. The study concluded that the cavitation technique for heat transfer enhancement is ineffective for higher flow rates, but showed considerable improvement for
controlled laminar flow regimes.

5.3   DEGRADATION OF METHYL VIOLET

High-intensity, low-frequency ultrasound is widely used to augment sonochemical
reactions. The sonochemical hotspots formed due to cavitational implosion lead
to the disintegration of water molecules into H+ and OH- radicals. These recombine to form hydrogen peroxide, molecular hydrogen, and a number of other radicals and species. Methyl violet dye is released as industrial waste and is extremely
harmful to plant and aquatic life owing to its highly toxic characteristics. The main
aim of this study was to boost the decolorization of methyl violet using ultrasound
(Dhanalakshmi, 2013). The experiments were conducted in both tank-type and
probe-type sonicators, and utilized frequencies ranging from 20 kHz to 1 MHz. The
main steps involved the preparation of Fenton's reagent, mixing with dye dissolved
in water, and irradiating with ultrasound for 150 minutes. The results showed that in
the presence of Fenton's reagent, ultrasonic irradiation of the contaminated sample
led to the near-complete removal of dye from the water. The effect of frequency
is interesting to note because as the frequency increases, decolorization of water
is faster, and reaches near-completion although the rate increase is not significant
when the frequency exceeds 430 kHz. At higher frequencies, the cavitation effect
decreases and is dominated by acoustic streaming; the large density of bubbles leads
to the higher gas-liquid interfacial area and hence greater contact with OH radicals, resulting in higher reaction rates. The rate of decolorization also increased with
increasing power.

5.4   SONO-FRAGMENTATION OF COAL

This study (Dhanalakshmi, 2013), focused on the breakage of coal particles suspended in a solution. Frequencies spanning from low (<100 kHz), intermediate

(100–400 kHz), and high (>400 kHz) were studied. The ground coal was mixed with three different reagents (hydrogen peroxide, acetic acid, and ethanol) separately and sonicated at the chosen power and different frequencies. The results obtained showed that particle breakage was higher at lower frequencies, while particle agglomeration set in at higher frequencies. A combined effect of both phenomena was observed at intermediate frequencies. Particle breakage was found to be a function of both frequency and time. A longer time of sonication, up to 5 minutes, led to reduced particle size, and particle breakage gradually decreased between 5 to 10 minutes of ultrasonic exposure. Particle size reduction with respect to frequency followed the order: 25 > dual > 132 >192 >470 >1000 kHz.

5.5   DESTRATIFICATION

Thermal stratification refers to the variation in temperature at different levels of a stored fluid. This is a challenge posed by cryogenic liquids due to boiling off losses associated with handling, storage, and transport. The work reported by Jagannathan (2010) underscores the potential effect of ultrasound on thermal destratification. Experiments were initially conducted in a storage tank with water and later extended to cryogenic liquids. This was done by heating the top layer of a tank containing water while the bottom layer remained at room temperature. This system was irradiated with ultrasound and the temperature was measured at all levels creating a complete temperature profile along the height of the vessel at different points of time. A range of frequencies starting from 25 to 470 kHz was used to study the destratification effect. Of all the frequencies chosen, the highest frequency of 470 kHz provided maximum breakage of the thermal layer and destratification. When the power input and the storage pressure were increased, the 470 kHz system showed a drastic increase in the Stratification Index compared to other lower frequencies. However, a dual frequency mode of operation such as 58/192 kHz, which combines the effect of cavitation and streaming, provides better mixing as opposed to using a high-frequency tank in isolation.

5.6   EFFERVESCENT ATOMIZATION

When an effervescent gas is mixed in with a liquid inside a chamber in such a way that bubbles are formed inside a nozzle, it is known as "effervescent atomization". This is used in the spraying of fuel as liquid droplets in the combustor and leads to instantaneous formation of vapor. The objective of this study (Jagannathan et al., 2011) was to produce very fine droplets from an effervescent atomizer, with the assistance of ultrasound. Ultrasound was used to break the bubbles formed by the aerator and distribute their size homogeneously. Experiments were carried out with 20 kHz frequency and showed that ultrasound helped in breaking up larger-sized bubbles into smaller ones. At lower power (of less than 40% input power), bubbles undergo shape change without fragmentation. Although acoustic streaming plays a role in the elongation of the bubble in the direction of the streaming force, the main mechanism involved in this process of bubble breakup is cavitation, and the spray obtained is qualitatively superior when compared to traditional methods employed for effervescent atomization.

5.7 COAL BENEFICIATION

The high percentage of sulfur and ash content found in Indian coals makes them suitable for techniques such as ultrasonic coal wash. This method, employed at different frequencies to remove ash and sulfur from Indian coals, forms the basis of the study reported by Ambedkar et al. (2011a, 2011b). Along with different frequencies, both aqueous- and reagent-based ultrasonic coal-washing methods were explored; 20, 25, 58/192 kHz dual-mode, and 430 kHz frequencies were used for the study. While the 20 kHz frequency was transmitted via a probe-type sonicator, the rest of the equipment was tank-type. For de-ashing experiments, the sieved coal was sonicated for 15 minutes and subjected to three-level decantation using a fabricated settling column. Ash, being heavier, settles faster than coal. Three layers are then separated and analyzed for ash content. The de-sulfurization experiments involved sonication of the sample, filtration, washing, drying, and analysis as their main steps. For de-ashing, the lowest frequency gave the best size reduction owing to cavitational bubble collapse and subsequent breakage of coal particles. Higher frequencies showed lesser rupture of particles due to streaming being the dominant mechanism in such systems. However, the ash removal percentage was optimum at 132 kHz and this behavior may be attributed to the simultaneous presence of streaming and cavitation producing a balanced effect. For de-sulfurization of coal, the lowest frequency systems showed excellent contact of oxidizing agents produced by ultrasound with fine coal particles, leading to higher removal of pyritic and organic sulfur compared to the high-frequency systems. The rate of removal was, however, more than 90% in all cases, proving the viability of ultrasonic coal wash as an option for de-sulfurization of coal.

Oil agglomeration after ultrasonic treatment was performed to enhance coal quality (Sahinoglu and Uslu, 2013). The effect of using sunflower oil on coal after treating with ultrasound of 20 kHz frequency was studied. It was found that ultrasonic treatment before oil agglomeration led to significant removal of ash and pyritic sulfur, and also increased the calorific value of coal. Sonication also leads to a reduction of the size of coal particles and a decrease of elements such as oxygen, while the carbon content in coal increases.

Another study involving exposure of coal to low-frequency ultrasound to remove sulfur and ash revealed interesting results (Saikia et al., 2014). Coal was mixed with NaOH and KOH in 1:1 ratio and exposed to 20 kHz frequency for a period of three hours. This proved to be a promising technique for coal beneficiation. This process also produced semi-ultra-clean coal with an ash content of $\leq 0.1\%$ and ash removal of up to 87.52%.

In a more recent study (Balakrishnan et al., 2015), two different coals were subjected to chemical and aqueous leaching under the effect of ultrasound to remove alkali content. The results were compared with stirring the sample without applying ultrasound. For water washing, the use of ultrasound did not enhance the removal of alkali elements. However, when the sample was washed using ammonium acetate, it showed enhanced leaching efficiency when ultrasound was used. It was concluded that sonication may be a promising method to contain fouling in power plants.

Table 5.1 summarizes the observed intensification effects of ultrasonic fields on various physicochemical phenomena. These will be explored in greater detail in subsequent chapters.

Some of the key insights regarding the use of ultrasonic and megasonic fields for process intensification include:

- When an acoustic field is coupled to a liquid medium, two phenomena are induced—cavitation and acoustic streaming—and two effects are created—physical and chemical. The interaction between these can be quite complex. However, some general insights emerge from a cause-and-effect analysis.

TABLE 5.1
Consolidated Representation of Process Intensification under the Influence of Ultrasound

No.	Process	Procedure	Frequencies Investigated	Observations
a	Heat-transfer enhancement in furnace tubes	Heat transfer characteristics of water flowing in a tube exposed to heat were studied in the presence and absence of ultrasound. The critical parameter was found to be the ratio of ultrasonic flow velocity and prevailing flow velocity.	20, 33, and 20+33 kHz	Order of enhancement was as follows: 33 kHz > 20+33 kHz combined > 20 kHz. While 20 kHz alone was a first-order effect, combination of 20 kHz and 33 kHz was found to have a second-order effect.
b	Degradation of Methyl Violet	Decolorization of methyl violet was boosted using ultrasound. Preparing Fenton's reagent, mixing with dye and irradiating with ultrasound were the main steps involved.	25, 132, 430 kHz, and 1 MHz	Decolorization increases with increasing frequency and reaches near completion at 430 kHz. At higher frequencies, though cavitation effect is reduced, there is a higher density of bubbles leading to better contact with OH radicals and higher reaction rates.
c	Sono-fragmentation of coal	Ground coal was mixed with three different reagents and sonicated at selected frequencies. Results were analyzed quantitatively using a laser particle counter and qualitatively in a Scanning Electron Microscope.	25, 132, 58+192, 190, 470 kHz, and 1 MHz	Lower frequencies showed effective particle breakage while particle agglomeration was enhanced at higher frequencies. Particle size reduction with respect to frequency (kHz) followed the order: 25 > 58+192 dual > 132 >192 >470 >1000.

(Continued)

TABLE 5.1 (*Continued*)
Consolidated Representation of Process Intensification under the Influence of Ultrasound

No.	Process	Procedure	Frequencies Investigated	Observations
d	Destratification	The effect of ultrasound on thermal destratification was studied. A tank filled with water is heated at the top, while the bottom is maintained at room temperature. The system is exposed to ultrasound and the temperature profile is obtained.	25, 40, 58, 68, 132, 192, 58+192, 172+192, and 470 kHz	470 kHz provided maximum breakage of thermal layer and destratification. When the power input and the storage pressure were increased, 470 kHz system showed a dramatic increase in Stratification Index. Dual frequencies, which combine effect of cavitation and streaming, provided better mixing compared to high or low frequencies.
e	Effervescent atomization	Very fine droplets were generated from an effervescent atomizer with the assistance of ultrasound. Ultrasound was used to break the bubbles formed by the aerator and distribute them homogeneously.	20 kHz	At a lower frequency, ultrasound helped in breaking up of larger sized bubbles into smaller ones. The main mechanism involved in bubble break-up is cavitation, and the spray obtained is qualitatively superior.
f	Coal beneficiation	Aqueous and reagent-based ultrasonic coal wash was investigated. For de-ashing, the sieved coal was subjected to sonication followed by a three-level decantation. De-sulfurization was carried out by sonication, followed by titration for analysis.	20, 25, 58+192, and 430 kHz	For de-ashing, the lowest frequency gave the best size reduction owing to cavitational bubble collapse and subsequent breakage of coal particles. For de-sulfurization of coal, the low-frequency systems showed excellent contact between coal constituents and oxidizing agents produced by ultrasound.

Source: Srivalli (2019).

- When first-order physical effects are pronounced–for example, in particle breakage and in aerosolization of liquids–low-frequency cavitational effects are dominant.
- However, when second-order physical effects prevail, such as diffusion and leaching, the combined effect of cavitation and acoustic streaming is superior to that produced by either mechanism acting alone.
- Convection and mixing may be intensified by either cavitation or acoustic streaming since they operate by different but equally effective mechanisms. While cavitational fields literally "throw" parcels of fluid around, streaming induces a unidirectional flow of the fluid away from the transducer, which can set up convection rolls that assist redistribution.
- Higher frequencies lead to a greater density of smaller-sized bubbles, thus increasing net effective interfacial area. Chemical reactions that rely upon surface contact between reacting species and radicals are therefore maximally enhanced at such frequencies at the higher end of the ultrasonic spectrum (or lower end of the megasonic range).
- When particle agglomeration is desirable, such as in the separation of contaminants from effluents, higher frequencies are ideal for promoting it, while lower frequencies will actually break up agglomerates.
- Acoustic field effects on transport phenomena—momentum, mass, and heat—are very similar when the dominant mechanisms are the same.
- In the case of chemical reactions, while the localized high temperatures and high pressures associated with cavitation can increase local rates, frequency and length of contact between reacting species can be more directly affected by streaming effects. Thus, a frequency-blended field may be more appropriate.
- In general, streaming enhancement factors scale as square of frequency, while cavitational-enhancement effects scale as $1/f^3$. Both scale linearly with input power (amplitude) until a lower threshold value at which the field collapses. When both effects are exerting a combined influence, the frequency exponent will range from +2 to -3, depending on which influence is dominant.
- It is apparent that the use of ultrasound for process enhancement is quite effective in almost all the areas of application.

We will return to an examination of ultrasonic/megasonic process intensification in all its glory, but for now, let us change tracks and explore other external fields in the next chapter.

EXERCISE QUESTIONS

1. Hydrodynamic cavitation is defined by Wikipedia as "the process of vaporization, bubble generation and bubble implosion which occurs in a flowing liquid as a result of a decrease and subsequent increase in local pressure". It can also potentially cause severe damage (Dular, 2016; Ferrari, 2017; Raikwar and Jain, 2017), or be harnessed for positive impact (Moholkar

et al., 1999; Sun et al., 2020; Wang et al., 2021). Compare the mechanisms of acoustic and hydrodynamic cavitation, and review their implications for process intensification and cost impact.

2. While the acoustic systems deployed in the studies reported in this chapter had high energy inputs (in the 500 W–1 kW range), many applications—ranging from dental cleaners to benchtop nebulizers to medical imaging—work with lower power input. Take these three as examples, and analyze the process intensification and cost impact aspects.

REFERENCES

Ambedkar, B., T.N. Chintala, R. Nagarajan, and S. Jayanti, "Feasibility of Using Ultrasound-Assisted Process for Sulfur and Ash Removal from Coal", *Chem. Eng. Process. Process Intensif.*, vol. 50, no. 3, 2011a, pp. 236–246 (b).

Ambedkar, B., R. Nagarajan, and S. Jayanti, "Ultrasonic Coal-Wash for De-Sulfurization", *Ultrason. Sonochem.*, vol. 18, no. 3, 2011b, pp. 718–726 (a).

Bakhtari, K., R.O. Guldiken, P. Makaram, A.A. Busnaina, and J.-G. Park, "Experimental and Numerical Investigation of Nanoparticle Removal Using Acoustic Streaming and the Effect of Time", *J. Electrochem. Soc.*, vol. 153, no. 9, 2006, pp. G846–G850.

Balakrishnan, S., V.M. Reddy, and R. Nagarajan, "Ultrasonic Coal Washing to Leach Alkali Elements from Coals", *Ultrason. Sonochem.*, vol. 27, 2015, pp. 235–240.

Cooper, D.W., "Particulate Contamination and Microelectronics Manufacturing: An Introduction", *Aerosol Sci. Technol.*, vol. 5, no. 3, 1986, pp. 287–299.

Dhanalakshmi, N.P., "Process Intensification via Power Ultrasound—Unifying Principles", Ph.D. Thesis, Indian Institute of Technology Madras, 2013.

Dular, M., "Hydrodynamic Cavitation Damage in Water at Elevated Temperatures", 16th International Symposium on Transport Phenomena and Dynamics of Rotating Machinery, April 2016, Honolulu, United States. ffhal-01879380.

Ferrari, A., "Fluid Dynamics of Acoustic and Hydrodynamic Cavitation in Hydraulic Power Systems", *Proc. R. Soc. A*, vol. 473, 20160345. http://dx.doi.org/10.1098/rspa.2016.0345, 2017.

Fisher, W.G., "Particle Interaction with Integrated Circuits", In *Particle Control for Semiconductor Manufacturing*, ed. R.P. Donovan, Marcel Dekker, Inc, New York, NY, 1990, pp. 1–8.

Ginsburg, E., M.D. Kinsley, and A. Quitral, "The Power of Ultrasound", *Adm. Radiol. J.*, vol. 17, no. 5, 1998, pp. 17–20.

Huth, S., O. Breitenstein, A. Huber, D. Dantz, and U. Lambert, "Localization and Detailed Investigation of Gate Oxide Integrity Defects in Silicon MOS Structures", *Microelectron Eng.*, vol. 59, 2001, pp. 109–113.

Jagannathan, T.K., R. Nagarajan, and K. Ramamurthi, "Effect of Ultrasound on Bubble Breakup Within the Mixing Chamber of an Effervescent Atomizer", *Chem. Eng. Proc.*, vol. 50, 2011, pp. 305–315.

Jagannathan, T.K., "Mechanistic Modeling, Visualization and Applications of Flow Induced by High-Intensity, High-Frequency Acoustic Field", Ph.D. Thesis, Indian Institute of Technology Madras, 2010.

Kuehn, T.H., D.B. Kittelson, Y. Wu, and R. Gouk, "Particle Removal from Semiconductor Surfaces by Megasonic Cleaning", *J. Aerosol Sci.*, vol. 27, Supplement 1, 1996, pp. S427–428.

Laborde, A., J.L. Hita, A. Caltagirone, and J.P. Gerard, "Fluid Dynamics Phenomena Induced by Power Ultrasounds", *Ultrasonics*, vol. 38, 2000, pp. 297–300.

Lindstrom, O. and O. Lamm, "The Chemical Effects Produced by Ultrasonic Waves", *J. Phys. Chem.*, vol. 55, 1951, pp. 1139–1146.

Mendicino, M.A., P.K. Vasudev, P. Maillot, C. Hoener, J. Baylis, J. Bennett, T. Boden, S. Jackett, K. Huffman, and M. Goodwin, "Silicon-on-Insulator Material Qualification for Low-Power Complementary Metal-Oxide Semiconductor Application", *Thin Solid Films*, vol. 270, 1995, pp. 578–583.

Menon, V.B., "Particle Adhesion to Surfaces: Theory of Cleaning", In *Particle Control for Semiconductor Manufacturing*, ed. R.P. Donovan, Marcel Dekker, Inc, New York, NY, 1990, pp. 359–382.

Moholkar, V.S., P.S. Kumar, and A.B. Pandit, "Hydrodynamic Cavitation for Sonochemical Effects", *Ultrasonics Sonochem.*, vol. 6, no. 1–2, March 1999, pp. 53–65.

Nagarajan, R., "Survey of Cleaning and Cleanliness Measurement in Disk Drive Manufacture", *Precision Clean. J.*, February 1997, pp. 13–22.

Nebenzahl, L., R. Nagarajan, L. Volpe, J.S. Wong, and O. Melroy, "Chemical Integration and Contamination Control in Hard Disk Drive Manufacturing", *J. IEST*, vol. 41 (September–October), 1998, pp. 31–35.

Raikwar, A.S. and A. Jain, "A Review Paper on Hydrodynamic Cavitation", *Int. J. Eng. Sci. Comp.*, vol. 7, no. 4, 2017, pp. 10296–10299.

Sahinoglu, E. and T. Uslu, "Increasing Coal Quality by Oil Agglomeration after Ultrasonic Treatment", *Fuel Process. Technol.*, vol. 116, 2013, pp. 332–338.

Saikia, B.K., A.M. Dutta, L. Saikia, S. Ahmed, and B.P. Baruah, "Ultrasonic Assisted Cleaning of High Sulphur Indian Coals in Water and Mixed Alkali", *Fuel Process. Technol.*, vol. 123, 2014, pp. 107–113.

Selwyn, G.S., C.A. Weiss, F. Sequeda, and C. Huang, "In-Situ Analysis of Particle Contamination in Magnetron Sputtering Processes", *Thin Solid Films*, vol. 317, 1998, pp. 85–92.

Srivalli, H., "Mechanistic Study of Leaching of Sodium and Potassium from an Indian Coal Using Ultrasonic and Additive Methods", Ph.D. Thesis, Indian Institute of Technology Madras, 2019.

Sun, X., S. Chen, J. Liu, S. Zhao, and J.Y. Yoon, "Hydrodynamic Cavitation: A Promising Technology for Industrial-Scale Synthesis of Nanomaterials", *Front. Chem.*, 15 April 2020, Sec. Nanoscience, https://doi.org/10.3389/fchem.2020.00259.

Suslick, K.S., "Ultrasound: Its Chemical, Physical and Biological Effects", VCH New York, 1988.

Suslick, K.S., Y. Didenko, M.M. Fang, T. Hyeon, K.J. Kolbeck, and W.B. McNamara, "Acoustic Cavitation and Its Chemical Consequences", *Phil. Trans. R. Soc. Lond. A*, vol. 357, 1999, pp. 335–353.

Wang, B., H. Su, and B. Zhang, "Hydrodynamic Cavitation as a Promising Route for Wastewater Treatment – A Review", *Chem. Eng. J.*, vol. 412, 15 May 2021, p. 128685.

Webster, E., "Cavitation", *Ultrasonics*, vol. 1, 1963, pp. 39–48.

Yoon, E.-S. and B. Bhushan, "Effect of Particulate Concentration, Materials and Size on the Friction and Wear of a Negative-Pressure Picoslider Flying on a Laser-Textured Disk", *Wear*, vol. 247, 2001, pp. 180–190.

6 Process Intensification Fields

Case Studies

6.1 MICROWAVE

Microwave technology utilizes electromagnetic radiation with wavelengths that vary from as short as 1 mm to as long as 1 m. The microwave electromagnetic spectrum ranges between infrared and radio frequency, leading to a frequency range from 300 MHz to 30 GHz. In comparison with conventional heating, microwave heating has several advantages, which make it an attractive proposition, e.g., instantaneous starting and stopping and the generation of rapid heating within the material (Ethaib et al., 2020a). Many microwave-assisted organic reactions are accelerated as a result of this rapid heating compared to that attained using conventional methods. In shorter reaction times, higher yields and selectivity of the target compounds can be achieved. Moreover, there is no direct contact between the energy source and the reactants as microwaves travel through the vessel wall. Usually, this vessel wall is almost transparent to microwaves, enabling direct interaction with reaction mixture components (Ethaib et al., 2017; Quitain et al., 2013).

6.1.1 Heating and Drying (Jones, Marion Mixers)

Microwave technology is an extremely efficient but under-utilized energy source for process drying. Applications where microwave processes prove beneficial include: dehydration, sterilization, pasteurization, tempering (thawing), blanching, and cooking. In dehydration, the main purpose is to remove water. With pasteurization and sterilization, microwave systems are designed to raise the product temperature to a certain level to destroy pathogens. Other applications include blanching where the product is heated then cooled rapidly or then maintained at an elevated temperature, as in cooking, tempering, or sintering. Microwave drying—especially when combined with agitation—is capable of reducing process time by over 50% in some cases and poses significantly less operator risk.

Microwave energy, alone or in combination with conventional energy sources, makes it possible to control the drying process more precisely to obtain greater yields and better quality products in the shortest possible time. The mechanism for drying with microwave energy is quite different from that of conventional drying. In conventional drying, heat is transferred to the surface of the material by conduction, convection, or radiation and into the interior of the material by thermal conduction. Moisture is initially flashed off from the surface and the remaining water diffuses to the surface. This is often a slow process and it is diffusion-rate limited, requiring high external

DOI: 10.1201/9781003283423-6

temperatures to generate the required temperature differences. The processing time is limited by the rate of the heat flow into the body of the material from the surface as determined by its specific heat, thermal conductivity, density, and viscosity.

Microwaves are not forms of heat but rather forms of energy that are manifested as heat through their interaction with materials. Microwaves initially excite the outer layers of molecules. The inner part of the material is warmed as heat travels from the outer layers inward. Most of the moisture is vaporized before leaving the material. If the material is very wet and the pressure inside rises rapidly, the liquid will be removed from the material due to the difference in pressure. This creates a pumping action forcing the liquid to the surface, often as vapor. This results in very rapid drying without the need to overheat.

The superior control offered by microwave systems also contributes to higher quality by making product temperature much easier to regulate. Microwaves can be used about 50% more efficiently than conventional systems, resulting in major energy cost savings. Thus, microwave technology has become a game-changer in heating and drying applications.

6.1.2 Applications in Food Industry (Shaheen et al., 2012)

In baking, an enhanced throughput is achieved by microwave acceleration where the additional space needed for microwave power generators is negligible. Microwaves in baking are used in combination with conventional or infrared surface baking; this avoids the problem of the lack of crust formation and surface browning. The main use of microwaves in the baking industry today is in the finishing when the low heat conductivity can lead to considerably higher baking times in the conventional process. Another process that also can be accelerated by the application of microwave heating is (pre-) cooking. Microwaves render the fat and coagulate the proteins by an increased temperature.

Thawing and tempering have received much less attention in the literature than most other food processing operations, but these can also be intensified by microwave technology. Conventional thawing and tempering systems supply heat to the surface and then rely on conduction to transfer that heat into the center of the product. Microwaves use electromagnetic radiation to generate heat within the food, which is more time-efficient and cost-effective. The widespread use of microwave-assisted pasteurization and sterilization has been motivated by the fast and effective microwave heating of many foods containing water or salts.

Blanching is an important step in the industrial processing of fruits and vegetables. It consists of a thermal process that can be performed by immersing vegetables in hot water (88–99°C, the most common method), hot and boiling solutions containing acids and/or salts, steam, or microwaves. Blanching is carried out before freezing, frying, drying, and canning. The main purpose of this process is to inactivate the enzyme systems that may cause color, flavor, and textural changes. Some of the advantages of microwave blanching compared with conventional heating methods include the speed of operation and the requirement of no additional water. Hence, there is lower leaching of vitamins and other soluble nutrients, and the generation of wastewater is eliminated or greatly reduced. Microwave blanching enables a finish blanching of the center sections more quickly and without being affected by thick or non-uniform sections.

Uniformity is also accomplished in microwave ovens of the continuous tunnel type in contrast to the customary non-uniformity in institutional or domestic ovens.

6.1.3 WASTE TREATMENT UNDER MICROWAVE IRRADIATION

Many industrial activities involve the creation and subsequent disposal of waste, which represents a considerable cost in terms of money and pollution. Moreover, sometimes waste materials are hazardous as well, e.g., materials containing asbestos or byproducts of a nuclear plant. The disposal of waste materials is now becoming a very serious problem since, in recent years, the great increase in their production has not been matched by a corresponding rise in the number of authorized dumps. Moreover, the existing regulation does not always allow all kinds of waste material to be recycled, especially if harmful or hazardous materials are involved. It is, therefore, necessary to study and develop alternative ways to treat and re-use the components of waste materials, for instance by converting them into secondary raw materials and, if possible, restoring them to accomplish the task they were initially meant for.

Waste, even if originated by the same manufacturing process, and thus belonging to the same category (e.g., ash, nuclear waste, asbestos-containing materials, etc.), can be regarded as a multi-component material having a wide range of compositions, and usually, it is the presence of only some of these components that makes the entire mixture a product to be disposed of. Thus, a process allowing selective treatment of the "unwanted" portion of the waste could represent an enormous advantage in terms of time and money, especially as far as materials presenting low thermal conductivity are concerned. Microwaves have emerged as an interesting candidate to fulfill the need for this kind of process (Shaheen et al., 2012), and this is particularly true if the matrix of the waste materials exhibits dielectric properties significantly different from those of the unwanted components.

Microwave treatment has been implemented as an inexpensive way to clean heavy metals from treated sewage (WaterWorld, 2019). The method removes three times the amount of lead from biosolids compared to conventional means and could reduce the total cost of processing by more than 60%. After treatment in a microwave, researchers were able to remove the heavy metals from biosolids with a lower dosage of treatment chemicals than traditional extraction requires. It is a technique that can be scaled up to facilities that serve a city or a region to give them a less expensive way to make biosolids safe.

Microwave radiation is also used as a heat source to treat medical waste. Microwave treatment units can be either on-site installations or transported in mobile treatment vehicles. The processing usually includes front-end shredding of the waste, both to increase the efficacy of the microwave treatment and to reduce the volume of the end waste for disposal. If the waste is dry, water is added and the wet waste is introduced to the microwave chamber. Typical operation is at 2450 Hz. While an autoclave provides heat from outside the waste, the microwave unit transmits energy as microwaves and that energy turns into heat inside the wet waste. Microwave disinfection works only when there is water in the waste since the radiation directly works on the water, not on the solid components of the waste. For this reason, treatment units are often supplied with a humidifier. Processing time is determined by the manufacturer and the experience of the operators, but about 20 minutes per batch is typical. Mechanical treatment is often positioned upstream of the microwave in order to make sure the waste pieces are small. Smaller pieces enhance the heating action as microwaves are able to penetrate to where infectious microbes are. With enough power, the water is converted to steam and makes all of the waste around 100°C. The entire process takes place within a single vessel. Treatment of medical waste through exposure to microwaves is less expensive than incineration (Malsparo, n.d.).

6.1.4 PYROLYSIS OF BIOMASS WASTE (ETHAIB ET AL., 2020A,B)

The utilization of biomass waste as a raw material for renewable energy is a global priority. Pyrolysis is one of the thermal treatments for biomass wastes that results in the production of liquid, solid, and gaseous products. The complex structure of the biomass materials matrix requires elevated heating to convert these materials into useful products. Microwave heating is a promising alternative to conventional heating approaches. Several operating parameters, such as microwave power and temperature, microwave absorber addition and its concentration, initial moisture content, initial sweep gas flow rate/residence time, etc., govern this process. Attractive products of microwave pyrolysis include synthesis gas, bio-char, and bio-oil. Generally, different conditions of microwave-assisted pyrolysis will lead to varying yields of products. The efficacy of microwave pyrolysis ultimately relies on the operating conditions of this process. The factors that have the most impact on product recovery in microwave pyrolysis are temperature rate, microwave absorber addition, and its concentration (e.g., metal oxides and sulfides, carbon-based materials, and silicon carbide), initial moisture content, and initial sweep gas flow rate/residence time (Ethaib et al., 2020a).

The ability of microwave heating to process a wide variety of biomass is of tremendous benefit when it comes to processing feeds with high moisture content. A characteristic feature of this technique is the ability to manipulate the heating rate of a sample without increasing the input power through the use of microwave absorbers, also known as susceptors (Prashanth et al., 2020).

From microwaves to lasers is a somewhat illogical progression, but as we shall see in the next section, they share the one trait that qualifies them for side-by-side treatment in this book—they are both process intensifiers of renown.

PIF/CIF ANALYSIS FOR MICROWAVE PROCESSING

Hasna (2011) has dealt with cost implications for industrial-scale microwaves. The purchase price of a microwave system is reported to be around $1500–5500 per kW depending on the system's power, frequency, and complexity. Maintenance costs of industrial microwave systems are moderate and, once established, they are very robust with the only "consumable" being the magnetron. On the assumption that a typical electrical unit costs a reference price of 11 cents per kilowatt hour, the running cost for 6,000 h is estimated to be approximately $1200. Industrial microwave systems are thus estimated to vary in cost between $4000 and $7000 per kilowatt.

Microwave-driven Process Intensification Factors (PIF) can be quite significant, based on the application.

Given the relatively low installation and operational cost of microwave systems, they appear to be an attractive process intensification option when suitable for the process under consideration.

6.2 LASER

Lasers represent one of the most important inventions of the 20th century. The first industrial application of the laser was to make holes in diamond materials which are extremely difficult to machine (Grum, 2007). The high-intensity CO_2-laser has proved to be extremely successful in various industrial applications from the point of view of both technology and economy. Some of the most notable advantages of laser sources are: energy savings in comparison to classical surface heat-treatment processes; and adjustable energy input over a wide range by changing laser-source power, with converging lenses having different foci, and by choosing different travel speeds of workpieces and/or laser beam.

6.2.1 SURFACE HARDENING (GRUM, 2007)

With the interaction of the laser light and its movement over the surface, very rapid heating of metal workpieces can be achieved, and subsequent to that, also very rapid cooling down or quenching. The cooling rate, which in conventional hardening defines quenching, has to ensure martensitic phase transformation. In laser hardening, the martensitic transformation is achieved by self-cooling, which means that after the laser light interaction, the heat has to be very quickly conducted into the workpiece interior. For this purpose, besides CO_2-laser, Nd-YAG and Excimer-lasers with relatively low power and a wavelength ranging between 0.2 and 10.6 μm have been successfully used. A characteristic of these sources is that, besides a considerably lower wavelength, they have a smaller focal-spot diameter and much higher absorptivity than CO_2-lasers.

A recent and very interesting process from the viewpoint of physics is laser-shock processing (LSP). This is the laser treatment of a surface with a pulsed beam of high power density. The effects of shock waves are increased microhardness and the presence of compressive residual stresses in the surface layer. The changes occurring in the material improve fatigue strength of the material and, consequently, fatigue resistance of the surface prepared in this way. The process is suitable for the application of high-precision parts of machines and devices where the fatigue resistance of the material is to be improved.

Chemical heat treatment is commonly applied to increase wear resistance, fatigue strength, and corrosion resistance of many metal surfaces. Nitriding is widely used for the surface hardening of iron-based metal alloys, as well as titanium-based alloys and molybdenum. The thickness of the nitride layer is of the order of several micrometers and increases with increasing heat-treating temperature (Alnusirat, 2019). The long exposure time at high temperatures results in considerable energy expenditure, and can also degrade the mechanical properties of the materials. The most appropriate way to increase the efficiency of the chemical thermal treatment (increasing the depth with much lower exposure time) is through the use of laser processing.

6.2.2 SURFACE CLEANING

Modern manufacturing processes for "hi-tech" devices, e.g., in microelectronics or data-storage industries, continue to demand higher resolution, smaller linewidths, miniaturization, and vanishingly small mechanical clearances; hence, there is an ever-increasing need to develop better techniques to remove minute particles or films of contaminants from a critical surface. A submicron particle can adhere to a surface with a relatively strong force over a million times its weight; hence, a cleaning technique targeted at submicron particles needs to be effective for particle removal but yet gentle to the surface so as not to cause surface damage or add dirt onto the surface. The possibility of using a "noncontact" photon beam to induce desired cleaning action on critical surfaces in an industrial environment has thus become a subject of great interest.

Laser cleaning effect can be produced on a surface using pulsed laser irradiation, with pulse duration typically 1–100 ns, fluence typically tens to hundreds of mJ/cm², and wavelengths typically in the 200–300 nm range. A thin liquid film can be used to enhance the removal efficiency of small particulates; this is called "steam laser cleaning" (Tam et al., 1998). Steam laser cleaning mode is very effective for the removal of particulates in the micron to the submicron range, although it may also remove films of contaminants. This mode relies on the fast superheating followed by rapid explosive vaporization of an mm-thick liquid film (typically water mixed with 10% of ethanol or isopropanol to enhance wetting) deposited on the surface just before the flash heating. A strong acoustic pressure pulse, originating preferably at the liquid/solid interface, is generated by the explosion and is responsible for the removal of particles. This steam laser cleaning mode is most effective when the incident laser radiation is strongly absorbed by the solid surface.

The dry laser cleaning mode relies on the laser-induced generation of acoustic shocks, photochemical and photothermal bond-breaking, or ablative removal of the contaminants on the surface; this mode is most effective for laser wavelengths in the ultraviolet. Dry laser cleaning is effective both for the removal of film contaminants, as well as particulates, although the efficiency for particulate removal is usually not as good as steam laser cleaning. In general, cleaning is often best done using a combination of suitable steam laser cleaning followed by dry laser cleaning.

Daurelio et al. (1999) have described the application of laser irradiation in removal of rust, paint, and oxide layers.

6.2.3 ABLATION

Laser ablation or photoablation is the process of removing material from a solid (or occasionally liquid) surface by irradiating it with a laser beam. At low laser flux, the material is heated by the absorbed laser energy and evaporates or sublimates. At high laser flux, the material is typically converted to plasma. Usually, laser ablation refers to removing material with a pulsed laser, but it is possible to ablate material with a continuous wave laser beam if the laser intensity is high enough. While relatively long laser pulses (e.g., nanosecond pulses) can heat and thermally alter or damage the processed material, ultrashort laser pulses (e.g., femtoseconds) cause only minimal material damage during processing due to the ultrashort light-matter interaction and are therefore also suitable for micromaterial processing (Chichkov et al., 1996). Excimer lasers of deep ultra-violet light are mainly used in photoablation; the wavelength of the laser used in photoablation is approximately 200 nm. The depth over which the laser energy is absorbed, and thus the amount of material removed by a single laser pulse, depends on the material's optical properties, the laser wavelength, and pulse length. The total mass ablated from the target per laser pulse is usually referred to as ablation rate.

The simplest application of laser ablation is to remove material from a solid surface in a controlled fashion. Laser machining and laser drilling are examples; pulsed lasers can drill extremely small, deep holes through very hard materials. Very short laser pulses remove material so quickly that the surrounding material absorbs very little heat; hence, laser drilling can be done on delicate or heat-sensitive materials, including tooth enamel (laser dentistry). Several workers have employed laser ablation and gas condensation to produce nanoparticles of metal, metal oxides, and metal carbides. Another class of applications uses laser ablation to process the material removed into new forms either not possible or difficult to produce by other means. A recent example is the production of carbon nanotubes.

Laser ablation can also be used to transfer momentum to a surface since the ablated material applies a pulse of high pressure to the surface underneath it as it expands. The effect is similar to hitting the surface with a hammer. This process is used in industry to work-harden metal surfaces and is one damage mechanism for a laser weapon. It is also the basis of pulsed laser propulsion for spacecraft.

One of the advantages of laser ablation is that no solvents are used; therefore, it is environmentally friendly and operators are not exposed to chemicals (assuming nothing harmful is vaporized). It is relatively easy to automate. The running costs are lower than dry media or dry-ice blasting, although the capital investment costs are much higher. The process is gentler than abrasive techniques, e.g., carbon fibers within a composite material are not damaged. The heating of the target is minimal.

Apart from acoustic, microwave and laser-based enhancers, processes can also be directly acted upon, as indicated earlier, by external "phoretic" fields that impart an additional velocity component. We will now look at a few of these is more detail.

PIF/CIF ANALYSIS

The average laser can cost anywhere from $50,000 to 250,000, which makes it a very significant investment for a business. It is clearly a niche process intensifier whose cost-effectiveness needs to be analyzed with care. It does speed up many processes to a considerable extent, offers precision and accuracy that are best-of-class, and is "green". However, because of their capital cost, laser machines tend to be one of a kind. Process disruption due to laser downtime can be significant unless utmost care is given to maintenance and upkeep. Trained operators would become necessary, and their replacement could become an issue. The benefits that the laser tool offers from a process intensification viewpoint have to be clearly substantial. Quantification of both PIF and CIF again becomes essential to ensure that the correct business decision is made.

6.3   THERMOPHORESIS

Heavy molecules and particles in a temperature gradient experience a force (usually in the direction from hot-to-cold, i.e., proportional to $-\mathbf{grad}(\ln T)$) which produces a net transport flux $\rho_i \mathbf{c}_{i,T}$ (see Chapter 4) which is termed "thermophoretic". This augmentation of convective motion is important in systems with large temperature gradients, such as aerosol systems encountered in combustion environments (e.g., soot and ash deposition on cooled heat exchanger surfaces, sampling targets, and turbine blades) (Rosner, 1986). Thermophoresis can enhance the capture rate of submicron particles on cooled surfaces by more than 200X. The corresponding particle capture efficiency, defined as the ratio of the number of surface-adhered particles to the number of surface-incident particles, is also enhanced by a factor of 200X or more. For a typical coal-burning power plant, this can result in a submicron particle deposition rate of 7.3 kg/year, compared to a rate of 0.035 kg/year if Brownian diffusion were to be the only transport mechanism prevailing.

The PIF in this case can be represented by F(suction) where:

$$F(\text{suction}) = \text{Pe}_{\text{suction}} / \left[1 - \exp\left(-\text{Pe}_{\text{suction}}\right) \right] \tag{6.1}$$

with the Peclet number, $\text{Pe}_{\text{suction}}$, being related to the thermophoretic velocity, $\mathbf{c}_{p,T}$, target diameter, d_w, particle diffusivity, D_p, and Nusselt number in the absence of thermophoresis, $\text{Nu}_{m,0}$ as:

$$\text{Pe}_{\text{suction}} = \left[\left(-\mathbf{c}_{p,T}\right) d_w \right] / \left[D_p \text{Nu}_{m,0} \right] \tag{6.2}$$

Particle thermophoresis, and its vapor species equivalent, thermal or Soret diffusion, are responsible for the intensification of transport phenomena in a variety of high-temperature applications. The following are some illustrative examples.

6.3.1   BULB BLACKENING (NAGARAJAN, 1986)

The oldest known light sources are of the incandescent type—the candle and the kerosene lamp, and the gas light that illuminated the London fog. In these, the high temperature was achieved by chemical reaction, and the light was emitted by soot particles brought to incandescence by the heat of the reacting gases. In 1879, Edison constructed a carbon filament lamp by electrically heating the filament. The high evaporation rate of carbon, however, meant that the bulb "blackened" in a short time, and the economic lifetime of the lamp was equally short.

Later, tungsten, with a much lower vapor pressure (one thousand times smaller at 2000 K) than carbon even at higher filament temperatures, was adopted by the lamp industry, and first used in vacuum lamps in 1906. While incandescent bulbs are making way for LCD and LED lighting, the ones still in use (nearly 30% of all bulbs in use worldwide in 2021) are gas-filled to reduce the evaporation rate of tungsten. The evaporated tungsten atoms from the filament collide with filler gas molecules, and many are reflected toward the filament, thereby reducing the effective evaporation rate. The net migration of tungsten atoms away from the filament now becomes diffusion-limited. Since it is essential that the tungsten is not chemically attacked by the filler gas, the gas employed is typically nitrogen or one of the inert gases.

Based on a model that could explain his measurements, Langmuir (1912) made variable-property heat and mass transfer calculations for a heated wire in a surrounding gas by assuming that they occur only through conduction and diffusion, respectively, in a layer of gas of a certain radius around the hot wire—the so-called "Langmuir layer". In this layer, the gas is considered completely stagnant, with convection onset only outside this layer. By keeping the Grashof number based on this layer dimension below its critical value for natural convection currents to set in, bulbs can be designed to have this stagnant layer stretch all the way from the filament to the bulb walls. This offers the considerable benefit of reducing mass loss from the tungsten filament by means of eliminating augmentation due to convection and turbulence. Apart from bulb blackening, filament thinning and breakage are potential failure mechanisms for incandescent bulbs, as may be observed from household use.

The introduction of halogens into the incandescent filament lamp was later conceived in order to enhance the transport back to the filament of evaporated tungsten by providing a suitably chemically reactive environment. Tungsten depositing on the bulb wall (which is at a lower temperature) forms a volatile tungsten halide compound, which circulates through the bulb and decomposes to tungsten and halogen at or near the hot filament. The reverse flux of tungsten carried by its halide opposes the outward flux of vaporized tungsten and its oxides. An idealized "perfect cycle" is realized when the two fluxes exactly balance each other, leading to a net "Zero Element Flux" (ZEF) condition of no net evaporation of the filament material, no bulb blackening, and eternal filament life. Oxygen in trace amounts enhances the reactivity of the system, and hence W-O-Br cycle lamps are in widespread use. Bromine can be conveniently added as HBr, CH_2Br_2, or CH_3Br.

In the simplest model of the tungsten-halogen cycle (Elenbaas, 1972), the reaction sequence may be written as:

$$W + X \rightarrow WX \left(\text{low temperature} \right) \tag{6.3}$$

$$W + X \leftarrow WX \left(\text{high temperature} \right) \tag{6.4}$$

In the absence of thermal diffusion, the ZEF condition becomes:

$$p_{(W),b} / p_{(W),f} = D_W / D_{WX} \tag{6.5}$$

where $p_{(W)}$ represents partial pressures of elemental tungsten, and D represents Fick diffusion coefficients of the tungsten-containing species, W and WX. This ratio is always <1 since the heavier molecule, WX will diffuse slower than W. If $p_{(W),b}$ is smaller than the value required for the ZEF condition, tungsten will be transported toward the bulb wall, and the wall will blacken, provided the condensate is thermo-dynamically feasible at the local states.

The effect of thermal diffusion in this case is to aid the diffusion process from the filament to the wall since that would be the direction down the temperature gradient. In order to compensate, this would require that the tungsten element partial pressure at the wall must be kept higher than predicted by Elenbaas. This can be a significant design consideration in selecting operating conditions such as fill gas, halide and oxide concentration, filament and bulb wall temperature, etc.

Incorporating thermal diffusion in the transport model, the mass diffusion flux on vapor species W in inert gas is given by:

$$j"_W = -\rho D_W \left(\mathbf{grad}\omega_W + \omega_W \alpha_{W-IG}/T\mathbf{grad}T \right) \tag{6.6}$$

where ρ is the mass density of the inert gas, ω_W the mass concentration of tungsten, D_W the Fick diffusivity of tungsten, and α_{W-IG} the temperature-independent thermal diffusion coefficient of tungsten in inert gas.

Equation (6.6) will also apply to the case of tungsten *element* diffusion in a halo-gen-cycle environment, with the diffusion coefficients being evaluated as weighted means of corresponding W-containing species diffusivities. This equation enables the estimation of tungsten deposition rates on bulb walls under non-ZEF conditions, and optimizes bulb fill and operating conditions to minimize this rate. Thermal diffusion, in this instance, is an intensifier for the failure mechanism and must be negated as an influence by recognizing it as such and designing appropriately.

6.3.2 HOT CORROSION

Nickel-base superalloys, used extensively for gas turbine engine components for their superior high-temperature mechanical properties, are generally susceptible to a form of environmental attack known as "hot corrosion". Hot corrosion is encountered when molten salts are deposited from combustion gases onto turbine rotor blades and stator vanes. The attack of molten salts and slag on boiler tubes in coal-fired power plants has been noted by Reid (1971). Hot corrosion has also been observed in technologies such as coal gasification, magnetohydrodynamics, fluidized bed com-bustion of coal, molten salt fuel cells, salt processing baths, solar energy storage by molten salts, and marine combustion turbines (Stearns et al., 1983).

Hot corrosion is a major life-limiting factor in many gas turbine applications. Protective oxide coatings have been developed to provide superior corrosion resis-tance compared to base metal alloys, but these are also soluble in molten salts. At locations where the oxide dissolution rate is sufficiently high, the oxide scale is par-tially or fully dissolved, and the metal substrate becomes vulnerable to hot corro-sion. If the melt becomes locally saturated with the oxide, further oxide dissolution will not occur, and hot corrosion will become self-limiting. However, this has not been observed in practice. Thus, a continuous fluxing mechanism must be present that supplies fresh solvent continuously to local sites and transports away the dissolved oxide. The dominant cause of this condensate layer flow is likely to be gas-aerodynamic shear exerted at the vapor/condensate interface. This process of molten salt deposition and condensate flow, coupled with oxide dissolution and ensuing hot corrosion, has been modeled by Rosner and Nagarajan (1987). "Hot corrosion" maps, once validated by field data, can be used as a design tool in providing varying levels of corrosion protection to exposes surfaces.

The mass flux of each species containing a condensable element—e.g., sodium— is given by: (Rosner et al., 1979)

$$j''_{i,w} = (D_i\rho)_e / L.F_i(\text{Soret}).F(\text{turb}).Nu_{m,i}.\left[(\omega_{i,e} - \omega_{i,w}) + \omega_{i,w}.\tau_i.F(\text{ncp})/F_i(\text{Soret})\right]$$

(6.7)

where D_i is the Fick diffusivity of species i, ρ the density of the combustion product gases, L is a characteristic target dimension, $Nu_{m,i}$ the mass-transfer Nusselt number of species i, ω_i represents mass concentrations of species i, τ_i is a thermophoretic parameter of species i, and subscripts e and w refer to the locations of gas mainstream (external to the boundary layer), and the deposition surface, respectively. F(turb), F(ncp), and F_i(Soret) are correction factors for turbulence, non-constant properties (across the boundary layer), and thermal (Sore) diffusion, respectively.

This model assumes a "chemically frozen" boundary layer within which no chemical reaction, condensation, nucleation, or stagnation occurs, and across which condensable species mass transport occurs via the mechanisms of Fick diffusion, thermal (Soret) diffusion, convection, and turbulence. The effect of thermal (Soret) diffusion is incorporated via the dimensionless thermophoretic factor τ_i, and via the correction factor, F_i(Soret):

$$F_i(\text{Soret}) = -B_{T,i} / \left[1 - \exp(B_{T,i})\right]$$

(6.8)

where $B_{T,i}$ is a "suction" parameter. Given the temperature gradient from the hot combustion gas to the cooled turbine blade surface, the effect of thermal diffusion in this system will always be an augmentation of the molten salt deposition rate, sometimes as much as 20–30%. It can also induce a "dew point shift" of 10–20 K whereby molten salts can condense at temperatures higher than their thermodynamic values (Rosner and Nagarajan, 1985). This can result in unexpected hot corrosion at locations where the surface temperature may have been deemed to be too high for molten salt deposition through vapor transport and condensation.

6.3.3 PARTICLE THERMOPHORESIS

The drive toward higher combustion turbine inlet temperatures (required for lower specific fuel consumption) has led to the development of more-efficient turbine blade cooling schemes, such as film cooling or full-coverage transpiration. This is usually accomplished by blowing a small portion of the compressed air into the external boundary layer of hot combustion product gases in the turbine stages. While transpiration is certainly advantageous from a heat-transfer (blade cooling) viewpoint, its thermophoretic effect on small particle (ash or salt) mass transport can severely exacerbate the fouling problem. Though transpiration blowing significantly reduces the Brownian diffusion rate of particles to the cooled surface, this effect is countered by the thermophoretic drift of small particles down the steep temperature gradient, made steeper by transpiration cooling. A quantitative estimate of this countereffect reveals the magnitude of this unexpected less-studied mechanism (Gokoglu and Rosner, 1984). The relative importance of thermophoresis was found to depend on both the blowing rate and the size of the depositing particles. In the numerical example considered by the authors, transpiration-cooling driven reduction in deposition rate for a particle size of 7 nm was calculated to be one order of magnitude (10X) less when thermophoresis was taken into account.

In recent years, interest in thermophoresis in the context of life sciences has grown. There are two new major applications for the effect: (1) The monitoring of protein binding reactions through the sensitivity of thermodiffusion to complex formation, and (2) accumulation of a component in microfluidic devices through a combination of thermophoresis and convection (Niether and Wiegand, 2019). Microscale thermophoresis has become an important and widely known application for thermophoresis. Although relatively new, it is already becoming a standard method for the quantitative characterization of biomolecular interactions and is used as a high-throughput screening method for drug discovery. It is based on the thermophoretic behavior of biomolecules and their sensitivity to non-covalent binding.

Thermogravitational columns were one of the first instruments used to measure thermal diffusion and also the first application of the effect. A temperature gradient is applied horizontally across a column filled with a gas or liquid mixture. A circular convectional flow down the cold side and up the warm side occurs and combines with the thermophoresis in a horizontal direction. If the thermophobic component also has a higher mass density, the combination with convection enhances the de-mixing effect of thermodiffusion: the thermophobic component is carried down on the cold side and the thermophilic one up on the warm side of the column. After some time, a steady-state is reached with a concentration gradient along the column from which the Soret coefficient can be calculated. Thermogravitational columns can also be used for isotope separation and were employed in the Manhattan project to enrich uranium (Reed, 2011), but have since been replaced by less energy- and cost-intensive methods. Nowadays, the strength of conventional thermogravitation columns lies in the investigation of ternary mixtures as this technique is superior to optical methods, given that the extracted sample can be investigated using state-of-the-art analytical methods which do not rely only on refractive index measurements (Köhler and Morozov, 2016). Similar geometries are promising for microfluidic applications. It has been shown that the same combination of convection and thermodiffusion which is used in thermogravitational columns can also be used in capillaries and pores to trap and enrich one component of a mixture.

Owing to the importance of soot in dictating radiative heat loads in industrial furnaces, fires, and turbine combustors, as well as post-combustion gas opacity and absorptivity, the number of papers on carbonaceous soot published in the scientific and engineering literature is considerable. Mainly, these investigations are concerned with the mechanisms of formation (nucleation, growth, coagulation) of soot particles and their (partial) burnout during hydrocarbon combustion. However, the factors governing soot particle transport to cold surfaces immersed in flame gases have not been extensively investigated, despite the importance of deposition in a variety of experiments (requiring sampling) and technologies, e.g., combustion turbines and carbon black synthesis. In particular, the dominance of thermophoresis as an intensifier of the particle transport process has frequently been neglected but has been highlighted by Eisner and Rosner (1985). They showed that, because of the dominance of thermophoresis, the transient response of a fine thermocouple inserted into a soot-laden combustion gas mixture can be used not only to infer local gas temperatures, but may also be used to estimate local soot particle volume fractions, without having to make rather restrictive assumptions about the prevailing particle size distribution, particle optical properties, or flame gas uniformity/symmetry.

The "Process Intensification Factor" in this case is the ratio of soot particle thermophoretic flux to diffusional flux, and may be estimated as (Eisner and Rosner, 1985):

$$\left[\text{Thermophoretic Flux} / \text{Diffusional Flux} \right]$$
$$\approx \left(\alpha_{T,p} \right)_e \left(T_w / T_g \right)^{-\frac{1}{2}} \left[1/(1+\kappa) \left\{ 1 - \left(T_w / T_g \right)^{1+\kappa} \right\} \right] \tag{6.9}$$

where, to a first approximation:

$$\alpha_{T,p} \approx 6 \left(n_g V_p \right) \left(l_g / d_p \right) \tag{6.10}$$

Here, $\alpha_{T,p}$ is the particle thermophoretic factor, estimated to be six times the product of the number of gas molecules present in a volume equal to the particle volume, and the prevailing particle Knudsen number. n_g (= p/kT) is the gas mixture number density, V_p is the soot particle volume, d_p is the soot particle diameter, and l_g is the prevailing gas mean free path. The Knudsen number is the ratio of d_p to l_g, and is assumed to be much smaller than 1. The subscripts e and w refer to locations in the gas mainstream and the deposition surface, respectively.

k in Equation (6.9) is the exponent appearing in the dependence of gas mixture thermal conductivity, k_g, on temperature, i.e.:

$$k = \left[d \ln k / d \ln T \right] \approx 0.84 \tag{6.11}$$

For a soot particle size of 0.05 mm (50 nm), atmospheric pressure, T_g of 1000 K and T_w of 1400–1300 K, soot particle mass flux has been estimated by Eisner and Rosner (1985) to be 500–1,000 times higher than Brownian diffusion fluxes in the same local environment—a substantial enhancement indeed!

Now on to another phoretic field worthy of consideration.

PIF/CIF ANALYSIS

Setting up a thermal gradient is a relatively simple intervention, and is unlikely to create a significant cost impact. Hence, if a process will respond positively to a thermal field, by all means, go for it. Frequently, however, temperature nonuniformities can set in that are not part of the design. Thermophoresis can have as many adverse consequences as beneficial. Temperature control to the degree required by the process is not always easy to achieve. Heat transfer occurs by myriad mechanisms. There are innumerable leak paths. No insulation is perfect. Reducing process sensitivity to temperature excursions is always advisable.

6.4 ELECTROPHORESIS

Electrophoresis refers to the motion of dispersed particles relative to a fluid under the influence of a spatially uniform electric field (Kruyt, 1952). It is caused by the presence of a charged interface between the particle surface and the surrounding fluid. It is the basis for analytical techniques used in Chemistry for separating molecules by size, charge, or binding affinity. Electrophoresis is used in laboratories to separate macromolecules based on size. The technique applies a negative charge so that proteins move toward a positive charge. It is used extensively in DNA, RNA and protein analysis, transport in micro-channels, precipitation of ash particles, and other applications. In every case, it provides an intensification effect that makes the process run better, faster, and cheaper.

6.4.1 BIOMOLECULAR SEPARATIONS

Nanoporous membranes have recently found increasing applications in therapeutic protein purifications, membrane chromatography, biosensors, and biomaterials. Much of the biotechnology revolution uses genetic modification of cells to express desired proteins for biomolecular treatments. However, the separation of a specific protein from the large mixture of physiological proteins in a living system is a complex process. Membrane-based protein separations have the potential for single-step continuous operation that would result in low-cost, high-speed, and high-throughput processes.

To accelerate biomolecule transport across a nanoporous membrane, one can apply an electric field (electrophoresis or electro-osmosis) or high pressure. In the case of applied pressure, membrane fouling is accelerated. Because proteins have different charge states, depending upon the buffer pH, an electric-field-induced transport of the biomolecule in nanopores can offer selectivity and acceleration. Nanoporous gold and alumina membranes with high protein selectivity have been reported using electrophoresis, and they offer promising new applications in bioseparations, biosensing, and biomedical drug delivery. Inorganic nanoporous membranes, such as carbon and alumina membranes, also show greatly enhanced chemical, thermal, and mechanical stability.

Graphitic carbon nanotube (CNT) membranes are a new class of membranes that potentially have an ideal geometry for protein separations: a non-interacting graphite core and functional chemistry at the tips, where carbon bonds are cleaved to open the membrane structure. The smooth graphitic cores allow for dramatic enhancements in flow velocity of 10,000-fold compared to conventional pores, and gatekeeper activity has been demonstrated. Recently, electro-osmosis and electrophoresis had been used to efficiently pump nicotine across CNT membranes for a programmable transdermal drug-delivery device and can thus be applied to the translocation of proteins (Sun et al., 2011).

The steady-state protein flux of the charged molecule at an electric field is given by (Srinivasan and Higuchi, 1990):

$$J_{pore} = -D\,dC/dx - zFDC/RT\,dU/dx + v_{eo}C \qquad (6.12)$$

where D is the effective diffusion coefficient, F is the Faraday constant, R is the gas constant, T is the absolute temperature, J_{pore} is the flux based on pore area, v_{eo} is the electro-osmotic velocity, z is the number of unit charges of solute, and dC/dx and dU/dx are the concentration and electric field gradient across the membrane, respectively.

The increase in protein flux with applied voltage can be understood by the electrophoretic term (second term on RHS) in Equation (6.12). The effective diffusion coefficient is given by the following Einstein relation between the effective mobility and diffusion constant:

$$D_{eff} = mRT/zF \qquad (6.13)$$

where m is the protein's effective electrophoretic mobility, and D_{eff} is an effective diffusion coefficient based on the observed electrophoretic mobility (Srinivasan and Higuchi, 1990). The associated electrophoretic enhancement in protein flux can be as high as 10–12 times.

6.4.2 Transport in Microchannels

The rise of micro- and nano-fabricated systems has been accompanied by an increased interest in the transport of Brownian particles and biomolecules through channels of periodically varying cross sections. When considering periodically varying microchannels, it is particularly important to focus on electrophoretic transport. Given the ubiquity of charged species in biology, electrophoresis is the most commonly employed animating agent used in bio-oriented micro- and nano-fluidic devices (Yariv and Dorfman, 2007). The driving force is provided by an electric field, which is excited by electrodes that are separated by many channel periods. The particle motion is therefore given by a combination of Brownian diffusion and deterministic migration, the latter driven by a spatially periodic electric field.

Tracer particles are often used in microfluidics, including in fluid flow visualization and velocimetry techniques. In pressure-driven flows, only one main driving force is usually present, while in electrokinetic flows, several forces can simultaneously act on the tracer particles. Micro-particle image velocimetry (micro-PIV) is often used to measure the velocity of suspended tracer particles in microfluidics, and it is no surprise that it has also been used to determine electrokinetic flow properties. In a straight microchannel filled with an electrolyte containing neutrally buoyant tracer particles, where an electric field is applied without any external pressure gradient imposed, and under steady-state conditions, the tracer particle velocity (observed velocity, u_{obs}) results from multiple contributions (Sadek et al., 2017):

$$u_{obs} = u_{eo} + u_{ep} + u_B \tag{6.14}$$

where the subscript "eo" refers to electroosmotic, "ep" to electrophoretic, and B to Brownian motion of particles. In most instances, Brownian motion can be neglected relative to the other two terms (after averaging the values among several particles, whereby the random motion component cancels out).

For Newtonian fluids, the electrophoretic velocity can be expressed as:

$$u_{ep} = \varepsilon \zeta_w / \mu \, E \tag{6.15}$$

where ε is the electric permittivity, ζ_w is the wall zeta potential, μ is solution viscosity, and E is the applied electric field.

For typical microfluidic devices and tracer particles, the electrophoretic velocity becomes fully developed in a time scale that is orders of magnitude faster than the electroosmotic velocity, and hence becomes the predominant driving force when an electric field is applied.

6.4.3 ELECTROSTATIC PRECIPITATOR

Electrostatic precipitators are important tools in the process of cleaning up flue gases. They are highly effective at reducing particle pollution, including those particles whose sizes approach 1 μm in diameter, and some precipitators can remove particles of 0.01 μm in diameter. In addition, they can handle large volumes of gas at various temperatures and flow rates, removing either solid particles or liquid droplets (Britannica.com). Precipitators function by electrostatically charging particles in the gas stream. The charged particles are attracted to and deposited on plates or other collection devices. The treated air then passes out of the precipitator and through a stack to the atmosphere. When enough particles have accumulated on the collection devices, they are shaken off the collectors by mechanical rappers. The particulates, which can be either wet or dry, fall into a hopper at the bottom of the unit, and a conveyor system transports them away for disposal or recycling.

The collection efficiency, η, of an electrostatic precipitator may be estimated by an empirical equation: (Weiner and Mathews, 2003)

$$\eta = 1 - \exp\left(-A v_d / Q\right) \tag{6.16}$$

where A is the total area of the collecting surface of the collection electrodes, Q is the flow rate of gas through the pipe, and v_d is the electrophoretic drift velocity.

The drift velocity is the velocity of the particles toward the collecting electrode and may be calculated theoretically by equating the electrostatic force on the charged particle in the electrical field with the drag force as the particle moves through the gas. The drift velocity is analogous to the terminal settling velocity, except that in the latter case the force acting in opposition to the drag force is gravitational rather than electrostatic. Drift velocity may be estimated by

$$v_d = 0.5d_p \qquad (6.17)$$

where d_p is the particle size (in µm). Drift velocities are usually between 0.03 and 0.2 m/s.

As the dust layer builds up on the collecting electrode, the collection efficiency may decrease. While electrostatic precipitators are very efficient collectors of very fine particles, the amount of dust collected is directly proportional to the current drawn. Hence, the electrical energy used by an electrostatic precipitator can be substantial, resulting in high operating costs. The (PIF/CIF) ratio analysis is highly relevant in this context.

Magnetic fields surround us without our ever fully becoming aware of them or giving them their due. When properly harnessed, they can enhance many processes to a considerable extent. We will see some examples in the next section.

PIF/CIF ANALYSIS

As with thermophoretic equipment, electrophoretic systems tend to run relatively low in cost but can produce a significant effect. As an investment option, these systems can be easily justified on the basis of techno-economic viability. Maintenance issues do need to be borne in mind when the field is applied in extreme environments.

6.5   MAGNETOPHORESIS

Magnetophoresis is a phenomenon where particles migrate in a magnetic field. The phenomenon can be further categorized as positive and negative magnetophoresis. Positive magnetophoresis is the migration of magnetic particles in a diamagnetic medium. Negative magnetophoresis is the migration of diamagnetic particles in a magnetic medium. Magnetophoresis occurs in the gradient of a magnetic field, a gradient of magnetization of the surrounding medium, or a combination of both. The overall objective of magnetophoretic manipulation is the efficient handling of a large number of sample particles over a short period of time. The magnetic permeability (μ), the magnetic flux density (B), and the susceptibility (χ) are the key parameters for designing and optimizing applications with magnetophoresis.

Magnetophoresis offers several advantages over other active methods for particle manipulation such as electrophoresis, thermophoresis, dielectrophoresis, optical trapping, and acoustophoresis. Electrophoresis requires an electric field with direct electrode contact that may cause Joule heating and electrolysis. Thermophoresis needs a temperature gradient across the sample. Dielectrophoresis uses an electric field to affect the trajectory of particles in a fluid flow and may not need direct electrode contact with the liquid sample. However, the alternative current (AC) field could polarize biological cells and changes their metabolic function. Optical actuation utilizes photon energy to move particles. The optical setup is usually complex and expensive. The heat generated by the focused laser beam could affect the behavior of sensitive biological particles and even kill them. Acoustophoresis utilizes the pressure field or acoustic streaming often induced by a surface acoustic wave (SAW) to manipulate particles. An intense acoustic field may cause heat that is harmful to biological particles such as cells.

Magnetophoresis is a contactless method for manipulation of particles (Munaz et al., 2018). Magnetic fields have been successfully utilized for mixing, positioning, transport, and separation of magnetic and non-magnetic objects. The migration of particles under an external magnetic field depends on the balance of forces acting on it. The main forces acting on the particle are the inertial force (m_p dv/dt), the magnetic force $\mathbf{F}_m$, and the fluid drag force $\mathbf{F}_d$. The force balance may be written as:

$$m_p d\mathbf{v}/dt = \mathbf{F}_m + \mathbf{F}_d \tag{6.18}$$

where m_p is the mass of the particle, and v is the particle velocity.

The magnetic force acting on a particle suspended in a fluid medium is:

$$\mathbf{F}_m = V\left[\left(\Psi - \Psi_m\right)/\mu_0\right](\Delta.\mathbf{B})\mathbf{B} \tag{6.19}$$

where V is the particle volume, Ψ is the magnetic susceptibility, Ψ_m is the susceptibility of the surrounding medium, μ_0 is the magnetic permeability of the air, and $\Delta.\mathbf{B}$ represents the field gradient of the magnetic flux density $\mathbf{B}$. Magnetic force is determined by the size of the magnet, its strength, its distance from the channel, and the orientation of the magnetization. Effective manipulation of particles requires a strong magnetic field gradient, which can be implemented using special magnet geometry, an array of magnets, integrated micro electromagnets, or ferromagnetic wire array.

6.5.1 SEPARATION

The sorting of microparticles has a great application prospect in the fields of oncology, stem cell research, gene sequencing, and so on (Du et al., 2020). Inertial microfluidics is often used in microparticle sorting by particle size. However, the inertial effect is not enough on its own to separate microparticles with similar sizes. Hence, a variety of approaches to increase the lateral migration of microparticles, by the use of sonophoresis, thermophoresis, dielectrophoresis, photophoresis, etc., have been employed in microparticle sorting. Magnetophoretic separation is an approach having wide application in biological medicine and chemical analysis to separate magnetic particles with various magnetic properties or sizes in a viscous fluid. Additionally, magnetic microparticles can also be sorted by their shapes under a properly applied magnetic field.

6.5.2 MIXING

Micromixers are important functional devices in microfluidic devices and play an important role in microreactors, concentration gradient generators, micro-heat exchangers, and lab-on-a-chip devices. However, since the flow in these devices is usually at low Reynolds numbers, the mixing of fluid streams in microfluidic devices cannot be greatly enhanced by turbulence. Diffusion is the primary mechanism for mixing in a low Reynolds number laminar flow, and hence, it would take a long time and require a long microchannel to complete the mixing process. Micromixers can be classified into both active and passive types. In general, passive micromixers rely entirely on molecular diffusion to induce mixing in long microchannels, whose geometry needs to be specifically designed to achieve sample mixing. Compared with passive micromixers, the required mixing time and channel length in active micromixers are shorter due to the introduction of external power sources such as electrokinetics, acoustic force, and magnetic force. Moreover, the active micromixers are more compact and have a substantially better mixing performance.

Magnetic actuation has received more and more attention owing to its unique advantages in microfluidics (Chen and Zhang, 2017). It is not affected by other parameters such as surface charges, pH, and ion concentration. It does not induce heating and does not require expensive external systems as compared to optical concepts.

In order to describe the concentration distribution of magnetic nanoparticles in the micromixer, a Convection/Diffusion equation may be applied as follows:

$$\delta c / \delta t + (\mathbf{v}.\Delta)c = D\Delta^2 C \qquad (6.20)$$

where c is the species concentration and D is the diffusion coefficient. $\mathbf{v}$ is the particle velocity that can be expressed as:

$$\mathbf{v} = \mathbf{u} + \mathbf{F}_m / (6\pi r \mu) \qquad (6.21)$$

where r is the radius of the magnetic nanoparticle, $\mathbf{u}$ is the fluid velocity, and μ the dynamic viscosity of the solution. As the magnetic force, $\mathbf{F}_m$, increases, mixing becomes more violent.

In microfluidic devices, the Peclet number, Pe (= uL/D, where L is a characteristic dimension) is usually very high (10^3–10^6) due to the convective term dominating over the diffusive term; for this reason, the stirring process must be accelerated with artificial vortices created by appropriate paths along the stirring channel (Affanni and Chiorboli, 2008). An externally applied magnetic field is able to accomplish this very well by inducing force $\mathbf{F}_m$, and an associated magnetophoretic velocity.

6.5.3 TRAPPING

On-chip sample processing and investigation require controlled capture and release of target molecules and cells. The target molecules need to be in the chip until they are exposed to the intended reagents. However, the concentrated beads may aggregate and potentially block the flow path. Particles can be trapped by both positive and negative magnetophoresis so that both paramagnetic and diamagnetic particles can be trapped inside a microchannel. Similar to mixing and separation, the effectiveness of trapping depends on the height and the width of the microchannel, the size and the strength of the magnets, the size of the beads and their susceptibility, and fluid viscosity and flow rates. If the magnetic force is dominant over the hydrodynamic drag force, trapping occurs.

Trapping with positive magnetophoresis occurs if the particle has a higher magnetic susceptibility than the surrounding medium. Utilizing positive magnetophoresis, diamagnetic cells can bind to magnetic beads to be trapped at field maxima. Targeted cells are often trapped on the side wall or a designated reservoir of a microchannel. As a consequence, a large bead cluster may form and trap unwanted particles (Munaz et al., 2018). Magnetophoretic trapping of red blood cells was demonstrated by Watarai and Namba (2002) under counter-current flow conditions in the capillary. Magnetophoretic trapping of microspheres flowing in polymeric fluid channels was demonstrated by James et al. (2004) under various buffer conditions. Both the mixing and the trapping exhibited excellent characteristics nearly independent of solution conductivity and pH. Surrogate solutions for bovine blood serum, urine, and milk, as well as strawberry juice and water solutions, were successfully tested.

Jain et al. (2020) have reported the trapping and coalescence of flowing diamagnetic aqueous droplets in a paramagnetic (oil-based ferrofluid) medium using negative magnetophoresis. Their study revealed that the trapping phenomenon is underpinned by the interplay of magnetic energy (E_m) and frictional (viscous) energy (E_f), in terms of magnetophoretic stability number, $S_m = (E_m/E_f)$. The trapping and non-trapping regimes are characterized based on the peak value of magnetophoretic stability number, S_{mp}, and droplet-to-channel size ratio, D*. The droplets were trapped when $D* < 0.7$ and the maximum ratio of the magnetic energy to viscous energy, i.e., S_{mp} exceeds 200.

And now, back to our first love—acoustic fields, which can also induce a phoretic field and an accompanying convective velocity....

PIF/CIF ANALYSIS

Compared to thermophoretic and electrophoretic systems, magnetophoresis is more of a niche application but is particularly effective where magnetic or magnetizable materials are involved. It is typically justifiable as a custom solution where cost becomes a secondary consideration. As with electrophoretic systems, the accumulation of magnetic material as contaminants on critical surfaces is an issue and one that drives up maintenance costs as well as downtime. Particular attention needs to be paid to this aspect.

6.6 ACOUSTOPHORESIS

Ultrasonic standing wave-based particle manipulation has been investigated in macroscale chamber models in the last five decades, and more recently it has entered the microfluidic arena, offering the integration of new cell and particle manipulation modalities in lab-on-a-chip and micro-total analysis systems. Recent advances have

shown that ultrasonic standing wave systems in microfluidic channels, i.e., acoustofluidics, match or outperform known methods of manipulating cells and particles
such as dielectrophoresis, magnetophoresis, and hydrodynamics.

The physical properties of particles/cells that affect the magnitude of the primary axial radiation force, F_{ax}, the major force component in acoustic standing wave
manipulation, are density, ρ, speed of sound, c, and the radius, R. These properties,
and the corresponding data for the suspending medium and the parameters of the
sound field constitute the equation for the acoustic radiation force, F_{ax}.

$$F_{ax} = 4\pi a^3 E_{ac} k \sin(2kz) \Phi \qquad (6.22)$$

where F_{ax} is the acoustic radiation force, E_{ac} is the acoustic energy density, a is the
particle radius, z is the distance from pressure anti-node in the wave propagation
axis, and Φ is the acoustic contrast factor (Lenshof et al., 2012).

Equation (6.22) is a special case of the primary acoustic radiation force as
obtained in a one-dimensional plane acoustic standing wave. The sign of the primary
radiation force is defined by the material properties of the particle/cell relative to
the suspending medium and is summarized by the acoustic contrast factor (Φ). The
contrast factor hence reveals whether a particle will move toward the pressure node
or the antinode in the acoustic standing wave field. A positive contrast factor results
in a movement toward the pressure node while a negative contrast factor results in a
movement to the antinode.

6.6.1 Continuous Flow Concentration of Cells and Particles

In continuous flow mode, the most straightforward acoustophoretic unit operation is
to concentrate particles in suspension. This is generally done by focusing particles
into a node and thereby depleting the surrounding medium of particles. The flow
is subsequently split into a fraction containing the concentrated particles while the
particle-free medium leaves the system in the other fraction.

One of the early microfluidic continuous cell concentrators was described by Yasuda
et al. (1997), which used a quartz chamber of a half wavelength width. In this case,
red blood cells were concentrated into a single band in the flow cell. Using this setup,
the release of intracellular components from ultrasonically focused red blood cells was
examined, and no damage to the cells was found. Cylindrical capillaries have been
used as intrinsic acoustic resonators to concentrate particles in continuous flow mode
as described by Goddard and Kaduchak (2005). Acoustic capillary focusing was also
proposed as an alternative to hydrodynamic focusing in FACS (fluorescence-activated
cell sorting) applications where the use of acoustic standing waves could replace the
sheath flow responsible for focusing the cells or particles into a single stream.

6.6.2 Filtration/Removal of Suspended Particles

Contrary to concentration, acoustic focusing can be used to remove the solid components, resulting in a particle-free solution. The same configuration as in the concentration application can be used with a flow splitter after a half-wavelength acoustic

focusing step. However, this basic configuration is only applicable at modest particle concentrations if reasonable particle clearance is to be obtained. What usually fails at high concentrations is the overloading of particles in the central pressure node and consequently loss of beads into the clarified medium fraction. To overcome this problem, Lenshof et al. (2012) developed a device where cells were sequentially removed from the dense stream of cells in the central region of the channel. This enabled the development of a plasmapheresis chip that was capable of handling undiluted whole blood, providing clarified blood plasma. The focused blood cells were removed through exit holes located in the center bottom of the channel where the density of cells was high. Sequential exit holes lowered the cell concentration gradually and at the end of the separation channel the final fraction of the remaining cells was removed through a flow splitter and blood plasma of low cell counts was obtained that fulfilled the requirements for clinical diagnostics.

6.6.3 Particle Separation, Mixing, and Focusing

Particle and cell separation using acoustic standing wave technology have been widely researched. Enhanced cell sedimentation is possibly the most common application and is currently also in industrial use, e.g., in fermentation processes (Pui et al., 1995). Sedimentation is accomplished by aggregating particles and cells into the pressure nodal planes of a standing wave. If the nodal planes are aligned with the direction of gravity, the aggregates can be allowed to settle to the bottom of the fluid container. The precondition for this is that the acoustic force resulting from the lateral (parallel to gravity) pressure gradient does not override gravity.

Sedimentation assays have not been widely adopted in the microfluidic domain, whereas aggregation of cells and particles is more commonly employed in the macroscale. Acoustic standing wave aggregation in microfluidic components is typically used in half wavelength resonators where the primary acoustic radiation force levitates the cells/particles into the nodal plane and subsequently the lateral gradient of the primary radiation force drives the cells into a dense cluster in the nodal plane.

In contrast to acoustic aggregation and trapping, continuous flow-based approaches in microfluidic systems can utilize two different modes of flow-based separation, i.e., free flow acoustophoresis or binary acoustophoresis. In free-flow acoustophoresis, the net acoustic force drives cells or particles (orthogonal to the direction of flow) toward a pressure node at the same time as they travel with the flow. An alternative mode of separation, which is also based on the magnitude of the primary acoustic radiation force, is the use of frequency switching such that the acoustic resonator is switched between different resonance modes at controlled time intervals. If the amplitude for the employed actuation frequencies and the actuation intervals are tuned accordingly, an alignment of two particle categories into the different pressure nodes of the two frequencies respectively can be obtained.

Binary acoustophoresis utilizes the fact that in a given buffer, two particle types may display significantly different acoustic contrast factors. In the case where the contrast factor for a particle type has an opposite sign (different polarity) relative to its counterpart, the primary acoustic radiation force acts in the opposite direction. Particles in aqueous systems with a density higher than water and a positive acoustic contrast factor will experience a radiation force toward the standing wave pressure node, whereas particles of lower density and a negative contrast factor will focus on the pressure antinodes.

Ultrasound (US) is a promising method to address clogging and mixing issues in microreactors (Dong et al., 2020). Another way to improve the energy efficiency of US is to utilize the acoustophoretic effect (Bruus et al., 2011) at high ultrasonic frequency (1–100 MHz). As the system here is operated at frequencies above the cavitation threshold, cavitation effects are normally not observed, whereas the effect of acoustic radiation force does manifest itself especially when a standing wave is formed in the fluid channel. The strong radiation force generated by the standing wave could drive particles to the pressure node. This principle has been successfully implemented in microreactors, whose channel size is in the range of the wavelength for high-frequency ultrasonics (15 µm–1.5 mm for 1–100 MHz ultrasound in water). Sriphutkiat and Zhou (2017) investigated the effect of standing surface acoustic waves on the accumulation and clogging of SiO_2 particles in a 100 µm microchannel at a frequency of 19.95 MHz and power input from 0.3 to 1 W. They found that the acoustophoretic effect reduced the accumulation area by 2–3.7 fold and thus delayed clogging.

Dong et al. (2019) have extended the application of acoustophoresis to material synthesis processes. They developed a high-frequency ultrasonic microreactor with a half-wavelength standing wave formed in both the depth and width direction of the microchannel, which focused the particles to the node point at the channel center. This focusing effect not only prevented solid deposition (and thus clogging) on the channel wall but also induced a more uniform velocity distribution for the particles, resulting in a narrower particle size distribution for both $CaCO_3$ and $BaSO_4$ synthesis processes. The reactor worked at a resonance frequency of 1.21 MHz and an input power range of 0.3–3.3 W. Due to the low power input, no active cooling was needed to control the reactor temperature.

In the past several sections, the effect of external fields in intensifying processes has been showcased, with a focus on the induced phoretic flow field. In what follows, we will revert to the most riveting one of them all—**acoustic**—and do a deep dive which will further illustrate the concepts of PIF and CIF in a quantitative context. We will start by looking at how ultrasonic fields are generated in coupled liquids, and how the resulting fields are harnessed.

PIF/CIF ANALYSIS

Samarasekera and Yeow (2016) point out that acoustophoresis while being a strong yet gentle method of process intensification, has typically involved high construction and labor costs for the resonant chambers. They have presented a new fabrication method that employs inexpensive materials and eliminates the need for cleanrooms or the special equipment typically found within them. The method utilizes a simple glass and polyimide sandwiching technique which addresses a major drawback—cost—that has prevented the widespread application of acoustophoretic technology. Such innovative solutions are de rigueur if this mode of process intensification is to be implemented successfully.

EXERCISE QUESTIONS

1. Microwaves are used at home in batch mode, but in industrial processing, lend themselves to continuous throughput (e.g., Nikam et al., 2017). Cost per unit product in continuous processing is not always lower, but it does lend itself to a higher manufacturing volume, and more consistent quality due to the elimination of batch-to-batch variations. The calculation of PIF and CIF in this case can lead to interesting insights into the two. Carry out this analysis for two different applications, and document your inferences. Also of concern are the harmful impacts on human health of microwave irradiation. This may require more mechanization and automation in place, and impact the cost of manufacturing.

2. Laser-induced and laser-controlled chemical processing is a well-known and long-practiced technology intervention (Knudtson and Eyring, 1974), for example, in laser isotope separation (Letokhov, 1980). Laser-induced chemical reactions in combustion allow lower process temperatures with decreasing energy expense and avoidance of high-temperature formation of byproducts (Wolfrum, 1986). The cost of laser processing depends on parameters such as laser type, wavelength, input power, spot size, beam polarization, processing time, etc. Select a specific application, and formulate a cost model. Integrate this with projections for "process intensification", and investigate the combined model with respect to implications for the viability of laser processing.

3. A thermophoretic sampler may be designed for the collection of particles smaller than 10 nm. The sampler used by Wen and Wexler (2007) is composed of heated and cooled surfaces separated by a gap of 0.1 mm; a bypass flow is introduced in the design to minimize the diffusional loss of nanoparticles in the upstream flow channel. Particles may be directly deposited on a 3 mm diameter TEM grid for chemical analysis, or on other substrates for other purposes. Carry out calculations at an inlet flow rate of 1.5 lpm and thermal gradient of 5×10^5 K/m to obtain the maximum collection efficiency for a particle diameter of 1 nm.

4. Sizing and costing of electrostatic precipitators is a crucial exercise for power plant operators involved in pollution abatement. Turner et al. (1988) describe the electrostatic precipitation process, types of ESPs by configuration and auxiliary equipment. They also review the electrostatic precipitation theory, including a discussion of electrical operation points, particle charging and collection, and sneakage and reentrainment. Estimation methods are given for specific collecting areas and pressure drop, followed by discussions of factors that affect ESP sizing and operation. Apply the methodology for a representative power plant, and draw conclusions regarding economic viability.

5. Magnetophoresis-based microfluidics is emerging as a major R & D direction (Alnaimat et al., 2018). Different sources for creating the magnetic field gradient are commonly employed in microfluidic devices. Positive- and negative-magnetophoresis are utilized for the manipulation of micro-scale

entities (cells and microparticles), and employed for operations such as trapping, focusing, separation, and switching of microparticles and cells. Investigate this integration of magnetophoresis and microfluidics with respect to process and energy efficiency.

6. Acoustic agglomeration technology has been adopted for the treatment of high-concentration particles in liquid and gaseous mediums in industries such as chemical processing, oil & gas, food production, and control of environmental pollution (de Sarabia et al., 2000; Riera et al., 2006). The focus has been on fine particles that make up a significant portion of particle emissions and are difficult to remove using other separation technologies. For instance, cyclone separators used in the control of industrial burning emissions have efficiencies of less than 40% for particles below 5 μm. By introducing acoustic pre-conditioning, fine particles form clusters that are large enough to be removed effectively through cascades of cyclone stages. Apart from industrial emissions control, there is also the potential for bio-aerosol removal, where typical filter efficiency is low for particles in the size range of 0.01–2 μm (Burge, 1995; Riley et al., 2002). The hypothesis is that fine PM and bioaerosols could be pre-conditioned by means of acoustic agglomeration to form larger clusters that can be removed more easily by filtration through the mechanisms of interception and inertial impaction in filtration systems, without the need to switch to higher efficiency filters that cause higher airflow resistance. Estimate the improvement in removal efficiency due to acoustic agglomeration in a typical application, making any necessary assumptions.

REFERENCES

Affanni, A. and G. Chiorboli, "Design of an Efficient AC Magnetohydrodynamic Stirrer", 16th IMEKO TC4 Symposium Exploring New Frontiers of Instrumentation and Methods for Electrical and Electronic Measurements, September 22–24, 2008, Florence, Italy.

Alnaimat, F., S. Dagher, B. Mathew, A. Hilal-Alnqbi, and S. Kashan, "Microfluidics Based Magnetophoresis: A Review", *Chem. Rec.*, vol. 18, no. 11, 2018, November, 1596–1612, doi: 10.1002/tcr.201800018. Epub 2018 June 11.

Alnusirat, W., "Application of Laser Radiation for Intensification of Chemical Heat Treatment", *Lasers Mfg. Materials Proc.*, vol. 6, 2019, pp. 263–279.

Bruus, H., J. Dual, J. Hawkes, M. Hill, T. Laurell, J. Nilsson, S. Radel, S. Sadhalg, and M. Wiklundh, "Acoustofluidics—Exploiting Ultrasonic Standing Wave Forces and Acoustic Streaming in Microfluidic Systems for Cell and Particle Manipulation", *Lab Chip.*, vol. 11, 2011, pp. 3579–3580.

Burge, H.A., Editor, *Bioaerosols*, 1st ed., CRC Press, Boca Raton, FL, USA, 1995, pp. 1–21.

Chen, X. and L. Zhang, "A Review on Micromixers Actuated with Magnetic Nanomaterials", *Microchim Acta*, vol. 184, 2017, pp. 3639–3649.

Chichkov, B.N., C. Momma, S. Nolte, F. von Alvensleben, and A. Tiinnermann, "Femtosecond, Picosecond and Nanosecond Laser Ablation of Solids", *Appl. Phys. A.*, vol. 63, 1996, pp. 109–115.

Daurelio, G., G. Chita, and M. Cinquepalmi, "Laser Surface Cleaning, De-Rusting, De-Painting and De-Oxidizing", *Appl. Phys. A.*, vol. 69, 1999, pp. S543–S546.

de Sarabia, E.R.F., J.A. Gallego-Juárez, G. Rodríguez-Corral, L. Elvira-Segura, and I. González-Gómez, "Application of High-Power Ultrasound to Enhance Fluid/Solid Particle Separation Processes", *Ultrasonics*, vol. 38, 2000, pp. 642–646. doi: 10.1016/ S0041-624X(99)00129-8.

Dong, Z., R.D. Fernandez, and S. Kuhn, "Acoustophoretic Focusing Effects on Particle Synthesis and Clogging in Microreactors", *Lab Chip.*, vol. 19, 2019, pp. 316–327.

Dong, Z., C. Delacour, K. McCarogher, A.P. Udepurkar, and S. Kuhn, "Continuous Ultrasonic Reactors: Design, Mechanism and Application", *Materials*, vol. 13, 2020, p. 344. doi:10.3390/ma13020344.

Du, J., L. Li, Q. Zhuo, R. Wang, and Z. Zhu, "Investigation on Inertial Sorter Coupled With Magnetophoretic Effect for Nonmagnetic Microparticles", *Micromachines*, vol. 11, 2020, p. 566. doi:10.3390/mi11060566.

Eisner, A.D. and D.E. Rosner, "Experimental Studies of Soot Particle Thermophoresis in Nonisothermal Combustion Gases Using Thermocouple Response Techniques", *Combust. Flame*, vol. 61, 1985, pp. 153–166.

Elenbaas, W., *Light Sources*, Philips Tech. Library, Crane, Russak & Co., Inc., New York, NY, 1972.

Ethaib, S., R. Omar, M.S. Mazlina, A.D. Radiah, and M. Zuwaini, "Evaluation of Solvent Level Effect on Sugar Yield During Microwave-Assisted Pretreatment", in *IOP Conference Series: Materials Science and Engineering*, IOP Publishing, Karbala, Iraq, Volume 871, 2020a, p. 012034.

Ethaib, S., R. Omar, S.M. Mustapa Kamal, D.R. Awang Biak, S. Syam, and M.Y. Harun, "Microwave-Assisted Pretreatment of Sago Palm Bark", *J. Wood Chem. Technol.*, vol. 37, 2017, pp. 26–42.

Ethaib, S., R. Omar, S.M. Kamal, D.R.A. Biak, and S.L. Zubaidi, "Microwave-Assisted Pyrolysis of Biomass Waste: A Mini Review", *Processes*, vol. 8, 2020b, pp. 1190–1207.

Goddard, G. and G. Kaduchak, "Ultrasonic Particle Concentration in a Line-Driven Cylindrical Tube", *J. Acoust. Soc. Am.*, vol. 117, 2005, p. 3440. doi: https://doi.org/10.1121/1.1904405.

Gokoglu, S.A. and D.E. Rosner, "Effect of Particulate Thermophoresis in Reducing the Fouling Rate Advantages of Effusion-Cooling", *Int. J. Heat Fluid Flow.*, vol. 5, no. 1, 1984, pp. 37–41.

Grum, J., "Comparison of Different Techniques of Laser Surface Hardening", *J. Achiev. Mater. Mfg. Eng.*, vol. 24, no. 1, 2007, pp. 17–25.

Hasna, A.M., "Microwave Processing Applications in Chemical Engineering: Cost Analysis", *J. Appl. Sci.*, vol. 11, no. 21, 2011, pp. 3613–3618.

Jain, S.K., U. Banerjee, and A.K. Sen, "Trapping and Coalescence of Diamagnetic Aqueous Droplets Using Negative Magnetophoresis", *Langmuir*, vol. 36, no. 21, 2020, pp. 5960–5966.

James, C.D., P.C. Galambos, M. Derzon, M. Hopkins, G.R. Anderson, P. Clem, T. Lemp, J. Martin, K. Rahimian, and L. Rohwer, "Magnetophoretic Bead Trapping in a High-Flowrate Biological Detection System", Sandia Report SAND2004-5466, November 2004.

Knudtson, J.T. and E.M. Eyring, "Laser-Induced Chemical Reactions", *Ann. Rev. Phys. Chem.*, vol. 25, 1974, pp. 255–274.

Köhler, W. and K. Morozov, "The Soret Effect in Liquid Mixtures—A Review", *J. Non-Equil. Thermodyn.*, vol. 41, 2016, pp. 151–197.

Kruyt, H.R., *Colloid Science*, Volume 1, Irreversible Systems, Elsevier, New York, NY, 1952.

Langmuir, I., "Convection and Conduction of Heat in Gases", *Phys. Rev.*, vol. 34, 1912, pp. 401–422.

Lenshof, A., C. Magnusson, and T. Laurell, "Acoustofluidics 8: Applications of Acoustophoresis in Continuous Flow Microsystems", *Lab Chip.*, vol. 12, 2012, pp. 1210–1223.

Letokhov, V.S., "Laser-induced Chemical Processes", *Phys. Today*, vol. 33, no. 11, 1980, p. 34. https://doi.org/10.1063/1.2913822.

Malsparo, "Medical Waste Management", n.d., https://www.malsparo.com/.

Munaz, A., M.J.A. Shiddiky, and N.-T. Nguyen, "Recent Advances and Current Challenges in Magnetophoresis Based Micro Magnetofluidics", *Biomicrofluidics.*, vol. 12, no. 3, 2018, p. 031501.

Nagarajan, R., "Theory of Multicomponent Chemical Vapor Deposition (CVD) Boundary Layers and Their Coupled Deposits", Ph.D. Thesis, Yale University, 1986.

Niether, D. and S. Wiegand, "Thermophoresis of Biological and Biocompatible Compounds in Aqueous Solution", *J. Phys.: Condens. Matter*, vol. 31, 2019, p. 503003.

Nikam, A.V., A.A. Kulkarni, and B.L.V. Prasad, "Microwave-Assisted Batch and Continuous Flow Synthesis of Palladium Supported on Magnetic Nickel Nanocrystals and Their Evaluation as Reusable Catalyst", *Cryst. Growth Des.*, vol. 17, no. 10, 2017, pp. 5163–5169.

Prashanth, P.F., M.M. Kumar, and R. Vinu, "Analytical and Microwave Pyrolysis of Empty Oil Palm Fruit Bunch: Kinetics and Product Characterization", *Bioresour. Technol.*, vol. 310, 2020, p. 123394.

Pui, P.W.S., F. Trampler, S.A. Sonderhoff, M. Groeschl, D.G. Kilburn, and J.M. Piret, "Batch and Semicontinuous Aggregation and Sedimentation of Hybridoma Cells by Acoustic Resonance Fields", *Biotechnol. Progr.*, vol. 11, no. 2, 1995, pp. 146–152.

Quitain, A.T., M. Sasaki, and M. Goto, "Microwave-Based Pretreatment for Efficient Biomass-to-Biofuel Conversion", in *Pretreatment Techniques for Biofuels and Biorefineries*, Springer, Berlin/Heidelberg, Germany, 2013, pp. 117–130.

Reed, C.B., "Liquid Thermal Diffusion During the Manhattan Project", *Phys. Perspect.*, vol. 13, 2011, pp. 161–188.

Reid, T.W., *External Corrosion and Deposits: Boilers and Gas Turbines*, American Elsevier Publishing, New York, NY, 1971.

Riera, E., J.A. Gallego-Juárez, and T.J. Mason, "Airborne Ultrasound for the Precipitation of Smokes and Powders and the Destruction of Foams", *Ultrason. Sonochem.*, vol. 13, 2006, pp. 107–116. doi: 10.1016/j.ultsonch.2005.04.001.

Riley, W.J., T.E. Mckone, C.K. Lai, and W.W. Nazaroff, "Indoor Particulate Matter of Outdoor Origin: Importance of Size-Dependent Removal Mechanisms", *Environ. Sci. Technol.*, vol. 36, 2002, pp. 200–207. doi: 10.1021/es010723y.

Rosner, D.E. and R. Nagarajan, "Transport-Induced Shifts in Condensate Dew-Point and Composition in Multicomponent Systems With Chemical Reaction", *Chem. Eng. Sci.*, vol. 40, 1985, pp. 177–186.

Rosner, D.E. and R. Nagarajan, "Vapor Deposition and Condensate Flow on Combustion Turbine Blades: Theoretical Model to Predict/Understand Some Corrosion Rate Consequences of Molten Alkali Sulfate Deposition in the Field or Laboratory", *Intl. J. Turbo Jet-Engines*, vol. 4, 1987, pp. 323–347.

Rosner, D.E., *Transport Processes in Chemically Reacting Flow Systems*, Butterworths, Stoneham, MA, 1986, pp. 465–467.

Rosner, D.E., B.K. Chen, G.C. Fryburg, and F.J. Kohl, "Chemically Frozen Multicomponent Boundary Layer Theory of Salt and/or Ash Deposition Rates from Combustion Gases", *Combustion Sci. Technol.*, vol. 20, 1979, pp. 87–106.

Sadek, S.H., F. Pimenta, F.T. Pinho, and M.A. Alves, "Measurement of Electroosmotic and Electrophoretic Velocities Using Pulsed and Sinusoidal Electric Fields", *Electrophoresis*, vol. 38, 2017, pp. 1022–1037.

Samarasekera, C. and J.T.W. Yeow, "Low-Cost Implementation of Acoustophoretic Devices", Proc. Volume 9705, Microfluidics, BioMEMS, and Medical Microsystems XIV; 97050C, 2016. doi: https://doi.org/10.1117/12.2223290.

Shaheen, M.S., K.F. El-Massry, A.H. El-Ghorab and F.M. Anjum, "Microwave Applications in Thermal Food Processing", in *The Development and Application of Microwave Heating*, ed. W. Cao, London, Intech Open, 2012. https://www.intechopen.com/ books/2226 doi: 10.5772/2619, Chapter 1.

Srinivasan, V. and W.I. Higuchi, "A Model for Ionophoresis Incorporating the Effect of Convective Solvent Flow", *Int. J. Pharm.*, vol. 60, 1990, pp. 133–138.

Sriphutkiat, Y. and Y. Zhou, "Particle Manipulation Using Standing Acoustic Waves in the Microchannel at Dual-Frequency Excitation: Effect of Power Ratio", *Sensors and Actuators A: Physical*, vol. 263, 2017, pp. 521–529.

Stearns, C.A., F.J. Kohl, and D.E. Rosner, "Combustion System Processes Leading to Corrosive Deposits", *High Temp. Corrosion,* NACE-6, 1983, pp. 441–450.

Sun, X., X. Su, J. Wu, and B.J. Hinds, "Electrophoretic Transport of Biomolecules Through Carbon Nanotube Membranes", *Langmuir*, vol. 27, no. 6, 2011, pp. 3150–3156.

Tam, A.C., H.K. Park, and C.P. Grigoropulos, "Laser Cleaning of Surface Contaminants", *Appl. Surf. Sci.*, vol. 127–129, 1998, pp. 721–725.

Turner, J.H., P.A. Lawless, T. Yamamoto, D.W. Coy, G.P. Greiner, J. McKenna, and W.M. Vatavuk, "Sizing and Costing of Electrostatic Precipitators", *JAPCA*, vol. 38, no. 5, 1988, pp. 715–726. doi:10.1080/08940630.1988.10466413.

Watarai, H. and M. Namba, "Capillary Magnetophoresis of Human Blood Cells and Their Magnetophoretic Trapping in a Flow System", *J. Chromatogr. A.*, vol. 961, no. 1, 2002 June 28, pp. 3–8. doi: 10.1016/s0021-9673(02)00748-3.

WaterWorld, 2019. https://www.waterworld.com/wastewater/treatment/article/14073748/ microwave-treatment-is-an-inexpensive-way-to-clean-heavy-metals-from-treated-sewage.

Weiner, R.F. and R.A. Mathews, *Environmental Engineering*, 4th ed., Butterworth Heinemann, Oxford, UK, 2003.

Wen, J. and A.S. Wexler, "Thermophoretic Sampler and Its Application in Ultrafine Particle Collection", *Aerosol Sci. Technol.*, vol. 41, no. 6, 2007, pp. 624–629.

Wolfrum, J., "Laser Induced Chemical Reactions in Combustion and Industrial Processes", *Laser Chem*, vol. 6, 1986, pp. 125–147.

Yariv, E. and K.D. Dorfman, "Electrophoretic Transport Through Channels of Periodically Varying Cross Section", *Phys. Fluids*, vol. 19, 2007, p. 037101.

Yasuda, K., S.S. Haupt, S.-I. Umemura, T. Yagi, M. Nishida, and Y. Shibata, "Using Acoustic Radiation Force as a Concentration Method for Erythrocytes", *J. Acoust. Soc. Am.*, vol. 102, no. 1, 1997, pp. 642–645.

7 Acoustic Fields Coupled to Liquids

Basics and Some Applications

A wave is defined as a disturbance that propagates through space and time, usually with transference of energy. Waves travel and transfer energy from one point to another, often with no permanent displacement of the particles of the medium (that is, with little or no associated mass transport); they consist instead of oscillations or vibrations around almost fixed locations. A sound wave is a mechanical pressure wave that propagates or travels through a medium due to the restoring forces it produces upon deformation. Ultrasound is an acoustic wave, but with a frequency higher than the upper limit of human hearing—the lowest ultrasonic frequency is normally taken as 20 kHz (20,000 cycles per second). The top end of the frequency range is limited only by the ability to generate the signals—frequencies in the gigahertz range (upwards of 1 billion cycles per second) have been used.

Ultrasound is being used for a large variety of applications (Jagannathan, 2010). The uses of ultrasound within this available frequency range may be divided broadly into two areas. The first area utilizes low-power (high-frequency) ultrasound and is concerned with the physical effect of the medium on wave propagation; applications include medical imaging and non-destructive evaluations. The second area involving high-intensity (lower-frequency) ultrasonic waves, known as "power ultrasound", lies between 20 and 100 kHz (Mason and Lorimer, 2002). Applications of power ultrasound include surface cleaning, cutting, machining, plastic welding, nano-material forming, sintering, disruption of biological cell walls, sonochemistry, etc. The extended range of frequency for sonochemistry is 20 kHz to 2 MHz. The ultrasound of frequency ranging from 400 kHz to 2 MHz, known as "megasonics", extensively serves semiconductor industries for non-contact acoustic cleaning of contamination-sensitive products—a process known as megasonic cleaning. Megasonics is most appropriate for heat- or chemical-sensitive substrates that cannot withstand the heat and pressure of transient cavitation and for applications requiring line-of-sight-dependent cleaning.

The usage of ultrasound in physicochemical processing is increasing due to its versatility and applicability in solving many engineering problems (Doraiswamy and Thompson, 1999; Lawrence, 1975; Nagarajan, 2006; Shoh, 1975). It is widely used to intensify many conventional processes involving momentum, heat, and mass transfer in chemical engineering. For example, ultrasonics and megasonics increase heat and mass transfer coefficients over conventional methods and enhance chemical reaction rates by several orders of magnitudes. The processing applications of ultrasound include mixing and emulsification, dissolution of solids, extraction, impregnation, filtration, crystallization, and atomization (Mason and Lorimer, 2002).

DOI: 10.1201/9781003283423-7

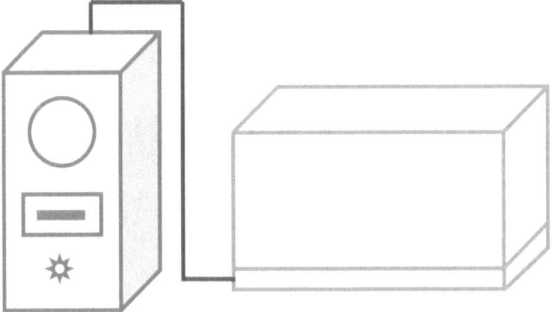

FIGURE 7.1 Tank-type ultrasonic system.

To produce ultrasonic waves in liquid, an ultrasound generator system is required. It essentially consists of a high-frequency generator and an ultrasonic tank or an ultrasonic horn. The schematic of two types of the ultrasonic system is illustrated in Figures 7.1 and 7.2. The generator converts a low-frequency electrical signal into a required high-frequency signal which actuates the horn or the transducers mounted, generally, at the bottom of the ultrasonic tank. The transducer creates vibrations at the bottom of the tank and produces ultrasonic waves propagating through liquids. In the other type, the horn tip vibrates and generates ultrasonic waves in the liquid present in a vessel. During propagation of wave, cavitation in liquids and bulk fluid motion due to streaming are formed.

An important aspect of ultrasonically induced flow is cavitation. Cavitation is the name given to the rupture of liquid, formation of cavity (filled up with gas or vapor or both), and associated dynamics of the cavities thus generated. Based on the cause of cavitation, four different types of cavitation are identified and the schematic overview is given in Figure 7.3. Hydrodynamic cavitation is produced by pressure variations in a flowing liquid, due to the geometry of the system. Acoustic cavitation is induced by the propagation of sound waves in a liquid. The sound waves cause regions of rarefaction and compression, and these pressure variations can set a tiny

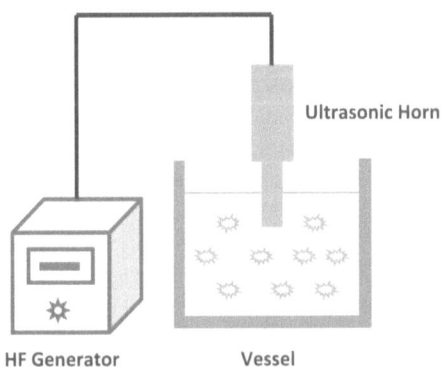

FIGURE 7.2 Probe-type ultrasonic system.

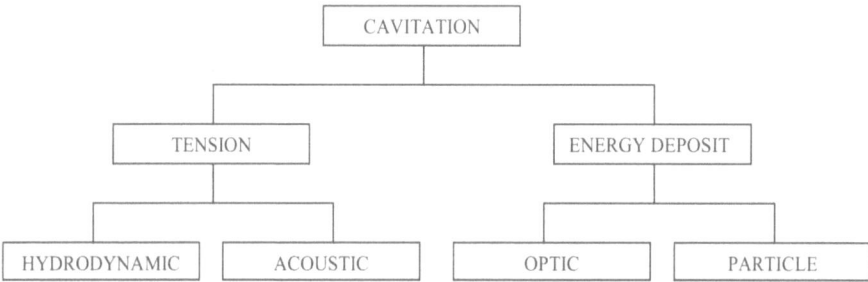

FIGURE 7.3 Classification of cavitation types.

bubble into motion. Optic cavitation is produced by light photons of high-intensity. Particle cavitation is produced by high-energy elementary particles such as a proton. In optic and particle cavitation, a high-energy pulse is transmitted from the elementary particle to liquid, producing rapid local heating and cavitation (Young, 1989).

Thus, when ultrasound having sufficient intensity is coupled with the liquid medium (water), it induces acoustic cavitation. Another important phenomenon induced by the propagation of ultrasound is known as "acoustic streaming", which is a time-independent fluid motion generated by a sound field. Hence, acoustic cavitation and streaming are the basic mechanisms that induce the fluid elements to flow when ultrasound is coupled with fluid.

Acoustic cavitation is the generation and action of cavities or bubbles in a liquid. As shown in Figure 7.4, when an acoustic wave propagates through a liquid medium (say, water), it creates regions of low and high pressures within the medium. The low-pressure areas are highly localized and changing all the time (for an ultrasonic standing wave, the time between lowest and highest pressure is typically 10–25 μs). When the liquid pressure momentarily drops below the vapor pressure of the liquid during the low-pressure portion of the acoustic wave (rarefaction), small evacuated areas, or cavities, are formed by tearing apart the liquid. These cavities quickly

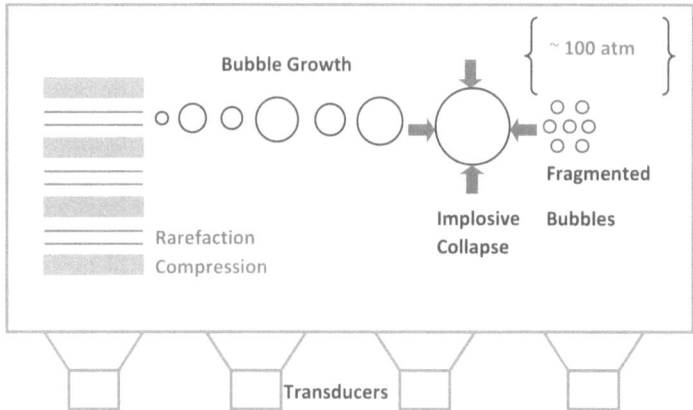

FIGURE 7.4 Bubble growth and collapse in acoustic cavitation.

become filled with gas (a foreign substance present in liquid such as dissolved oxygen, air) and/or vapor. These tiny bubbles are set in motion by the acoustic wave. The cavities formed undergo growth due to further reduction in pressure during rarefaction. When the bubble reaches a size that can no longer be sustained by its surface tension, the bubble will expand and then collapse, or implode, which is an important component of the cavitation phenomenon. The bubble implosion momentarily creates immense temperatures—thousands of Kelvin, and pressures—hundreds of bars (Suslick, 1988). Imploding bubbles generate shock waves which are transmitted to the surrounding liquid layers and induce flow.

Cavitation implosion force varies with the size and contents of the bubble. Larger bubbles are unstable and implode with larger force; smaller bubbles are stable and collapse with less force. Vapor collapses more quickly, resulting in a larger implosion force, whereas gas cushions and slows the collapse, resulting in a smaller implosion force. The typical flow induced by low-frequency ultrasound via cavitation process is shown in Figure 7.5.

Acoustic streaming is a time-independent fluid motion generated by a sound field. It may result from either vibration of a solid body adjacent to a fluid at rest, or from an acoustic standing wave generated in a fluid adjacent to a solid wall. Lord Rayleigh (1884) pioneered the analysis of acoustic streaming in a uniform duct. The work by Nyborg (1958) concludes that the velocity of streaming is determined by the coefficient of sound absorption, no matter from what source the sound may arise. The

FIGURE 7.5 Typical surface condition of an ultrasonically cavitated liquid.

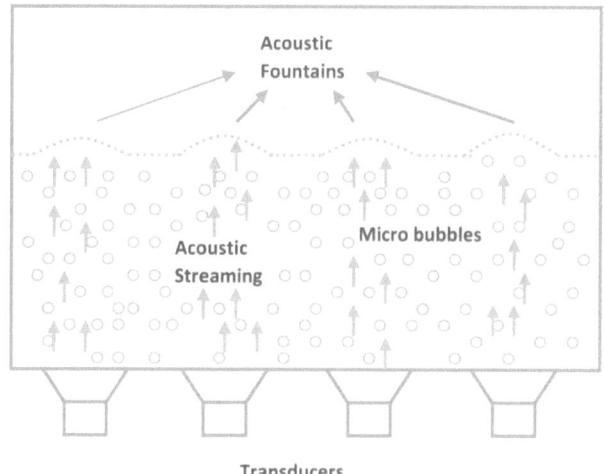

FIGURE 7.6 Acoustic streaming flow pattern in a megasonic tank.

work by Lighthill (1978) revealed that the momentum flux that causes the streaming motion becomes available due to the attenuation of acoustic energy flux.

The typical acoustic streaming that occurs in high-frequency ultrasonic and megasonic tanks is shown in Figure 7.6. The formation of acoustic fountains is observed at the center of the transducer locations. The microbubbles generated also follow the fluid path and initiate microstreaming. Acoustic streaming velocity is a function of acoustic energy intensity, geometry, energy absorption, liquid density and viscosity, and sound speed in the liquid. Streaming velocity has been found to increase linearly with acoustic intensity (power). It also increases with square of the frequency, and decreases with distance from the source, due to attenuation.

Acoustic streaming comprises several important effects: (i) Bulk motion of the liquid, (ii) microstreaming (Eckert streaming), and (iii) streaming inside the boundary layer (Schlichting streaming). The primary effect of acoustic streaming is the bulk motion of the liquid, the strong localized flow of liquid which generates shear force. The second effect of acoustic streaming is microstreaming. It is also known as Eckart streaming and occurs near oscillating bubbles or any compressible substance in the liquid. Microstreaming occurs at the substrate surface, or outside the boundary layer, due to the action of bubbles as acoustic lenses that focus sound power in the immediate vicinity of the bubble. This is a powerful type of streaming in which the bubbles scatter sound waves and generate remarkably swift currents in localized regions. The currents are most pronounced near bubbles that are undergoing volume resonance and are located along solid boundaries. Microstreaming aids in dislodging particles and contributes to megasonic cleaning.

Most of the flow induced by acoustic streaming occurs in the bulk liquid outside the boundary layer. However, there is a third effect of acoustic streaming, Schlichting streaming, which is associated with cavitation collapse. Schlichting streaming occurs inside the boundary layer and is characterized by very high local velocity and vortex (rotational) motion. The vortices are of a scale much smaller than the acoustic wavelength. Schlichting streaming results from interactions with a solid boundary. Steady viscous stresses are exerted on the boundaries where this type of rotational motion occurs, and these stresses may contribute significantly to the removal of surface layers. Acoustic streaming, both inside and outside the boundary layer, enhances cleaning and other chemical reactions. The strong currents and small boundary layer thickness that result from acoustic streaming enhance transport properties.

Now that we understand the basics of ultrasonic/megasonic fields coupled to liquids, let us segue in the remaining sections in this chapter to some fascinating applications in mixing, atomization, and size reduction, to be followed in the next chapter by applications in transport phenomena and chemical reactions.

7.1   DESTRATIFICATION/MIXING

Stratification is the existence of fluid at different temperatures within a storage container. The temperature gradient may be set up due to heat leak, intermolecular forces between molecules, etc., resulting in one part of the fluid layer becoming warm and rising to the top of the container. In order to keep the fluid in the vessel at a uniform temperature, this stratification layer must be broken.

Cryogenic liquids are generally defined as liquids that have boiling points below the normal boiling point of nitrogen (77 K). Cryogenic liquids such as liquid nitrogen (LN_2), liquid oxygen (LO_2), and liquid hydrogen (LH_2) have found many uses in areas such as space applications, medical field, and laboratory and industrial applications. LH_2 is primarily used as rocket fuel and LN_2 serves as a refrigerant. These cryogens are stored in tanks of various sizes and shapes. Even though the tanks are generally super-insulated, heat leak into these tanks occurs due to the large temperature difference between the ambient and the cryogen. As the fluid gets heated, natural convection currents are set up, leading to thermal stratification within the fluid. Thermal stratification increases the vessel pressure to a value corresponding to the temperature of the warmer upper layers of the liquid, and the vapor hence formed has to be vented to maintain the tank pressure. The stratification of the cryogen leads to reduced lock-up times, and the warmer stratified column of the cryogen can also cause cavitation in cryo-pumps (Khurana et al., 2006). Sherif et al. (1997) point out that boil-off losses associated with the storage, transportation, and handling of liquid hydrogen can consume up to 40% of its available combustion energy. One of the causes of boil-off losses is stratification which must be destroyed to improve the cost-effectiveness of cryogen storage.

Hence, effective in-situ destratification methods are required to reduce vapor loss and improve the cost-efficiency of the operation. Destratification also prevents the onset of hydrodynamic cavitation from the warmer stratified liquid layers that can occur while transferring the liquid, thereby affecting the flow and causing damage to the turbo pumps. Conventional ways of decreasing boil-off losses due to stratification include employing high-conductivity plates (conductors) installed vertically in the storage vessel. The plates produce heat paths of low resistance between bottom and top of the vessel and can eliminate temperature gradients and excessive pressures. Efficient magnetic refrigerators that pump the heat out and maintain the liquid at sub-cooled or saturated conditions are also known to be used for storage purposes (Sherif et al., 1997). Theoretical studies by Khurana et al. (2006) show that mixing of fluid layers at the walls of the container using ribs as roughness elements results in 30% less stratification.

Application of ultrasonics to counter stratification is a classic case-study of process intensification to reduce cost. The combined effect of cavitation and acoustic streaming on the destratification of thermally stratified liquid layers in insulated, pressurized storage containers has been investigated by Jagannathan (2010). Initially, studies were carried out in a water storage tank since water, as a medium, is easier to handle compared with cryogens. Later, it was extended to cryogenic storage systems.

The schematic representation of experimental setup used is shown in Figure 7.7. It consists of an ultrasonic system and a storage container. The ultrasonic system consists of an ultrasonic tank (US tank) made of stainless steel and a high-frequency generator. The bottom of the US tank is mounted with a transducer which generates sound waves of required frequency. The high-frequency generator converts the electric power of low frequency to required high-frequency. The storage container is also made of stainless steel with an inner diameter of 20 cm and height of 56.6 cm.

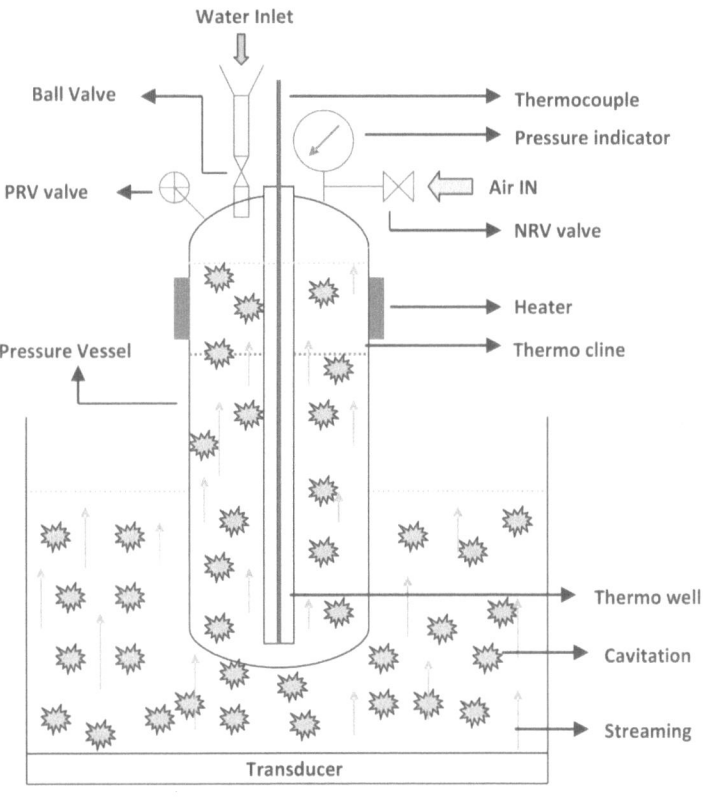

FIGURE 7.7 Experimental set-up for the study of ultrasonic destratification.

It is fitted with water inlet and outlet ports, pressure gauge, non-return valve (NRV), and pressure relief valve (PRV). The container is pressurized by pumping air through NRV. The total volume of water taken in the storage container is 15 liters for the studies. The top layer of water in the vessel is heated with the help of ceramic band-type heater (21 cm ID × 12 cm height) with a power input of 3500 W. The temperature of the water within the vessel is monitored by T-type (copper-constantan) thermocouples with an accuracy of ±0.5–1°C. The temperature was monitored at nine ports along the height. The distance between two adjacent ports is 5 cm. Ports 1 and 9 correspond to bottom and top of the tank, respectively. The vessel is kept inside a US tank with the help of a support. Clearance of 2–3 cm is ensured between bottom of the vessel and surface of the transducer. The vessel is insulated above the surface of water level in US tank in order to minimize the heat loss to the surroundings.

Stratified regions have been created within the vessel by heating the top layer of water in the vessel. The temperature of water at the bottom of the tank was at room temperature. The temperature difference between hot and cold water ($\Delta T_{initial}$) was maintained within the range of 32–40°C at the start of the experiment (at t = 0) in all cases. It is assumed that the temperature of the water at any point in a particular cross-section of the vessel is the same. Due to the density difference, the warmer layer remains at the top of the vessel. Ultrasound of desired frequency and power was introduced to break up strata and the temperature of the water was monitored at all ports. Subsequently, the temperature profile along the height of the vessel was obtained at different time instants and the destratification parameters were analyzed. Experiments were carried out at different storage pressures at a fixed power level of ultrasound in order to study the effect of pressurization on ultrasonic destratification.

Coupling of an ultrasonic field with water produces regions of low pressure and high pressure. When the pressure exceeds the cavitation threshold, it causes the formation of bubbles at low-pressure regions. Bubbles formed to follow a certain path to reach high-pressure regions where they implode. During travel, the bubble may grow and oscillate about some equilibrium radius, thus imparting momentum to the liquid medium surrounding it. This causes the mixing of water layers exhibiting different temperatures and destroys the stratification region. On the other hand, at higher frequencies having megasonic characteristics, acoustic streaming also breaks up thermally stratified regions within the vessel.

In order to analyze the destratification process in detail the following three parameters are defined—Stratification Index, Mixing Time, and Mixing Effectiveness.

Stratification index (SI) is a dimensionless parameter that indicates the extent of mixing in the storage tank at any instant of time. It is defined as

$$SI = \frac{A_t}{A_0} \tag{7.1}$$

where A_t and A_0 are areas under temperature versus height curve with respect to T_{mix} (uniform temperature of water at complete mixing) at any time t and time t = 0, respectively. Figure 7.8 illustrates the area under the curve of temperature versus height with respect to T_{mix} at time t = 0.

SI scales between 0 and 1. SI value of one indicates no mixing or existence of strong thermally stratified layers, which is obviously at initial time t = 0, whereas

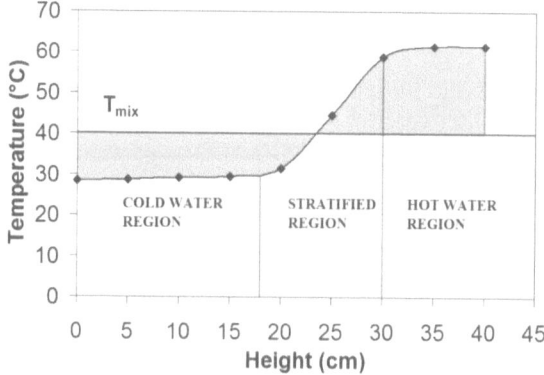

FIGURE 7.8   Area under the curve of temperature versus height.

zero refers to complete mixing or uniform temperature within the container, which indicates the achievement of destratification. Ideally, the temperature versus height curve must become flat and must coincide with T_{mix}-line when SI becomes zero.

Mixing time, t_{50}, is the time required for the completion of 50% of mixing, i.e., the time corresponding to an SI value of 0.5. It is converted into dimensionless form given as $t_{50}/t_{50,ref}$, by choosing $t_{50, ref}$ as mixing time in the case of natural mixing conditions without ultrasonic field.

Mixing effectiveness (ME) is defined as the ratio of the reduction in temperature difference between hot and cold layers of water after an hour of operation to the initial difference between the hot and cold layers of water in the vessel. It is given by

$$ME = \frac{\left(\Delta T_{initial} - \Delta T_{t=60\ min}\right)}{\Delta T_{initial}} \tag{7.2}$$

where $\Delta T = (T_{hot} - T_{cold})$

T_{hot} – Temperature of hot layer of water

T_{cold} – Temperature of cold layer of water

Experiments were first carried out in the absence of an ultrasonic field in order to study the natural mixing of hot and cold layers of water without ultrasound. The absolute pressure within the vessel is maintained at 1 atm. It is apparent that it will take longer time for the removal of stratified regions without any action of external forces. Figure 7.9 shows the temperature distribution within the vessel at different times for natural mixing conditions without ultrasound. The stratified region exists between 20 and 30 cm in height. Here, the principal mechanism of heat transfer is conduction between hot and cold layers of water which reduces temperature gradient to some extent. Heat loss to the surroundings was minimized by using insulation.

Figure 7.10 shows the decay of SI with time for natural mixing. Mixing time, t_{50}, is calculated as 190 minutes ($t_{50, ref}$) for natural mixing at a storage pressure of 1 atm. Ideally, under natural mixing conditions, it will take infinite time to get a uniform temperature of the liquid within the vessel, if the system is adiabatic.

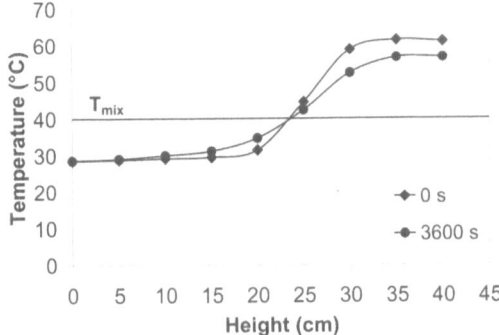

FIGURE 7.9 Temperature profile in the absence of ultrasonic field.

Figures 7.11(a)–(j) show the SI decay with time at different storage pressures for various frequency settings of the acoustic field with a power input of 500–900 W, and storage pressures ranging from 1 to 3 atm.:

These data at 1 atm storage pressure are summarized in Figure 7.12.

In low-frequency regimes, 68 kHz provides good mixing. In the intermediate frequency regime, 132 kHz provides better mixing, but on the whole, these intermediate frequencies showed poor mixing behavior since the net forces resulting from cavitation implosions and streaming are the lowest. Dual frequencies provide better destratification compared with single frequencies due to the higher energy input and combined destratification mechanisms associated with them. Among dual frequencies, the combination of low and intermediate frequencies (58/192 kHz) shows better results at 1 atm. Among all the frequencies studied in the range of 25 to 470 kHz, the highest frequency (470 kHz) breaks up the thermal strata faster, thereby providing the most effective destratification. This indicates that acoustic streaming, which elicits macroscopic, unidirectional fluid motion, is much more effective at the mixing of stratified layers compared to acoustic cavitation. Higher storage pressures, however, suppress acoustic streaming effects.

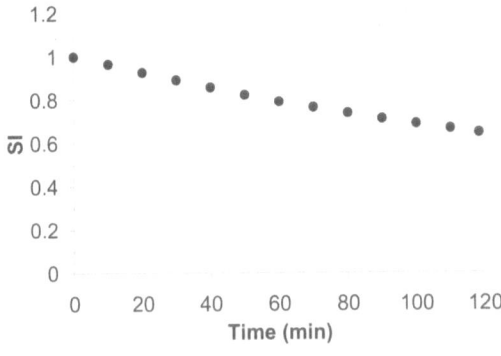

FIGURE 7.10 Time evolution of stratification index in the absence of ultrasonic field.

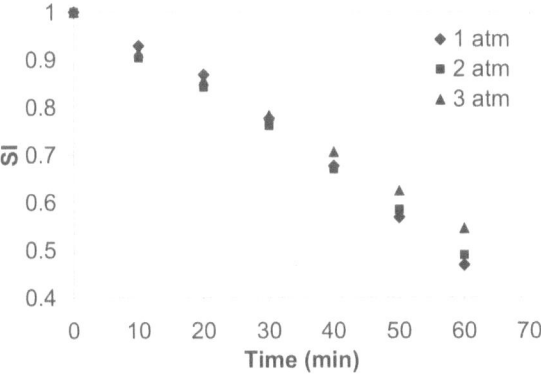

FIGURE 7.11a Stratification decay at 25 kHz, 500 W.

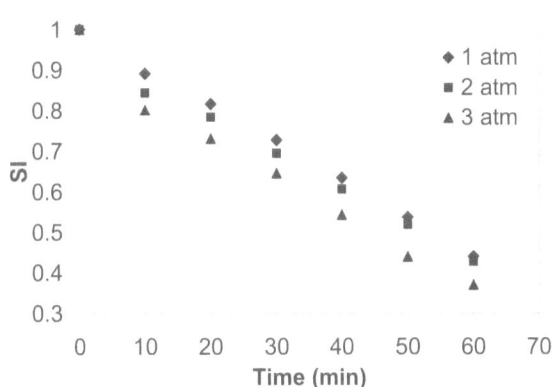

FIGURE 7.11b Stratification decay at 40 kHz, 500 W.

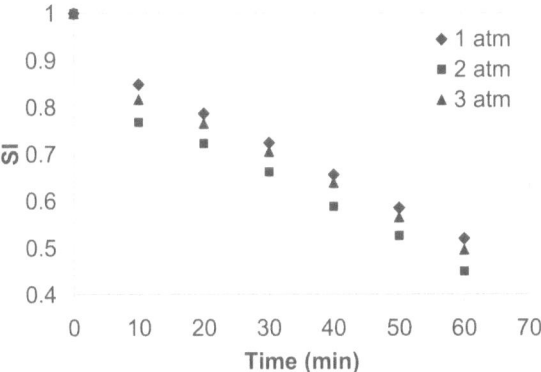

FIGURE 7.11c Stratification decay at 58 kHz, 500 W.

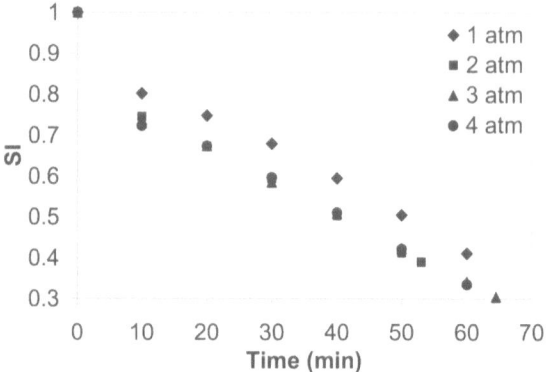

FIGURE 7.11d   Stratification decay at 68 kHz, 750 W.

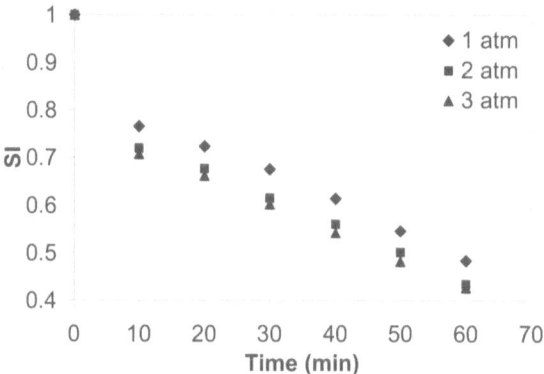

FIGURE 7.11e   Stratification decay at 132 kHz, 500 W.

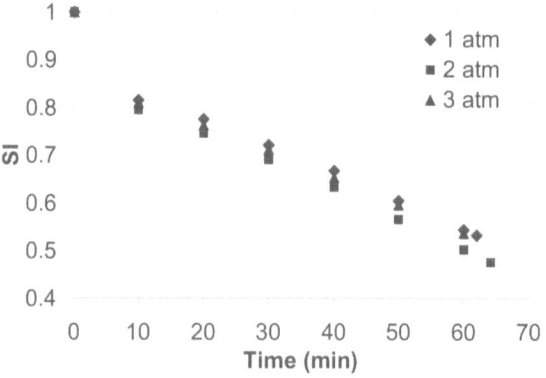

FIGURE 7.11f   Stratification decay at 172 kHz, 500 W.

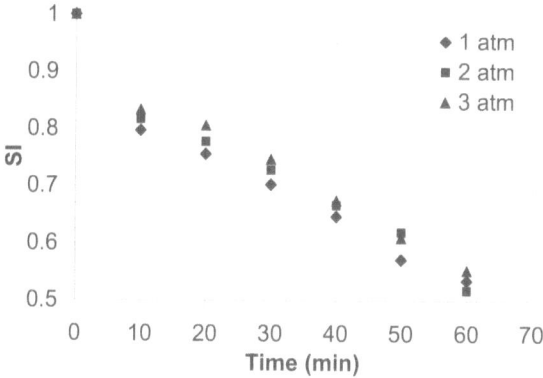

FIGURE 7.11g   Stratification decay at 192 kHz, 500 W.

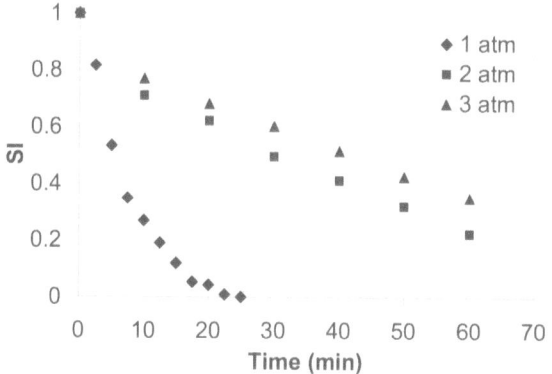

FIGURE 7.11h   Stratification decay at 470 kHz, 900 W.

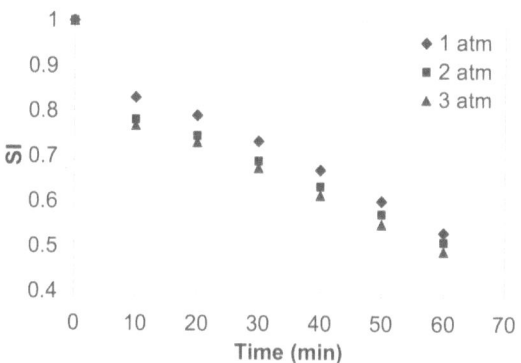

FIGURE 7.11i   Stratification decay at 58/192 kHz (dual frequency), 600 W.

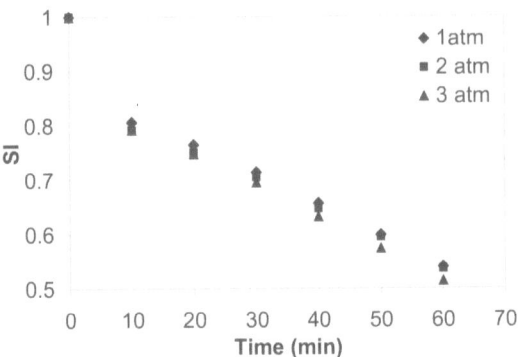

FIGURE 7.11j Stratification decay at 172/192 kHz (dual frequency), 600 W.

At lower frequencies, and cavitation-dominated fields, higher storage pressure results in better mixing. In acoustic streaming-dominated regimes (higher frequencies), lower storage pressure (close to atmospheric pressure) is preferable for faster destratification. At a particular storage pressure, the mixing time decreases initially (low-frequency regime), then increases (intermediate-frequency regime where total acoustic forces are minimum), and later decreases quite rapidly (streaming-dominated high-frequency regime) with the frequency of ultrasound. The use of higher power levels (amplitude) guarantees better destratification at all frequencies of ultrasound, irrespective of the storage pressure. Higher power levels uniformly result in faster destratification.

Figures 7.13 and 7.14 show how ultrasonic frequency influences Mixing Effectiveness at various power levels and storage pressures.

Mixing Effectiveness increases when storage pressure is increased in the low-frequency regime and displays no significant improvement at intermediate frequencies. As storage pressure increases, its value drops from 100% to 58% at 3 atm, for 470 kHz but it still holds better destratification abilities compared with all other

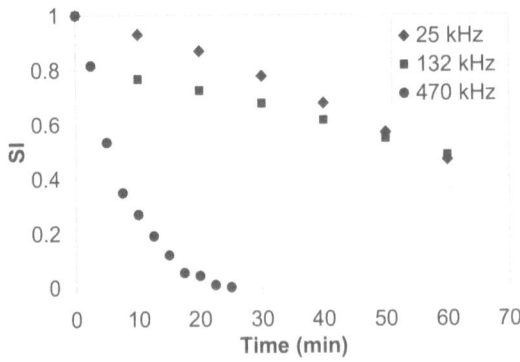

FIGURE 7.12 Effect of ultrasonic frequency on stratification decay at 1 atm storage pressure, 100% input power.

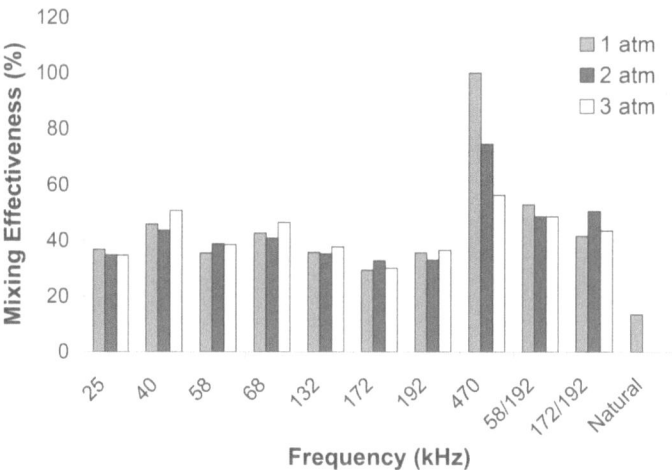

FIGURE 7.13 Effect of frequency on mixing effectiveness at storage pressures of 1–3 atm (100% power level).

frequencies. When power level is decreased, a corresponding decrease in mixing effectiveness was observed at all frequencies. A total of 13.5% mixing was obtained without ultrasonic field during 1 hour of natural mixing conditions. The introduction of 470 kHz acoustic field gives 100% mixing within 15 minutes, clearly demonstrating the suitability of high-frequency ultrasonics for thermal destratification under atmospheric-pressure storage conditions.

The mixing efficiency obtained by Vetrimurugan (2007) for different cavitation intensities is shown in Figure 7.15. From the results for a 68 cm cylinder, it can be observed that the mixing efficiency is increasing with cavitation intensity.

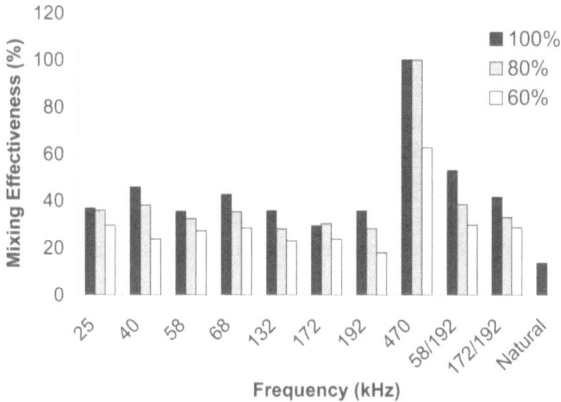

FIGURE 7.14 Effect of frequency on Mixing Effectiveness at input power levels of 60–100% (1 atm storage pressure).

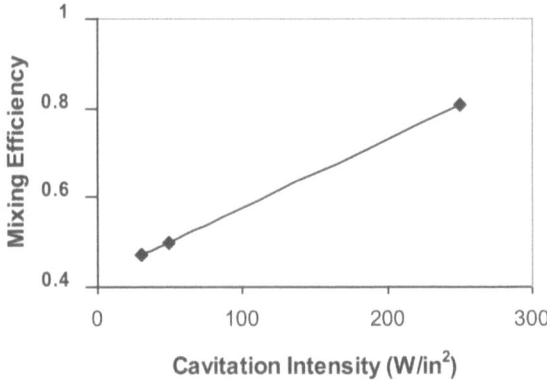

FIGURE 7.15   Effect of cavitation intensity on mixing efficiency in a 68 cm FRP cylinder.

CIF/PIF ANALYSIS

The cost associated with boil-off loss of cryogenic fuels during on-ground storage is high enough that any feasible solution will get serious consideration. The data demonstrate that inducing an ultrasonic/megasonic field is effective in destroying stratification. Retrofitting every fuel storage container with the required acoustic field does entail cost, although "pulsing" is an option that can reduce operating expenses by 10–20%.

7.2   EFFERVESCENT ATOMIZATION

Atomization is a process for the conversion of bulk liquid into a spray or mist (i.e., collection of drops), by subjecting the liquid to pressure variations. The process can also be described as nebulization. In internal combustion engines, fine-grained fuel atomization is instrumental to efficient combustion. In spray drying, fine droplets of the liquid feed are required for better product size and quality. The fineness of the droplet size of the spray determines the quality of the spray. Ultrasonic vibrations in a thin film of liquid can disintegrate the film and break it up into fine droplets. Many ultrasonic nebulizers are available commercially to get fine sprays.

Effervescent atomization is a process that utilizes the effervescent gas to create bubbles (weak spots) in the liquid in a mixing chamber. The two-phase mixture is then allowed to discharge through a nozzle orifice, where the bubbles rapidly expand and shatter the liquid into ligaments and droplets (Sovani et al., 2001). The quality of the spray mainly depends on the type of flow existing in an orifice. Literature suggests that the existence of fine bubbly flow or mist-like flow in an orifice gives a better quality spray.

Another application for effervescent atomization is driven by the need for cheap and plentiful energy that has increased the usage of coal-based fuels. Coal is lique-fied by suspending microscopic coal particles in water or some other suitable liquid

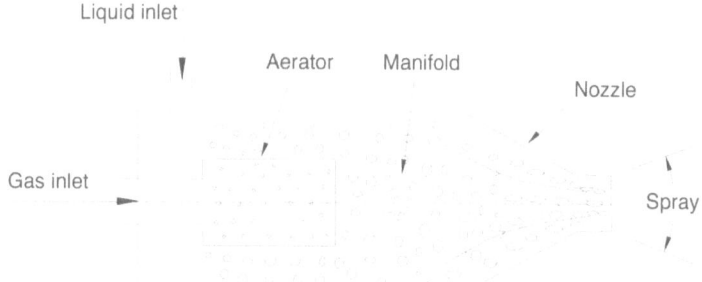

Liquid inlet

Aerator   Manifold

Nozzle

Gas inlet

Spray

FIGURE 7.16 Schematic of effervescent atomizer.

for its easier transportation, thereby facilitating its use in pipelines, tankers, and the combustion zones of gas turbine and internal combustion engines. A fine spray of coal-water slurry is required in the combustion chamber for efficient burning of the fuel. But the majority of atomizers are incapable of efficiently atomizing coal-water slurries since they are sensitive to slurry rheological properties. Also, the coal particles make it abrasive and prone to clogging orifices. In order to overcome these problems, an effervescent atomization method has been developed.

Figure 7.16 illustrates the functioning of an atomizer (Jagannathan, 2010). A two-phase flow, consisting of fine bubbles in a liquid, is expanded through an orifice to form a spray of fine liquid droplets. An aerator issues gas bubbles in the flowing liquid. A mixing chamber or manifold mixes the gas bubbles with the stream of liquid. The two-phase mixture is discharged through an orifice into the combustor. The gas bubbles in the two-phase mixture grow rapidly on being discharged from the orifice and shatter the surrounding liquid into ligaments and droplets.

Typically, the nozzle orifice dimensions are of the order of millimeters in diameter in order to ensure the formation of a mist-like spray and sufficient penetration of spray in the combustion chamber. In order to get a bubbly flow through such a small discharge orifice, the sizes of the gas bubbles in the spray need to be extremely small. The generation of such tiny bubbles calls for fine pores in the aerator, and therefore, substantial pressure drop in the aerator. Better means of generating fine bubbles uniformly dispersed in the flow through the discharge orifice is necessary. The need, then, is to obtain finely dispersed gas bubbles in the mixing chamber of the effervescent atomizer, thereby producing a spray consisting of fine liquid droplets.

The application of ultrasound in improving the performance of an effervescent atomizer, by breaking up larger bubbles produced by an aerator into tiny bubbles, is another representation of acoustic intensification of a process and has been investigated by Jagannathan (2010). The schematic diagram of the test setup is depicted in Figure 7.17. The test setup contains three main components, namely, air supply unit, water supply unit, and a test article. The air supply unit issues air to the test article from the high-pressure air cylinder at desired pressure level using a pressure regulator. The flowrate of air was controlled by a needle valve and measured using a calibrated air rotameter. The injection pressure of the air was obtained from the

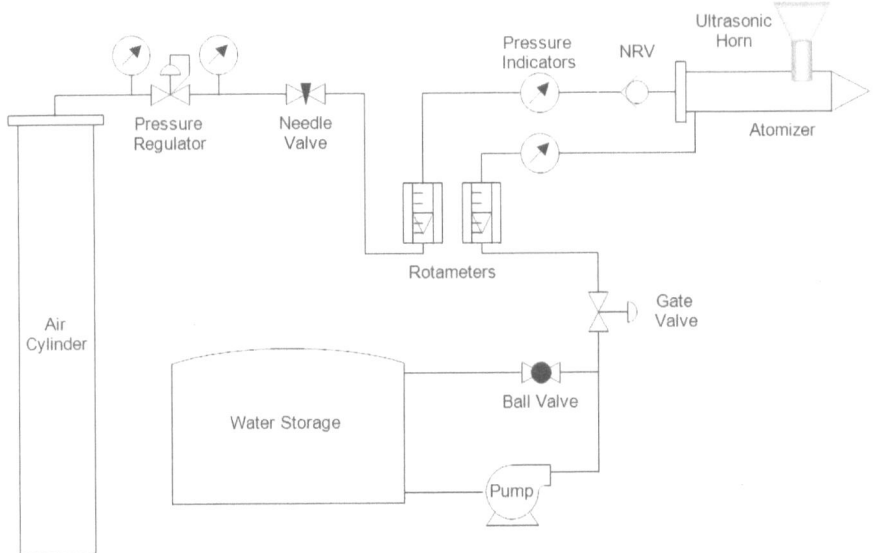

FIGURE 7.17   Schematic of experiment setup.

pressure indicator before it enters into the atomizer via a non-return valve. The water supply unit issues water from a storage tank with a capacity of 50 liters, using a pump having a maximum capacity of 900 liters per hour with 0.9 m head. The flowrate of water entering the atomizer was controlled by a control valve, as well as by the return loop. The flowrate of water was monitored by a calibrated water rotameter. The test setup comprises an atomizer and ultrasonic probe system.

The atomizer is made of acrylic sheet which has a rectangular mixing chamber of length 180 mm with a cross-section of 20×20 mm^2 and a conical converging section of length 35 mm. The nozzle has an exit orifice of 4 mm diameter and 7 mm length. The mixing chamber was made rectangular in order to avoid curvature effects while capturing images. Air was admitted into the mixing chamber through an aerator tube of length 120 mm and diameter 10 mm, having a hole of 0.5 mm diameter at the center of the circular end.

A 20 kHz ultrasonic horn was positioned in the mixing chamber such that ultrasound propagates perpendicular to the flow direction at a distance of 32 mm before the converging section. The details of the atomizer are given in Figure 7.18(a). This test configuration was used to carry out initial experiments to find out the effect of ultrasound on bubble breakup. Based on the results obtained, the atomizer configuration was modified and re-fabricated.

The second atomizer was designed in such a way that the ultrasound propagates parallel to the direction of liquid flow. Air was introduced directly into the mixing chamber via a 0.5 mm hole drilled in one of the side walls of mixing chamber, and mixed with the liquid flow without the need for a separate aerator. The atomizer was slightly shortened to a length of 130 mm but has the same cross-section of 20×20 mm^2. The converging section was made on the wall of the mixing chamber

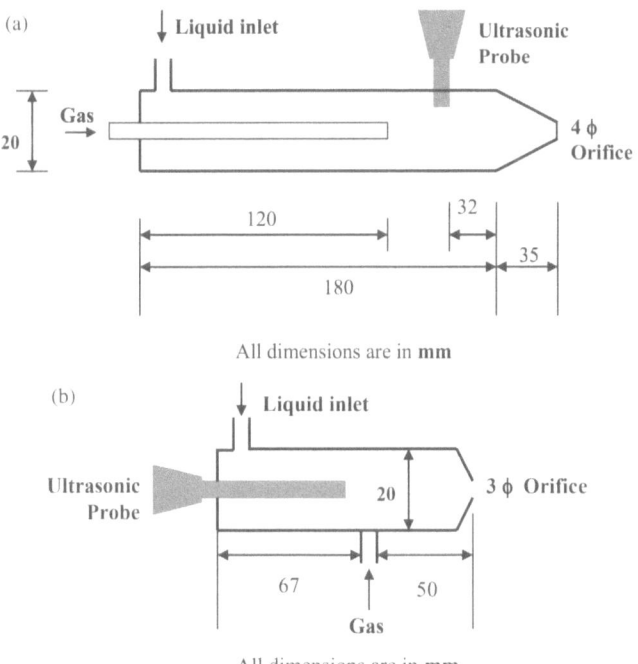

FIGURE 7.18   Ultrasonic configurations in mixing chamber.

itself on outlet side. The converging section is 10 mm in length and has an exit orifice of 3 mm diameter and 3 mm length. The details of the atomizer are shown in
Figure 7.18(b).

The ultrasonic probe system consists of an ultrasonic horn made of titanium
rod and a high-frequency generator. The high-frequency generator converts the
low-frequency electrical signal to a high-frequency electrical signal which actuates the ultrasonic horn at its excitation frequency. In the present study, a 20 kHz
ultrasonic probe with 750 W power input was used since bubble breakup is generally higher due to intense acoustic cavitation encountered at lower ultrasonic
frequencies.

Water was supplied first at desired flowrate to the test setup, and then the air
was admitted into the mixing chamber of the atomizer at desired flowrate. This
forms a two-phase flow in which a train of bubbles flows along with the liquid. The
mixing chamber was visualized initially without ultrasound to characterize initial
bubble sizes. Then, the ultrasonic field was turned on with known power input, and
again images were captured of bubbles present in the mixing chamber. A high-speed
CMOS camera was used to observe bubble motion and breakup inside the mixing
chamber in the absence and presence of the ultrasonic field. Diffused light emanating
from a translucent acrylic sheet was used for backlighting the atomizer. Similarly,
pictures of mixing chamber and spray were captured with and without ultrasound
at a frame rate of 1000 fps using the second atomizer configuration. The effect of
ultrasonic power input on bubble breakup and spray formed was also studied in the

same way as described earlier. All the experiments were carried out at an air injection pressure of 0.05 MPa.

The interface between water and air bubble in the captured images was first identified by tracing the interface manually. The trace lines were extracted using imageJ software. This gives the details of the number and the corresponding size distribution of bubbles present in an image. The size of bubble was estimated based on the projected area of bubble on the image acquired. In this manner, the number and size of the bubbles within the test section were estimated from randomly picked 20 images for a particular experimental run and bubble size distribution was calculated.

The effect of the presence of an ultrasonic field on breaking up larger bubbles into smaller was studied. An ultrasonic field of 20 kHz frequency was used at different power levels. Thus, the effect of ultrasonic power input was also studied. The effect of orientation of the ultrasonic field was also studied by applying the ultrasonic field in the directions parallel and perpendicular to the liquid flow direction.

Figure 7.19 shows the bubbles present in a mixing chamber with and without an ultrasonic field. In the absence of the field, the bubbles were of the order of 5 to 8 mm in diameter and moved along the liquid without breakup. Bubble coalescence also took place, making the bubble size even larger. When ultrasound is turned on, the larger bubbles introduced by an aerator were broken up into tiny bubbles of diameter ranging from 2 mm to sub-micrometer.

Figure 7.20 displays the bubble size distribution (BSD) with and without ultrasound for a GLR (gas-to-liquid mass flowrate ratio) of 0.063%; the data clearly demonstrate the ability of ultrasound in increasing the number of tiny bubbles in liquid by breaking up bubbles of larger size. An ultrasound power of 750 W was used (100% of available power).

Figure 7.21 shows the effect of ultrasonic power input on breaking up bubbles present in the mixing chamber. At lower power input (40% of maximum), the bubbles produced are relatively larger and BSD shifts toward the larger size. However, 80% of the bubble population is less than the orifice diameter of 3 mm and 97% of bubbles are less than the orifice diameter of 4 mm. Generally, higher power levels are preferable since they produce higher acoustic pressure amplitudes which would result in more intense bubble breakup and produce many finer bubbles. In the present study, at a 40% power level, the desired bubble breakup was obtained. When power was lowered further below 40%, ultrasound did not break up bubbles; instead, bubbles were pushed away from their flow path toward the direction of sound propagation and underwent shape changes without disintegrating.

It was observed that disintegrated bubbles often coalesce before reaching the exit orifice once they move away from the field, especially when ultrasound propagates across the flow. This can be avoided by placing the ultrasonic probe as close as possible to the exit orifice, and/or by making the ultrasound field propagate in the direction of liquid flow. Hence, the direction of sound propagation was changed parallel to the flow direction in the second configuration, and experiments were repeated. This configuration also has the advantages of augmenting the movement of tiny bubbles after breakage and adding momentum to liquid flow apart from preventing coalescence.

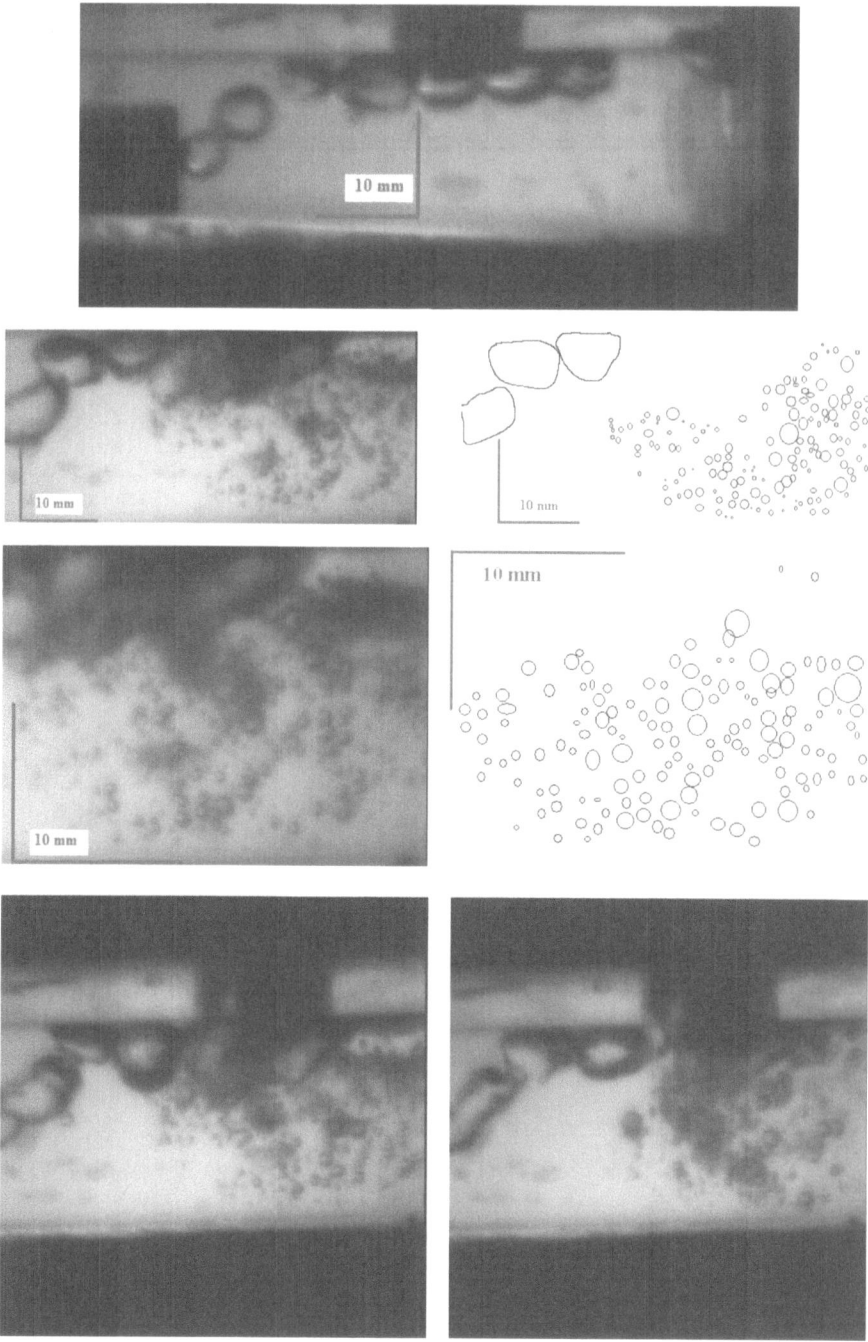

FIGURE 7.19 Bubbles within mixing chamber (Configuration-1).

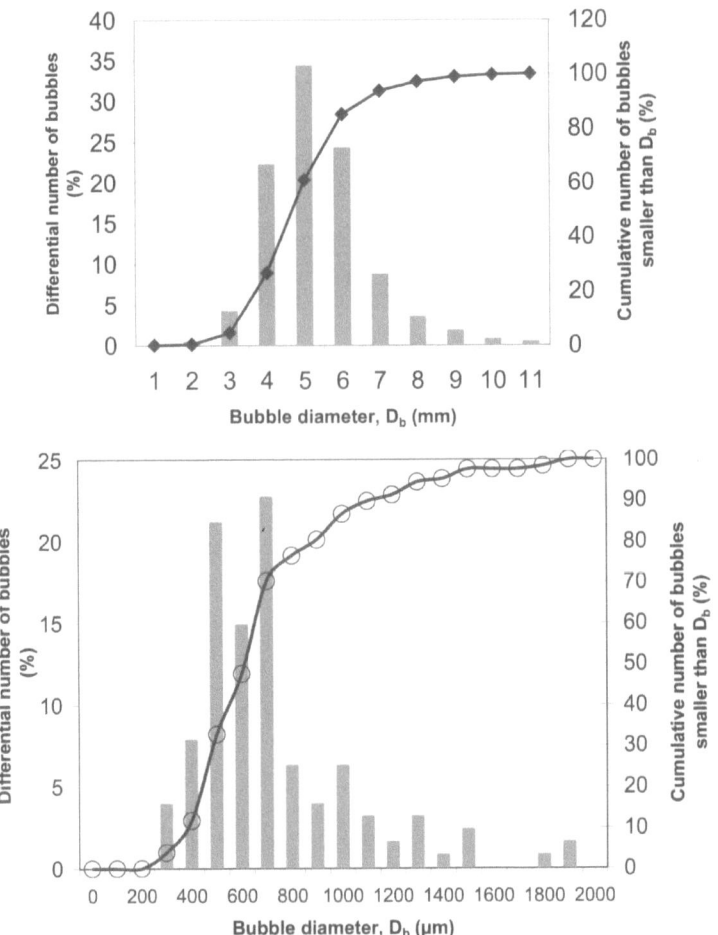

FIGURE 7.20   Bubble size distribution with and without sonication.

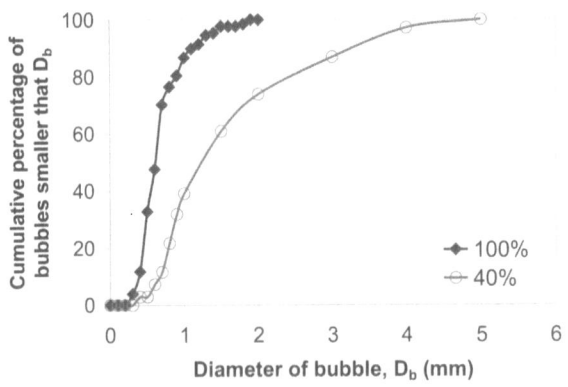

FIGURE 7.21   Effect of ultrasonic power input on BSD.

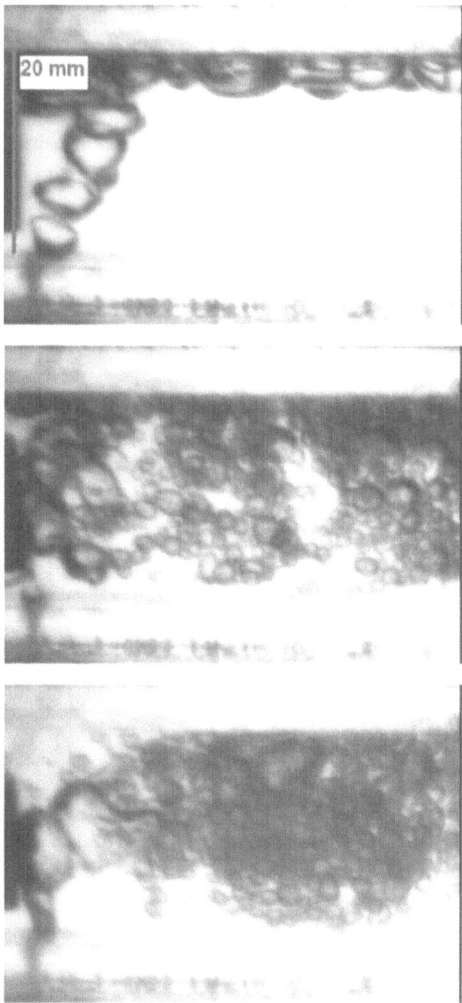

FIGURE 7.22   Bubbles within mixing chamber: Ultrasonic Configuration-2.

Figure 7.22 shows the images of mixing chamber of the second configuration taken at 1000 fps in which air is admitted through a 0.5 mm diameter hole from the bottom of the chamber, and ultrasound propagates parallel to the liquid flow. It is evident that bubbles remain larger in size when ultrasound is absent. In this configuration, the larger bubbles disintegrated as soon as they enter the mixing chamber and produce many smaller bubbles. When the ultrasound (20 kHz) is present, bubbles are uniform in size and present throughout the mixing chamber, and a fine bubbly two-phase mixture was produced.

Figure 7.23 illustrates the BSD in the mixing chamber of Configuration-2. The bubbles were more uniform in size after breakup, and 100% of bubbles are less than 3 mm, the diameter of the orifice, with 60% power level.

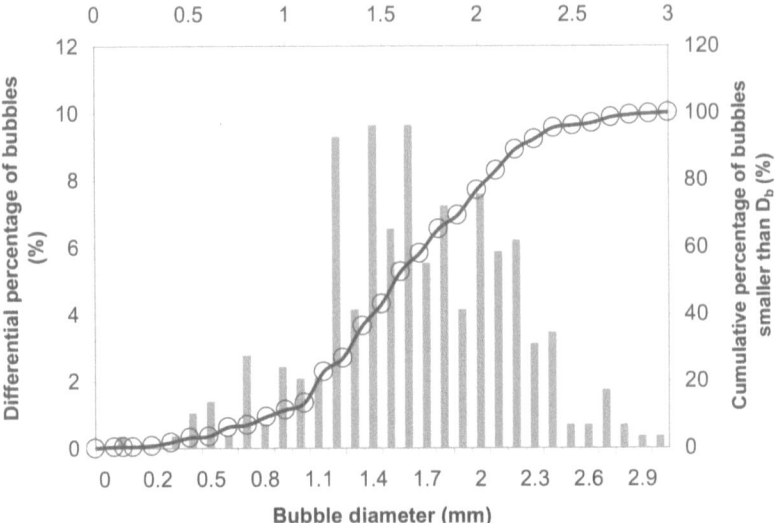

FIGURE 7.23   BSD after break-up in Configuration-2.

Figure 7.24 shows photographs of the movement of the gas bubble and its breakup into a large number of small bubbles while passing through the ultrasonic flow field. It was observed from the experiments that the gas bubbles spontaneously break up into a cloud of fine bubbles by cavitational collapse. In addition, acoustic streaming causes a sudden drag force experienced by the bubbles moving through the ultrasonic flow field, due to which the bubbles tend to elongate in the direction of the streaming force. Acoustic streaming augments bubble breakup and distribution throughout the mixing chamber. Hence, ultrasound can be used effectively as an enhancer for effervescent atomization. Its use can even be considered in stand-alone mode. Indeed, ultrasonic atomizers are available as off-the-shelf products.

It is evident from the results that the gas bubbles introduced by an aerator in the mixing chamber are broken up into smaller bubbles using 20 kHz ultrasound, and then distributed uniformly in the mixing chamber, thereby ensuring that a fine bubbly flow takes place though the discharge orifice. Bubbles disintegrate as soon as the ultrasonic field interacts with the bubble. The break-up of larger bubbles into fine bubbles is intensified in presence of ultrasonic field, provided

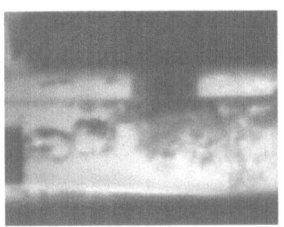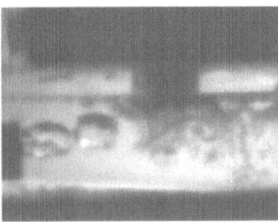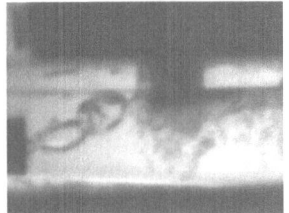

FIGURE 7.24   Bubble moving in a mixing chamber in presence of the ultrasonic field.

sufficient acoustic pressure amplitude (power input) was supplied. Larger bubbles of size 5–10 mm in diameter were broken up into 1 mm to sub-micrometer bubbles at 20 kHz, 750 W power. At lower power amplitude, bubbles undergo shape change without fragmentation. The main reason for bubble disintegration was identified as the deformation of bubble due to instabilities associated with higher acoustic stresses. Ultrasound can, therefore, be used effectively as an "intensifier" for bubble breakup in the atomizer, resulting in enhancement of spray quality produced.

PIF/CIF ANALYSIS

Ultrasonic/megasonic atomization results in smaller droplet sizes compared with other means of atomization; the resulting enhancement in the efficiency of combustion is, for the most part, sufficient to justify the technology. The logistics of locating the probe and aligning it optimally with fluid flow are sometimes challenging, and any retrofitting required for the process equipment can be an added cost factor.

7.3 NANOPARTICLE SYNTHESIS

Nanoparticulate technology is really a boon to humanity provided safe practices are followed. "Nano" refers to the size regime of nanometers or 10^{-9} m. Manipulation of matter at atomic level is possible. Much may be learned from nature. For example, human bones, rat teeth, and spider webs are made up of nano-structured material, and hence exhibit higher strength. The properties of materials change drastically as they reach nano-dimension. For example, the color of alumina changes from white to gray as it reaches sub-100 nm range. Surface area increases, hence resulting in higher surface energies, and hence in more number of atoms on the surface and a higher level of interaction at atomic levels. There is a growing demand for nano-engineered products in various disciplines such as (a) semiconductor nano-crystallites for use in microelectronics, (b) ceramics for use in demanding environments, (c) polymers with enhanced functional properties, (d) transparent coatings with ultraviolet/infrared absorption properties, (e) catalysis, (f) personal care products (e.g., sunscreen) and pharmaceutical applications, (g) enhanced heat-transfer fluids, (h) ultrafine polishing of, e.g., rigid memory disks, optical lenses, etc.

With the emergence of nano-technology from its cocoon of nano-science, large-volume production of nano-dimensional materials has increasingly occupied the minds of manufacturers. Nanoparticle synthesis may be broadly classified into two techniques—bottom-up and top-down. In the former, smaller molecules are assembled into larger agglomerates of nanometer dimension, whereas in the latter, larger, typically micron-sized, particles are fragmented into the smaller nanometer dimension. Acoustic fields have an "intensification" role to play in both scenarios.

7.3.1 BOTTOM-UP METHODS

Nanoparticles are fundamental building blocks of nano-technology. They form a starting point for "bottom-up" approaches for preparing nano-structured materials and devices. Their synthesis is an important research component and often quite challenging. There are several routes for nanoparticle synthesis, e.g., colloidal processes (Bognolo, 2003), liquid-phase synthesis (Grieve et al., 2000), gas-phase synthesis (Kruis et al., 1998), vapor-phase synthesis (Swihart, 2003), etc.

In case of *colloidal process route*, nanoparticles are produced directly to the required specifications and assembled to perform a specific task. It involves the use of surface-active agents, e.g., cadmium sulfide 50 nm particles produced by mixing two solutions containing inverted micelles of sodium bis(2-ethyl hexyl) sulfosuccinate in heptane and antiferromagnetic nanoparticles of Fe_2O_3 by decomposition of $Fe(CO)_5$ in a mixture of decaline and oleyl sarcosine. Nanoclusters are produced by co-ordinating ligands. Surfactant plays a major role in this process.

Liquid-phase synthesis is effectively used for the preparation of "quantum dots" (semiconductor nanoparticles). Dispersions can be stabilized indefinitely by capping particles with appropriate ligands. "Sol-Gel" is a method used to synthesize glass, ceramic, and glass-ceramic nanoparticles. It is performed in the aqueous or alcohol-based medium. It involves the use of molecular precursors, mainly alkoxide or alternatively metal formats. The mixture is stirred until gelation. The gel is dried at 100°C for 24 hours over a water bath, then ground to a powder. The powder is heated gradually (5°C/min), then calcined in air at 500–1200°C for 2 hours. This allows the mixing of precursors at the molecular level and better control of particulate size. This process has some very good advantages, e.g., high purity, low sintering temperature, high degree of homogeneity, and suitability for nano-sized multi-component ceramic powders.

Vapor phase synthesis route is another route for nanoparticle production. The vapor phase mixture is rendered thermodynamically unstable relative to the formation of desired solid materials. Particles nucleate homogeneously if the degree of supersaturation is sufficient and reactions/condensation kinetics are initiated. Once nucleation occurs, remaining supersaturation is relieved by condensation or reaction of vapor-phase molecules, resulting in particles. This initiates the particle growth phase. Rapid quenching after nucleation stops particle growth. This occurs in two ways: (a) by removing the source of super saturation, and (b) by slowing the kinetics. Coagulation rates are proportional to the square of number concentration and have a weak dependence on particle size. At high temperatures, particles coalesce (sinter) rather than coagulate. Spherical particles are produced during this process. At lower temperatures, loose agglomerates with open structures are formed. At intermediate temperatures, partially sintered, non-spherical particles are formed. Control of coagulation and coalescence is critical in case of nanoparticle generation. In general, nanoparticles generated in the gas phase always tend to agglomerate. This is one significant limitation of this process. Loosely agglomerated particles can be redispersed with reasonable effort, but hard (partially sintered) agglomerates cannot be fully re-dispersed.

In case of *gas-phase synthesis,* supersaturation is achieved by vaporizing materials into a background gas, then cooling the gas. Methods using solid precursors are: (a) inert gas condensation, (b) pulsed laser ablation, (c) spark discharge generation, (d) ion sputtering. Methods using liquid or vapor precursors are (a) chemical vapor synthesis, (b) spray pyrolysis, (c) laser pyrolysis/photothermal synthesis, (d) thermal plasma synthesis, (e) flame synthesis, (f) flame spray pyrolysis, and (g) low-temperature reactive synthesis.

Inert gas condensation route is suited for the production of metal (e.g., Bi) nanoparticles. Reasonable evaporation rates are required at attainable temperatures. The solid precursor is heated into gas and they are mixed with background gas to reduce temperature. Reactive gas (e.g., O_2) is introduced into the cold gas stream to prepare compounds (e.g., oxides). Controlled sintering after particle formation is used to prepare composite nanoparticles (e.g., PbS/Ag, Si/ln, Ge/ln, Al/Pb).

In *pulsed laser ablation* route, plumes of materials are vaporized using the pulsed laser in a tightly confined and spatial ambience. This process is suitable for producing only small amounts of nanoparticles. However, materials that cannot be easily evaporated can be vaporized in this route. For example, synthesis of Si, MgO, titania, hydrogenated silicon nanoparticles is possible in this route. In *ion sputtering* route, the solid precursors are sputtered with a beam of inert gas ions, e.g., magnetron sputtering of metal targets. Low pressure (appr. 1 m Torr) is required for this process. Here again, the key limitation is the further processing of nanoparticles in aerosol form.

In case of *spark discharge generation* route, electrodes made up of metal to be vaporized are charged in presence of inert gas until breakdown voltage is reached. Arc formed across electrodes vaporizes a small amount of metal, e.g., Ni. This route yields a small amount of nanoparticles but in a reproducible manner. Reactive gas (e.g., O_2) can be used to make compounds (e.g., oxide). Background gas can be pulsed between electrodes as the arc is initiated.

In case of *chemical vapor synthesis*, vapor phase precursors are brought into a hot-wall reactor under nucleating conditions. Vapor phase nucleation of particles is favored over film deposition on surfaces. This process is very flexible and can produce a wide range of materials. Precursors can be solid, liquid, or gas under ambient conditions, but they are delivered to the reactor as vapor (using bubbler, sublimator, etc.), e.g., oxide-coated Si nanoparticles for high-density non-volatile memory devices, and W nanoparticles by decomposition of tungsten hexacarbonyl. Another advantage of this process is that it allows the formation of doped or multi-component nanoparticles by the use of multiple precursors. For example, europium doped in Si nanoparticles and zirconia doped with alumina can be produced. One material encapsulated with another (e.g., metal in metal halide) can prevent agglomeration.

In the *spray pyrolysis* technique, a nebulizer is used to inject very small droplets of the precursor solution. Also known as *aerosol decomposition synthesis*, this involves direct droplet-to-particle conversion. The reaction takes place in solution in the droplet, followed by solvent evaporation, e.g., preparation of TiO_2 and Cu nanoparticles. In case of *laser pyrolysis/photothermal synthesis*, precursors are heated by absorption of laser energy. This allows highly localized heating and rapid cooling. Infrared (CO_2) laser is used for heating and this energy is absorbed by precursors, or by inert photosensitizers (SF_6), e.g., Si from silane, MoS_2, SiC. The significant advantages of pulsed laser techniques are that they shorten reaction time, and allow the preparation of even smaller particles. In case of *thermal plasma synthesis,* precursors are injected into thermal plasma ambience. Precursors are generally decomposed fully into atoms, which then react or condense to form particles when cooled by mixing with cool gas or expansion through nozzle. This technique is mainly used for production of SiC and TiC nano-phase hard coatings.

Flame synthesis is the most commercial and successful approach to produce quite a large number of fine particles. For example, millions of metric tons per year of carbon black and metal oxides are produced. Particle synthesis is done within the flame. Heat is produced in-situ by combustion reactions. The main limitation is the complexity of process and it is quite difficult to control. This is primarily used for making γ-Fe_2O_3 nanoparticles and sintered agglomerates such as titania and silica. The particle size is influenced by application of direct current electric field.

In *flame spray pyrolysis* technique, liquid precursors are sprayed directly into the flame. This method allows the use of low-vapor-pressure precursors. A typical application is the synthesis of silica particles from hexamethyldisiloxane. In case of *low-temperature reactive synthesis*, vapor phase precursors react directly without addition of heat and also without any significant heat generation due to the process. For example, ZnSe nanoparticles are produced from dimethylzinc-trimethylamine and hydrogen selenide. This is done by mixing in a counter-flow jet reactor at room temperature. The heat of reaction is sufficient to allow particle crystallization.

In case of *sonochemical synthesis* of nanoparticles, molecules undergo a chemical reaction enhanced by power ultrasound (20 kHz–10 MHz). Ultrasound induces acoustic cavitation, during which very large numbers of cavities are produced and subsequently collapse due to implosion effect. Here, as the bubble implodes, very high temperatures (5,000–25,000 K) are realized for a few nanoseconds; this is followed by very rapid cooling (10^{11} K/s). High cooling rates hinder product crystallization, and hence, amorphous nanoparticles are formed. This process is superior for (a) preparation of amorphous products ("cold quenching"), (b) insertion of nano-materials into mesoporous materials by acoustic streaming, (c) deposition of nanoparticles on ceramic and polymeric surfaces, (d) formation of proteinacious micro and nano-spheres, and (e) for very small particle generation. For example, nano-metal particles such as gold, Co, Fe, Ni, Au/Pd, nanophased oxides (titania, silica, ZnO, ZrO_2, MnOx, magenetic Fe_2O_3, TiO), nanotubes (carbon, TiO_2, HC), nanorods (CdS, ZnS, Pbs, WS_2, WO_2, Fe_2O_3), and nanowires of Se can be effectively generated via this route. Liang et al. (2003) have synthesized pure zirconia nano-powders via sonochemical method using ultrasonics at room temperature. They proved that the ultrasonic process offered a rapid, controllable way to synthesize nanostructured materials with uniform shape compared to the conventional methods such as sol-gel, hydrothermal, co-precipitation, and surfactant templating methods. They have shown that ultrasound dramatically reduces the temperature and makes the reaction conditions very easy to maintain in synthesis of nano-powders.

7.3.2 TOP-DOWN METHODS

Fragmentation by conventional attritional means such as grinding or milling is reported widely in literature, but contamination from the milling material as well as energy efficiency are drawbacks; comminution by high-intensity ultrasound has also been studied but to a lesser extent.

Comminution, especially ultra-fine comminution, is an energy-intensive stage in the overall process to provide materials in fine size ranges for the required properties of the final product. High energy consumption and inefficiency in comminution technology for such materials as minerals, cement, pigment, chemicals, and food have long been regarded as an impediment for producing particles below micron sizes. Conventional devices (mainly, tumbling ball mills) have been used for comminution for many years, but the basic problem is that the power consumed is limited by the centrifugation occurring at speeds above the critical, and the grinding media could not be made very small, for the impact energy of each ball would otherwise be insignificant. The various studies in this area have been focused on the effect of operational conditions on efficiency of particle breakage. Chemicals or grinding aids used in comminution generally increase grinding energy efficiency, lower the limit of grinding, prevent agglomeration or aggregation of ground particles, avoid grinding media coating, and improve the rheology of material flow as well. Kass et al. (1996) related the extent of size reduction to intensity of ball milling. In case of high-energy ball-milling, there is a limit on the smallest size that can be achieved, and that depends on the size of the grinding media used. Typically, the smallest size that can be achieved is about 1/1,000th the grinding media size. An increase in rotation/vibration intensity only affects the kinetics of the process but has no bearing on the smallest size achievable.

Sonication of suspensions of particles provides a number of significant benefits, not the least of which is better dispersion. Ultrasonics substantially reduces the particle size of ultrafine suspensions by one-tenth the time of traditional ball milling methods. In addition, one can expect disaggregation and de-agglomeration of clumps (particle size reduction), degassing of the carrier liquid, enhanced slurry flow properties, higher homogeneity, as well as denser casting, sintering, and packing. The principal process intensification parameters are the intensity and frequency of ultrasound, contact time, and temperature. The extent of breakage increases as the intensity of sound increases. The reduction in size is also enhanced by increasing the contact time but the instantaneous breakage rate, which is particle-size dependent, decreases with time. The relation with temperature is somewhat complicated as it depends on a complex interplay between surface tension, viscosity, and vapor pressure of the fluid. The breakage increases up to a certain temperature and then falls. This is because of the combined effect of the increase in the number of cavities (which dominates initially) and the decrease in the cavitation intensity (which dominates at higher temperatures).

Another property that has been studied is the concentration of solids in the suspension, which has been found to have no effect if the concentration is kept low. Suslick (1993) stated a theory explaining that fragmentation can occur either by fracture or erosion. Erosion refers to particle size reduction due to the loss of primary particles from the surface of the agglomerate, whereas fracture is the partitioning of the original agglomerate into several smaller agglomerates. Which breakage mechanism dominates may depend on the applied ultrasonic intensity, but it is certainly a function of material properties. For erosion to take place, the primary particles have to be freed from the surface of the agglomerates. This means that the cavitation pressure must be larger than the cohesive strength with which the surface primary particles are bound together.

Friedman (1972) proposed a theory and mechanism which explains the effects of dispersion and coagulation in a cavitating liquid medium based on the interaction mechanism of a cavitation bubble and a particle of the material medium. Experimentally, it was shown that for more efficient dispersion, a combination of ultrasonic cavitation and mechanical dispersion is preferable. Jing et al. (2007) studied the significant role of ultrasonic irradiation over conventional mechanical stirring in the preparation of micron-sized irregular polyaniline (PANI) particles. They noticed that as the reaction proceeds, the primary nanofibers grow and agglomerate into irregular-shaped PANI particles in a mechanical stirring system; this has been prevented in ultrasonic irradiation. They reported that the use of ultrasound in synthesis is one of the facile and scalable approaches in synthesizing PANI nanofibers in comparison with other conventional approaches.

Gaete-Garreton et al. (2000) carried out experiments to study the fine grinding of hard materials, based on a high-compression roller mill with ultrasonics. They tested the high-compression roller mill by producing powder from fine to hard rocks, with and without ultrasonic activation. They found that the application of ultrasonic energy diminishes the torque required, the stress over the shafts and the total energy consumed for the same grinding results, and results in a reduction in the erosion of the grinding surfaces. Their results show that the ultrasonic device produces high-quality ground products with a lower energy and material consumption than the conventional systems. They also stated that ultrasonic grinding is a very promising new technology, as it may help improve efficiency and broaden the capabilities of conventional grinding systems.

Some other advantages of sono-fragmentation as cited in the literature are:

- Comminution (Teipel et al., 2004) by an ultrasonic field in an aqueous phase is ideally suited for hard, brittle materials. The limiting factor for ball-mill comminution is the size of the grinding medium.
- Cohesive forces between the mill charge and medium increase with decreasing media size. Below a certain size limit, the charge adheres to the medium and comminution does not occur. This effect can be overcome through mechanical agitation. An alternative technique for overcoming this limiting factor is through the application of ultrasonication during the milling operation. In addition, the morphology of the ultrasonically treated particles was considerably less angular than the untreated particles (Kass et al., 1996).

Gasgnier et al. (2000) carried out experiments to study sono-fragmentation of various materials by subjecting them to ultrasound after mixing in water and other

dilute acids such as acetic acid. They observed a decrease in particle size and formation of unexpected and unknown compounds on a mesoscopic scale. They also found
that the magnetic properties, i.e., susceptibility and effective magnetic moment, were
slightly changed in case of iron and iron oxides.

Kusters et al. (1994) performed experiments to show the effect of agglomerate
concentration and suspension volume on fragmentation. They found that the fragmentation rate was independent of solid concentration up to some extent, giving
ample evidence that this type of fragmentation results from particle-cavity rather
than particle-particle interactions. The reduction ratio increases with time faster at
the small than at the large suspension volume for equal power input.

Raman and Abbas (2008) investigated the effects of intensity of ultrasound on
particle breakage in liquid medium and also the effect of sonication power, temperature, and contact time on particle breakage. Three different flow rates were studied: 1.0 l/min, 1.6 l/min, and 2.2 l/min. Breakage was more predominant at lower
flow rates corresponding to larger values in residence time. As the residence time
increases, the particles spend more time in contact with the breakage forces of the
high-intensity ultrasound (HIU) field.

Temperature has a significant effect on the cavitation phenomenon, which, in
a liquid medium, is affected by its surface tension, viscosity, and vapor pressure.
Increasing temperature results in the liquids cavitating at lower intensities. This can
be attributed to the increase in vapor pressure of the liquid, decrease in surface tension, and reduced viscosity of the liquid medium. The decrease in viscosity decreases
the magnitude of the natural cohesive forces acting on the liquid, and thus, decreases
the magnitude of the cavitation threshold. Lower cavitation thresholds translate into
ease of cavity formation, thereby making higher temperatures more favorable for
particle breakage. This is the reason for an increase in the breakage of particles as
temperature is varied from 10°C to 25°C. As the temperature is increased beyond
25°C, a decrease in particle breakage is observed. This is primarily caused by the
cushioning effect of increased cavity internal vapor pressure at higher temperatures.
Due to this cushioning effect, the intensity of the collapse, and subsequently the
breakage, decreases above 25°C.

There are, thus, two opposing factors that are at play during particle breakage; the
first being the increase in the number of cavitation events with an increase in temperature due to which particle breakage increases, and the second being the cushioning effect of the cavity internal vapor pressure, which has a suppression effect on the
cavitation intensity and subsequently on particle breakage.

Published literature only addresses particle size reduction to the 1 μm size range.
Ultrasonic fragmentation of silica particles (Kusters et al., 1994) to produce a fine
fraction of crystalline materials in the 10 μm size range, and of zirconia (Franco et al.,
2004) in the 1–2 μm size range are examples of studies reported in the literature.
When literature data are extrapolated, they show that sub-micron fractions could be
achieved with higher sonication power as displayed in Figure 7.25. By application of
20 kHz ultrasonic field at 500 W input power, it is possible to achieve sub-100 nm size
range. Nano-ceramic particles such as alumina, zirconia, etc. can be fabricated in an
energy-efficient way, when compared to the conventional grinding process, by using
low frequency (20–35 kHz), high-intensity power ultrasound (Kusters et al., 1994).

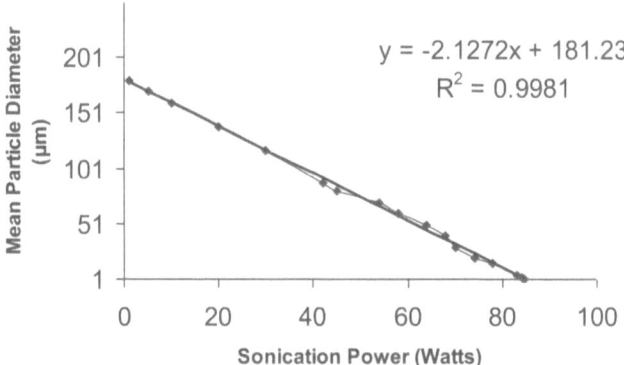

FIGURE 7.25   Extrapolation from literature data on sono-fragmentation.

In ultrasonic fragmentation of powders, the energy consumption per unit mass is less compared to ball milling for eroding particles (such as silica and zirconia). Ultrasound induces acoustic cavitation, during which very large numbers of cavities are produced and subsequently collapse due to the implosion effect, which, in turn, helps to fragment the micron-sized ceramic particles into submicron and nano-dimensional particles. The purity obtained by sono-fragmentation is significantly better when compared to other dry processes for nanoparticle fabrication. Kusters et al. (1994) developed a sono-fragmentation model for agglomerate powder as shown in Equation (7.3).

$$FR = \frac{2.0 \times 10^{-3} \; \in^{0.38} \; v_p^{\frac{1}{3}}}{V_{Tot}} \tag{7.3}$$

where $\in$ is the ultrasonic power input, V_{Tot} is the total suspension volume, and v_p the particle volume.

Since there was no mathematical expression reported in the literature with respect to sono-fragmentation of particles, Gopi (2007) developed his own semi-empirical models. In general, the sono-fragmentation process is less complex when compared to sonochemical process, since chemical reactions are not involved.

The susceptibility of materials to ultrasonic damage is complex. Kass (1996, 2000) had done some work on the fragmentation of alumina particles and the effect of sonication time and power level on the fragmentation process. Pe´rez-Maqueda (2005) also studied alumina fragmentation showing that the powder had significantly finer mean and median particle sizes, narrower particle-size distributions, and less angular morphologies than powders that were conventionally wet-ball milled.

Suslick (1990) has identified mainly two mechanisms responsible for particle fragmentation in the presence of ultrasonic cavitation (Figure 7.26):

1. Particle-particle interaction (due to shock wave generation) when particles are too small to perturb the ultrasonic field. It has been found that these intense shock waves can accelerate small particles to more than 500 km

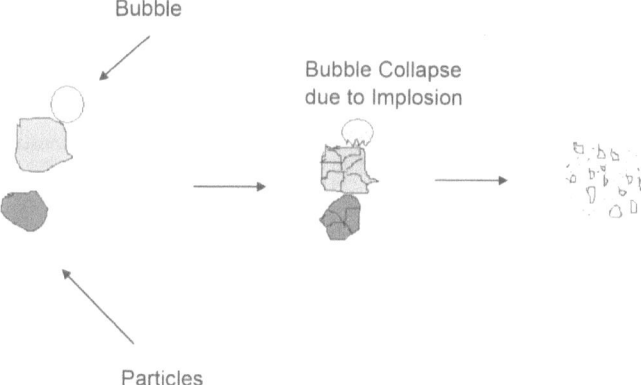

FIGURE 7.26   Schematic representation of particle fragmentation.

per hour in liquids; hence, particles traveling at these speeds can collide with other particles, causing profound structural and chemical changes to the solid.

2. Particle-cavity interaction (due to liquid jet generation) when particles are large enough, e.g., agglomerates. The liquid jets move with velocities of roughly 400 km per hour and hit the surface with nearly the same force as solid projectiles. The jet can cause severe damage at the point where it hits the surface.

Friedman (1972) observed that during sonication, the constituents of the medium disperse in some cases, in others they coagulate. He attempts to clarify the observed phenomena by looking at the interaction mechanism of a cavitation bubble with particles of the material medium. On the basis of the calculations and the analysis of a large number of films, the following hypothesis regarding the mechanism of cavitation dispersion and coagulation has been made by Friedman.

With the collapse of a bubble, the resulting shock wave primarily disrupts the large particles lying on the bubble surface. If the diameter of the particle is less than 2 μm, then they accumulate at a distance from the bubble's surface. Consequently, the disruptive effect of the shock wave on them is less, while their accumulation in the vicinity of the bubble leads to an increase in the concentration of the particles (coagulation). Therefore, there is a limit to the size reduction of particles using ultrasonication.

Kusters et al. (1993, 1994) deal with the fragmentation of ceramic agglomerates in a suspension. They have concluded that it is most likely that the intense pressures generated in the vicinity of imploding cavitation bubbles are the primary cause of particle degradation. A population balance model was developed describing quantitatively the ultrasonic fragmentation of agglomerate powders. Expressions were derived based on the assumption that agglomerates break up during ultrasonication by their interaction with collapsing cavities in the liquid.

Kusters et al. (1994) also report an experimental study pertaining to the effect of agglomerate concentration and suspension volume on fragmentation. The following are some conclusions from the study:

- Fragmentation rate was found to be independent of solid concentration up to some extent, giving ample evidence that this type of fragmentation results from particle-cavity rather than particle-particle interactions.
- The reduction ratio increases with time faster at the small than at the large suspension volume for equal power input.

Some conclusions pertaining to energy consumption are:

- Energy consumption increases with power input for the same amount of size reduction. This is attributed to the indirect fragmentation mechanism of ultrasonic dispersion.
- Lowering the power input helps prevent agglomeration and hence improves the efficiency of size reduction.

The size distribution of silicon particle (SiC) feed particles obtained by Gopi (2007) is shown in Figure 7.27. Corresponding size distribution data of SiC Particles after exposure for 30 minutes to 33 kHz ultrasonics at 500 W power input is displayed in Figure 7.28. Clearly, the sub-micron to nanometer size fraction has increased substantially due to the sonication process. It was observed that sono-fragmentation rate increases with increasing applied sonic power, as shown in Figure 7.29, and with a decrease in ultrasonic frequency, as shown in Figure 7.30. In both, the vertical axis represents a fragmentation factor—the ratio of mean particle size post-sonication ($d_{p,\ final}$) to the mean size prior to sonication ($d_{p,initial}$). At a lower

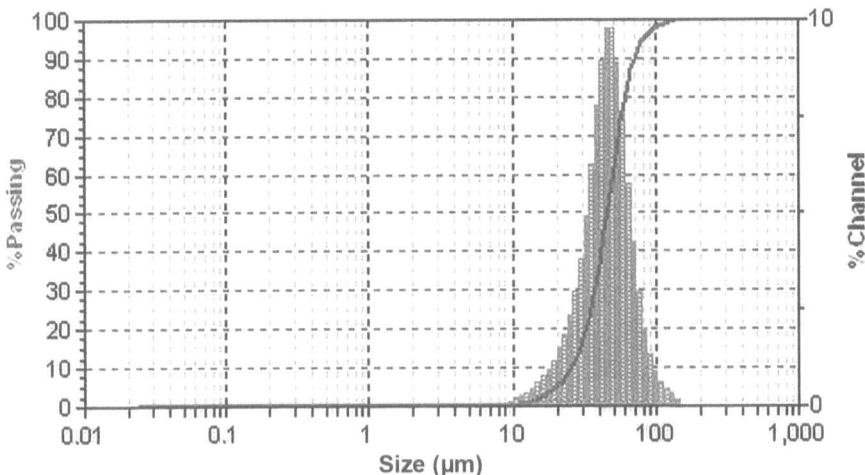

FIGURE 7.27 Particle size distribution of SiC feed particles.

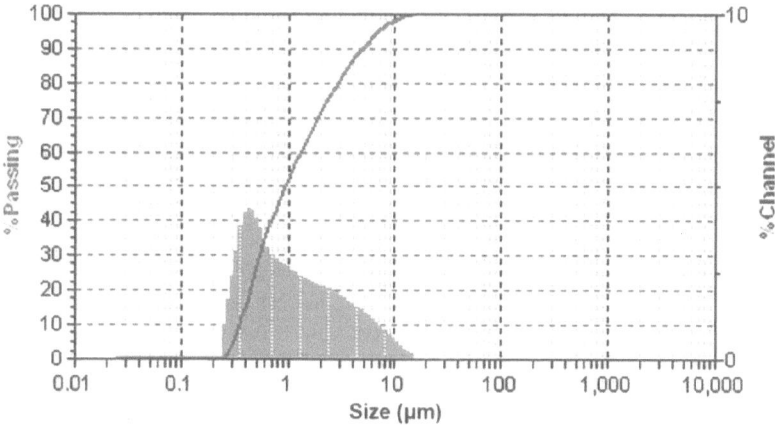

FIGURE 7.28 Particle size distribution of SiC particles after exposure to 33 kHz, 500 W sonication.

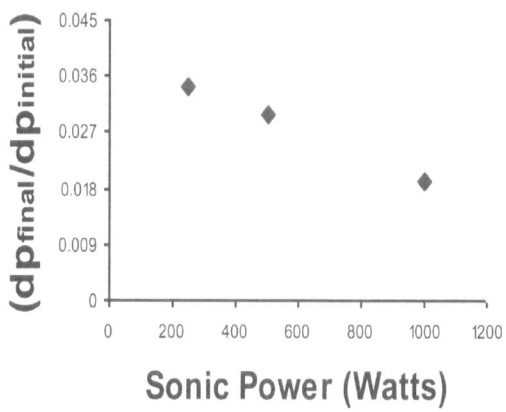

FIGURE 7.29 Effect of ultrasonic power input on particle fragmentation.

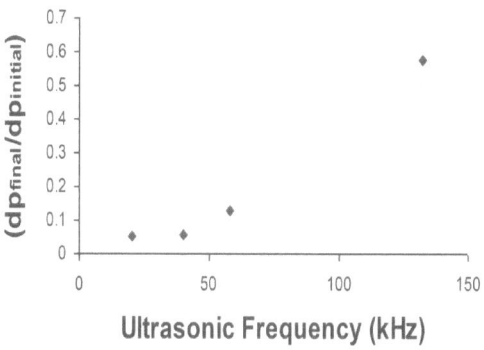

FIGURE 7.30 Effect of ultrasonic frequency on particle fragmentation.

frequency, acoustic cavitation phenomena are predominant, whereas at a higher frequency, acoustic streaming is predominant. Particle breakage is only facilitated by the cavitation effects of acoustic fields.

Low-frequency ultrasonic fields result in higher size reduction compared to dual, intermediate and megasonic frequency systems. The spectrum of average particle size distribution achieved at low frequencies typically lies to the left (smaller size) of the virgin sample particle size spectrum. As frequency is increased, the spectrum shifts to the right of original spectrum of unprocessed sample. This implies that particle breakage is higher at low frequency, agglomeration of particles occurs at higher frequency, and intermediate frequencies see the net effect of both mechanisms.

Particle breakage extent and rate are a function of frequency and time of sonication, and a small particle size implies a large surface area per unit mass. This reduction in original size can enhance transport processes and chemical reaction rates associated with the particles, including those that play the role of a catalyst. Hence, sono-fragmentation of particles has great potential for use in industrial applications.

PIF/CIF ANALYSIS

Top-down methods are clearly cost-advantageous over bottom-up methods for nanoparticle production at scale. For synthesizing nano-dimensional particles from a feed population in the micrometer size range, the only viable methods are high-energy ball milling and sonication. Both are energy intensive, but the sono-process offers the advantage of taking less time to reach a finer size distribution. It also enables continuous processing, which can potentially realize production quantities of tons per day. Such systems are commercially available, and the technology is proven. Sono-fragmentation systems provide a ready-made, possibly superior alternative to other processes for synthesizing nanoparticles.

EXERCISE QUESTIONS

1. Sonoluminescence was first observed from water in 1934. As with sono-chemistry, sonoluminescence derives from acoustic cavitation. There are two classes of sonoluminescence: multiple-bubble sonoluminescence (MBSL) and single-bubble sonoluminescence (SBSL) (Suslick and Flannigan, 2008). Cavitation is a nucleated process, and liquids generally contain large numbers of particulates that serve as nuclei. As a consequence, the "cavitation field" generated by a propagating or standing acoustic wave typically consists of very large numbers of interacting bubbles, distributed over an extended region of the liquid. If this cavitation is sufficiently intense to produce sonoluminescence, then we call this phenomenon "multiple-bubble sonoluminescence" (MBSL). For rather specialized but easily obtainable conditions, a single, stable, oscillating gas bubble can be forced into such large amplitude pulsations that it produces sonoluminescence emissions on each (and every) acoustic cycle, a phenomenon called single-bubble

sonoluminescence (SBSL). The intensity of sonoluminescence can thus be used as a measure of cavitation intensity. There are commercial probes available that are based on this premise. While sound pressure monitors can measure acoustic pressure as a whole, they do not enable the distinction of cavitation pressure from the prevailing total pressure. Sonoluminescence probes do and are therefore widely employed in process industries to measure and control cavitation. Based on your understanding of cavitational fields inside a water-filled tank with transducers mounted at the bottom, sketch how cavitation energy would be distributed in the tank, and contrast this with your projection of acoustic pressure distribution. Streaming is an effect that comes into play at higher, non-cavitational frequencies. Are streaming pressure fields more likely to be aligned with the total pressure field, and if so, why?

2. Micromixing is the backbone of microfluidic devices for the applications of reagents and analytes detection, drug screening, diseases diagnosis, and various fluidic operations at the microscale. The micromixing characteristics of microfluidics are contrary to mechanical mixing in which the latter falls in a turbulence regime, whereas the former is associated with the laminar regime as a consequence of miniaturized feature of microchannel having hydraulic diameters of less than several hundred microns. Micromixing can be divided into two distinguishable mixing principles, namely active and passive mixing. Of many active mixing mechanisms, ultrasound and thermal energies are the two common modes of inductive external source for creating fluid disturbance in micromixers. Ultrasound has been reported as an effective means to enhance mixing performance in microfluidics. This is because the dissipated ultrasound energy induced acoustic streaming from the formation of intensified macroscopic liquid flow. The mechanical effect due to acoustic streaming was shown to be effective in improving mixing at significantly higher flow rates. Similarly, thermal energy was found effective in enhancing micromixing as the fluid viscosity reduces with the increase in temperature, thereby promoting greater molecular diffusion (Mahmud et al., 2021). For a representative test case, compare thermal and ultrasonic mixing processes for their effectiveness and cost impact. Under what circumstances would one be preferred over the other?

3. In the definition of atomization efficiency, the interfacial energy, E_L, required to atomize the liquid may be taken as the product of surface tension (σ_L) and the area change dA between the initial hydrodynamic structure and the spray droplets, expressed as (Panão, 2021):

$$E_L = \sigma_L \, dA$$

Atomization efficiency may then be defined as (E_L/E_A), where E_A is the energy available for atomization. Based on this definition, a lower input of energy that leads to the same level of atomization is clearly desirable. Would ultrasonic atomization fit the bill, or are the more traditional (mechanical, vibrational) methods preferable? Identify relevant parameters that can enable this judgment to be made on a quantitative basis.

4. While breakage of micron-sized particles to the nano-size range is clearly demonstrated with cavitational fields, the breakage is not 100% in one pass, and it will require multiple passes through the sonicator to achieve near-complete size reduction. A continuous-throughput sonicator can be installed and supplemented with classification by size. If the liquid volume can be decanted at various levels to draw out particles of narrow size ranges, and these can be fed back into the processing volume in a phased manner, the overall efficiency of the sono-fragmentation process may be improved significantly. Derive a strategy to achieve this, assuming a feed particle size distribution of 10–50 μm, and a desired product size distribution of 50–100 nm. Make any necessary assumptions regarding particle shape, possibility of hindered settling, etc.

REFERENCES

Bognolo, G., "The Use of Surface-Active Agents in the Preparation and Assembly of Quantum-Sized Nanoparticles", *Adv. Colloid Interf. Sci.*, vol. 106, no. 1–3, December 2003, pp. 169–181.

Doraiswamy, L.K. and L.H. Thompson, "Sonochemistry: Science and Engineering", *Ind. Eng. Chem. Res.*, vol. 38, 1999, pp. 1215–1249.

Franco, F., L.A. Perez-Maqueda, and J.L. Perez-Rodriquez, "The Effect of Ultrasound on the Particle Size and Structural Disorder of a Well Ordered Kaolinite", *J. Colloid Interf. Sci.*, vol. 274, 2004, pp. 107–117.

Friedman, V.M., "The Intersection Mechanism between Cavitation Bubbles and Particles of the Solid and Liquid Phases", *Ultrasonics*, vol. 10, 1972, pp. 162–172.

Gaete-Garreton, L.F., Y.P. Vargas-Hermandez, and C. Velasquez-Lambert, "Application of Ultrasound in Comminution", *Ultrasonics*, vol. 38, 2000, pp. 345–352.

Gasgnier, M., L. Beaury, and J. Derouet, "Ultrasound Effects on Metallic (Fe and Cr); Iron Sesquioxides (alpha-, gamma-Fe2O3); Calcite; Copper, Lead and Manganese Oxides as Powders", *Ultrasonics Sonochem.*, vol. 7, 2000, pp. 25–33.

Gopi, K.R., "Application of Sono-Fragmentation, Siono-Blending and Sono-Cavitation Testing in Development of NanoStructured Composite Materials", Ph.D. Thesis, IIT Madras, 2007.

Grieve, K., P. Mulvaney, and F. Griser, "Synthesis and Electronic Properties of Semiconductor Nano-particles/Quatum Dots", *Curr. Opin. Colloid. Interf. Sci.*, vol. 5, no. 1–2, 2000, pp. 168–172.

Jagannathan, T.K., "Mechanistic Modeling, Visualization and Applications of Flow Induced by High-Intensity, High-Frequency Acoustic Field", Ph.D. Thesis, Indian Institute of Technology Madras, 2010.

Jing, X., Y. Wang, D. Wu, and J. Qiang, "Sonochemical Synthesis of Polyaniline Nanofibers", *Ultrasonics Sonochem.*, vol. 17, 2007, pp. 75–80.

Kass, D., "Ultrasonically Induced Fragmentation and Strain in Alumina Particles", *Mater. Lett.*, vol. 42, 2000, pp. 246–250.

Kass, D.M., J.O. Kiggans, and T.T. Meck, "Ultrasonic Modification of Alumina Powder During Wet Ball Milling", *Mater. Lett.*, vol. 26, 1996, pp. 241–243.

Khurana, T.K., B.V.S.S.S. Prasad, K. Ramamurthi, and S.S. Murthy, "Thermal Stratification in Ribbed Liquid Hydrogen Storage Tanks", *Inter. J. Hydrogen Energ.*, vol. 31, 2006, pp. 2299–2309.

Kruis, F.E., K. Nielsch, H. Fissan, B. Rellinghaus, and E.F. Wassermann, "Preparation of Size-Classified PbS Nanoparticles in the Gas Phase", *Appl. Phys. Lett.*, vol. 73, 1998, pp. 47–549.

Kusters, K.A., S.E. Pratsinis, S.G. Thoma, and D.M. Smith, "Energy—Size Reduction Laws for Ultrasonic Fragmentation", *Powder Technol.*, vol. 80, 1994, pp. 253–263.

Kusters, K.A., S.E. Pratsinis, S.G. Thoma, and D.M. Smith, "Ultrasonic Fragmentation of Agglomerate Powders", *Chem. Eng. Sci.*, vol. 48, 1993, pp. 4119–4127.

Lawrence, C.L., "Industrial Applications of Ultrasound-A Review II. Measurements, Tests, and Process Control Using Low-Intensity Ultrasound", *IEEE Transact. Sonics Ultrasonics*, vol. U-22, 1975, pp. 71–101.

Liang, J., X. Jiang, G. Liu, Z. Deng, J. Zhuang, F. Li, and Y. Li, "Characterization and Synthesis of Pure ZrO$_2$ Nanopowders via Sonochemical Method", *Mater. Res. Bullet.*, vol. 38, 2003, pp. 161–168.

Lighthill, J., "Acoustic Streaming", *J. Sound Vib.*, vol. 61, 1978, pp. 391–418.

Mahmud, F., K.F. Tamrin, S. Mohamaddan, and N. Watanabe, "Effect of Thermal Energy and Ultrasonication on Mixing Efficiency in Passive Micromixers", *Processes*, 2021, 9, pp. 891–894; https://doi.org/10.3390/pr9050891.

Mason, T.J. and J.P. Lorimer, *Applied Sonochemistry*, Wiley-VCH, Weinheim, 2002.

Nyborg, W.L., "Acoustic Streaming Near a Boundary", *J. Acoust. Soc. Am.*, vol. 30, 1958, pp. 329–339.

Panão, M.R.O., "Interpreting Liquid Atomization Efficiency", *Int. J. Eng. Technol. Inform.*, vol. 2, no. 5, 2021, pp. 121–124. ISSN: 2769-6146.

Pérez-Maqueda, L.A., A. Duran, and J.L. Perez-Rodriquez, "Preparation of Submicron Talc Particles by Sonication", *Appl. Clay Sci.*, vol. 28, 2005, pp. 245–255.

Raman, V. and A. Abbas, "Experimental Investigations on Ultrasound Mediated Particle Breakage", *Ultrasonics Sonochem.*, vol. 15, 2008, pp. 55–64.

Rayleigh, L., "On the Circulations of Air Observed in Kundt's Tubes and on Some Allied Acoustical Problems", *Philos. Transc. Roy. Soc. Lond.*, vol. A175, 1884, pp. 1–71.

Sherif, S.A., N. Zeytinoglu, and T.N. Veziroglu, "Liquid Hydrogen: Potential, Problems and a Proposed Research Program", *Inter. J. Hydrogen Energ.*, vol. 22, 1997, pp. 683–688.

Shoh, A., "Industrial Applications of Ultrasound-A Review I. High Power Ultrasound", *IEEE Transact. Sonics Ultrasonics*, vol. U-22, 1975, pp. 60–71.

Sovani, S.D., P.E. Sojka, and A.H. Lefebvre, "Effervescent Atomization", *Progr. Energy Combust. Sci.*, vol. 27, 2001, pp. 483–521.

Suslick, K.S., "Sounding Out New Chemistry", *New Sci.*, vol. 1702, 1990, pp. 50–53.

Suslick, K.S. and D.J. Flannigan, "Inside a Collapsing Bubble: Sonoluminescence and the Conditions During Cavitation", *Annu. Rev. Phys. Chem.*, vol. 59, 2008, pp. 659–83; 10.1146/annurev.physchem.59.032607.093739.

Suslick, K.S., "Ultrasound: Its Chemical, Physical and Biological Effects", VCH New York, 1988.

Suslick, K.S., *Ultrasound: Applications to Materials Chemistry*, Encyclopedia of Materials, 1993.

Swihart, M.T., "Vapor-Phase Synthesis of Nanoparticles", *Curr. Opin. Coll. Int. Sci.*, vol. 8, 2003, pp. 127–133.

Teipel, U., K. Leisinger, and I. Mikonsaari, "Comminution of Crystalline Material by Ultrasonics", *Int. J. Miner. Process.*, vol. 74, no. 4, 2004. doi: 10.1016/j.minpro.2004.07.011.

Vetrimurugan, R., "Investigation of High-Frequency, High-Intensity Ultrasound for De-stratification of Liquids Stored in Insulated Containers", Ph.D. Thesis, Indian Institute of Technology Madras, 2007.

Young, F.R., *Cavitation*, McGraw Hill, London, 1989.

8 Intensification of Transport Phenomena and Chemical Reactions

8.1 HEAT TRANSFER

Heat transfer enhancement techniques can be broadly classified into active enhancement techniques and passive enhancement techniques. Some of the basic techniques used for passive enhancement include flow disruption, secondary flows, surface treatments, and entrance effects.

8.1.1 SURFACE ROUGHNESS

One passive technique is to alter the characteristics of the heated surface. This method reduces the thermal boundary layer thickness and also aids in the early transition into turbulent flow. The conventional way to alter the surface is to increase its roughness. The effective roughness ratio is increased to create a boundary layer influence. Kandlikar (2002) and Kandlikar et al. (2003) studied the effect of surface roughness in a minichannel flow. They determined that the roughness ratio has a greater effect in a smaller diameter channel than the same ratio in a conventional channel.

8.1.2 FLOW DISRUPTIONS

The inclusion of flow interruptions is an attractive technique to incorporate. The flow disruptions provide increased mixing and can also serve to trip the boundary layer, causing flow transition. In conventional-sized passages, the flow interruption can be achieved using flow inserts, flow disruptions along the sidewalls, and offset strip fins.

Figure 8.1 shows sidewall obstacles in a microchannel. The triangular obstruction seen in Figure 8.1(a) would serve to cause some swirl flow. The steps shown in Figure 8.1(b) would break up the boundary layer and change the thermal and hydraulic gradients. Another possibility is using flow obstacles in the bulk area of the microchannel. Figure 8.2 shows some channel flow obstructions. Figure 8.2(a) shows some obstructions with simple rectangular geometries. Figure 8.2(b) shows some obstructions with circular geometries. The heights of these structures could vary to increase the secondary flows in the flow field. As dimensions of interest, such as the achievable critical dimensions in lithography, are reduced, a more refined circular profile can be achieved. These geometries can be optimized and intermixed

DOI: 10.1201/9781003283423-8

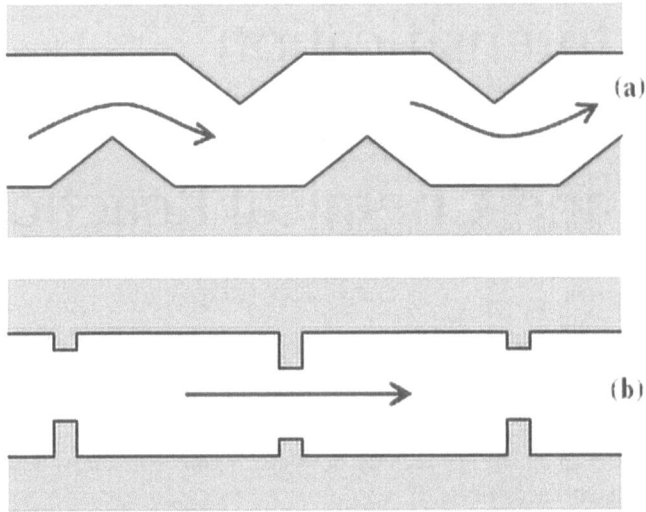

FIGURE 8.1 Sidewall flow obstructions in a microchannel: (a) triangular obstructions; (b) square obstructions.

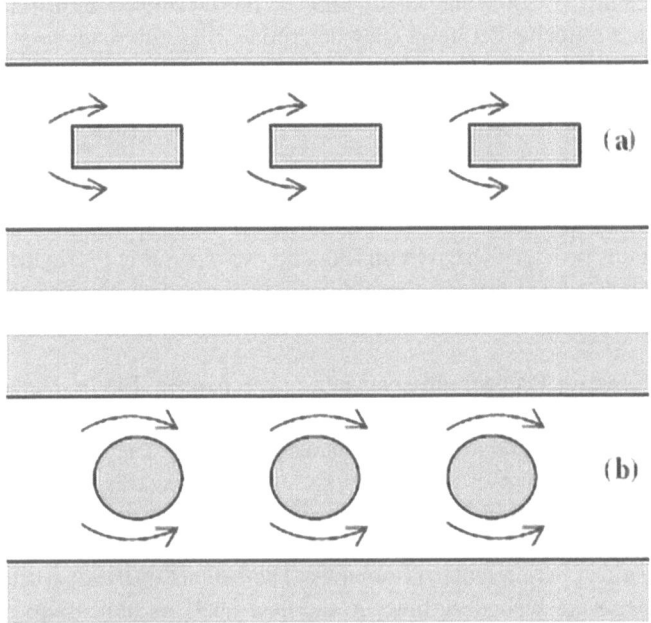

FIGURE 8.2 Flow obstructions in the channel: (a) simple rectangular geometry; (b) circular profile.

to achieve the maximum amount of heat transfer enhancement with the lowest pressure drop penalty.

8.1.3   CHANNEL CURVATURE

Several researchers have demonstrated that heat transfer enhancement can be achieved by having a curved flow path. The traditional parabolic velocity profile is skewed due to the additional acceleration forces. This causes the angle between the gradients to decrease and facilitate enhancement. Sturgis and Mudawar (1999) demonstrated the enhancement in a curved channel. The radius of curvature was 8.3 mm and the channel had a cross-section of 5.0 mm × 2.5 mm. The resulting hydraulic diameter was 3.33 mm. The enhancement reached as much as 26% for the curved channel versus a straight channel.

This technique is not really practical in a large-sized conventional passage. The application is a possibility in minichannels such as in compact heat exchangers. However, the greatest potential lies in a microchannel. The radius of curvature can be of the order of a few millimeters to centimeters but still large compared to the channel diameter. The compact nature of the microchannel flow network could allow for serpentine flow channels to utilize the curvature enhancement. This concept can be seen in the work involved in fabricating a microsystem gas chromatography column.

8.1.4   RE-ENTRANT OBSTRUCTIONS

The entrance region of channels can also provide heat transfer enhancement. A few researchers have reported the enhancement gained in the entrance region of a microchannel, e.g., Gui and Scaringe (1995). This technique could also find application in minichannels. However, the short lengths and low Reynolds numbers found in microchannel flow seem to be more appropriate. The authors reported heat transfer enhancement in microchannel heat sinks. The hydraulic diameters ranged from 221 to 388 mm. They suggested that the high heat transfer coefficients resulted from the decreased size, entrance effects, pre-existing turbulence at the inlet, and wall roughness.

The short lengths in a microchannel could allow a design that incorporates entrance spaces in the flow network. The sudden expansions and contractions would generate entrance effects. This would cause the flow to be in a perpetual state of development and allow for heat transfer enhancement. Figure 8.3 demonstrates such an arrangement. The cavities could also be used for pressure measurements and possible mixing sites.

Another possibility in utilizing entrance effects is shown in Figure 8.4. In this case, a re-entrant structure is incorporated in the channel. The structure could eventually be optimized to provide the maximum entrance effect with minimal pressure drop. The structure can be added several times in the channel, thereby causing a continual developing flow. Similar structures could be included in a minichannel as well. However, the design of those structures might be limited to simple geometries such as orifices. The pressure drop penalties in these apparatuses need to be carefully evaluated.

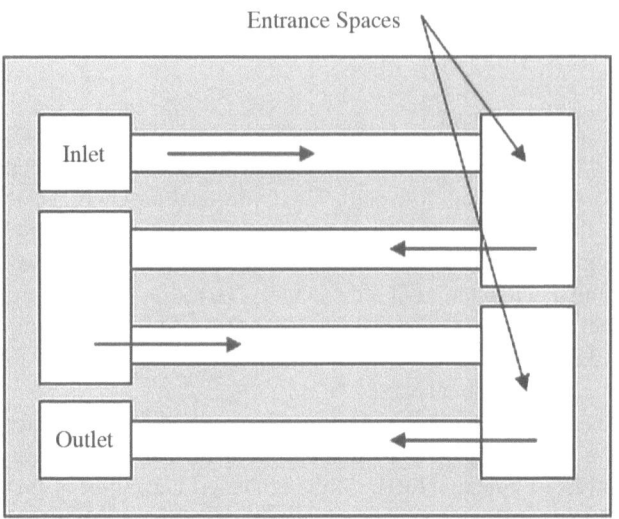

FIGURE 8.3   Entrance spaces in a microchannel flow.

8.1.5   SECONDARY FLOWS

Many researchers have demonstrated that secondary flows within the flow field provide enhancement. This can be seen in conventional channels. This technique can also be applied in minichannels using offset strip fins and chevron plates, and further refinement for minichannels is possible. The optimal shape and pitch of such devices can continue to improve the heat transfer enhancement.

The generation of secondary flows or swirl flows also has potential in microchannels. The geometry of the microchannel can be manipulated to produce secondary flow. Figure 8.5 shows a simple geometry that can generate secondary flows. Smaller channels are added between the main flow channels. The secondary flow will move from one channel to another via these channels. A second method for generating secondary flow comes from a conventional device. A Venturi can be manipulated to generate secondary flow without external power and a major increase

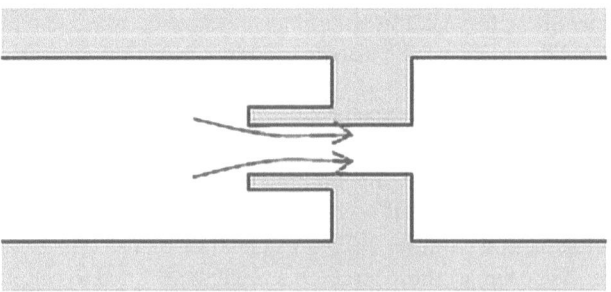

FIGURE 8.4   Incorporating re-entrant structures.

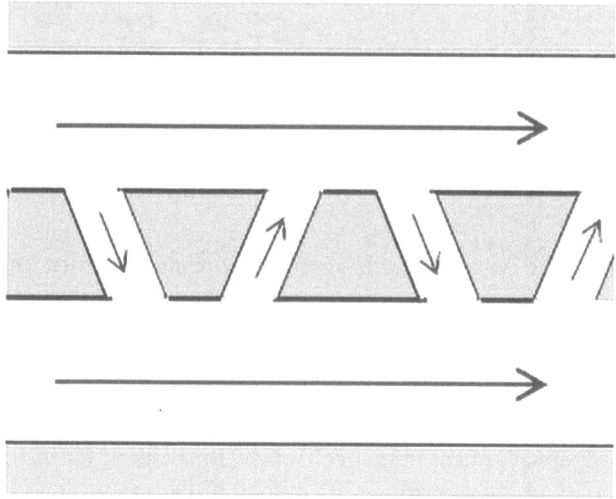

FIGURE 8.5   Secondary flow channels.

in pressure drop. Figure 8.6 shows a Venturi-based secondary flow apparatus. The throat area is connected to the larger area section of an adjacent microchannel. The reduction in pressure at the throat area seen from the reduction in area will draw flow in from the larger area of the adjacent channel. This technique could also be utilized to increase fluid mixing or the addition of another flow stream to the main

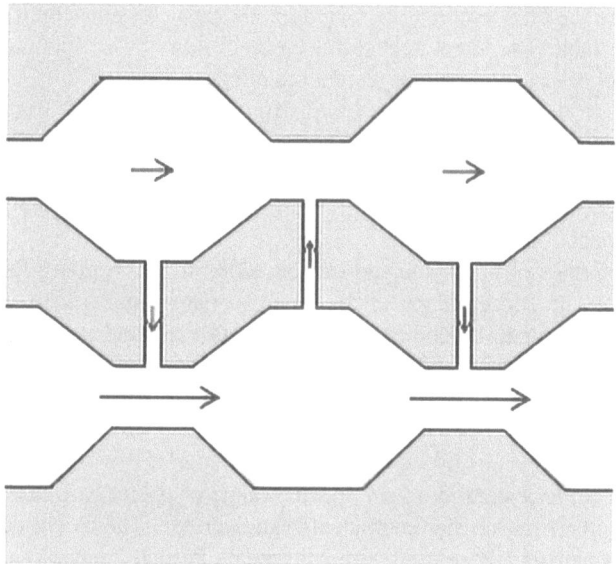

FIGURE 8.6   Venturi-based secondary flow.

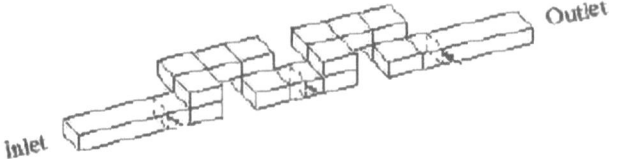

FIGURE 8.7   Three-dimensional twisted microchannel.

flow stream without the need for secondary flow pumping power. Once again, the pressure drop penalties could be a limiting factor for these devices (Vinay, 2007).

8.1.6   OUT-OF-PLANE MIXING

A technique being developed to increase binary fluid mixing could also be applied to heat transfer enhancement in microchannels. Bondar and Battaglia (2003) have studied the effect of out-of-plane or 3D mixing of two-phase flows in microchannels. They have achieved a high degree of fluid mixing. An example of a three-dimensional twisted microchannel is shown in Figure 8.7. This work could provide a path to follow for single-phase heat transfer. The rotation of the fluid will promote mixing and therefore change the hydrodynamic and thermal gradients.

8.1.7   FLUID ADDITIVES

The addition of small particles to the fluid can sometimes provide heat transfer enhancement. The use of small particles containing a phase change material (PCM) to achieve heat transfer enhancement has been studied. The particles begin in the solid state. As the fluid temperature increases, the particles reach their melting point and begin to melt. The latent heat of fusion involved with the melting of the PCM creates an enhancement. In other words, the effective heat capacity of the fluid has changed due to the presence of the PCM. Hu and Zhang (2002) studied the effect of microcapsules containing a PCM. The radius of the particles used was 50 μm in a 1.57 mm radius duct. This method works well in conventional channels. Smaller particles are being developed for use in minichannels and may possibly be extended to microchannels.

Small quantities of another liquid can be added to the working fluid to achieve enhancement. The concentration of the secondary substance affects the amount of enhancement obtained. A good example is seen in Peng and Peterson (1996). They used water as the working fluid and methanol as the additive. Technically, this would be considered a two-phase mixture. However, the authors reported the results in terms of single-phase heat transfer enhancement. The mixing of these additives typically does not achieve a perfect mixture. Therefore, the use of this technique can provide heat transfer enhancement in a very specific range of operations. Due to the imperfect mixture, the heat transfer coefficients can decrease. The recent development of nanoparticles, such as those used with Microscale Particle Image Velocimetry, provides some possibilities for microchannel heat transfer enhancement. The particles could be included to augment the heat transfer without causing a major clogging issue.

Active enhancement techniques require additional, external inputs into the system. These could be in the form of power, electricity, RF signals, or external pumps.

8.1.8 Vibration

Vibration in the fluid or surface is an active technique that has been applied to conventional channels. The tubes in some conventional heat exchangers can vibrate and provide heat transfer enhancement. This technique could easily be applied to a minichannel heat exchanger. The smaller, more compact nature of the tube bundles would allow for easier access for tube vibration. Go (2003) has studied the effect of microfins oscillating due to flow-induced vibration. The working fluid was air at velocities of 4.4 and 5.5 m/s. A microfin array was fabricated on a heat sink. As the fluid moves over the microfins, a vibration is induced that causes heat transfer enhancement.

8.1.9 Electrostatic Fields

The enhancement achieved from exposing a flow to an electrical field has been studied by a large number of researchers. They have demonstrated the enhancement in conventional-sized heat exchangers as well as minichannel flows. Allen and Karayiannis (1995) present a review of the literature on electrohydrodynamic enhancement. The governing equations, working mechanisms, and existing correlations are presented with some experimental work. It is concluded that corona wind and electrophoresis contribute the most to single-phase heat transfer enhancement. This technique could also be applied to microchannel flows (Figure 8.8). In a conventional or a minichannel application, a small insert electrode is present in the flow field. A potential is applied between the insert probe and the channel surface. The electric field that results will provide a moving corona effect and enhance the heat transfer. Figure 8.9 shows this arrangement for all three channel sizes: conventional, minichannel, and microchannel.

The possibility of incorporating the electrodes in the walls of a microchannel is quite attractive. The doped region, which forms the electrode, could easily be formed

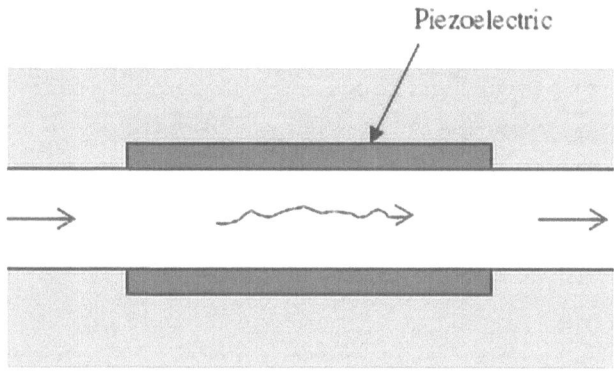

FIGURE 8.8 Piezoelectric-enhanced micro-channels.

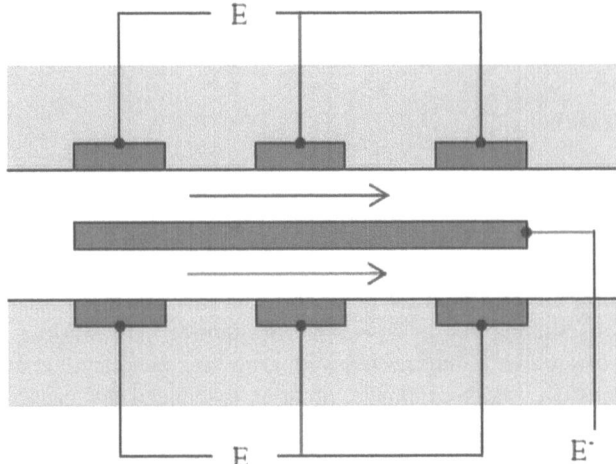

FIGURE 8.9 Electrostatic forces with insert probe for conventional channel, minichannel, and microchannel.

directly in the wall. The electrical connection can be achieved through the substrate of a cover wafer. Figure 8.10 shows an integrated electrostatic apparatus that incorporates the electrodes in the microchannel walls.

8.1.10 FLOW PULSATION

The variation of the mass flow rate through the channel can also provide heat transfer enhancement. Several researchers have demonstrated the mixing enhancement

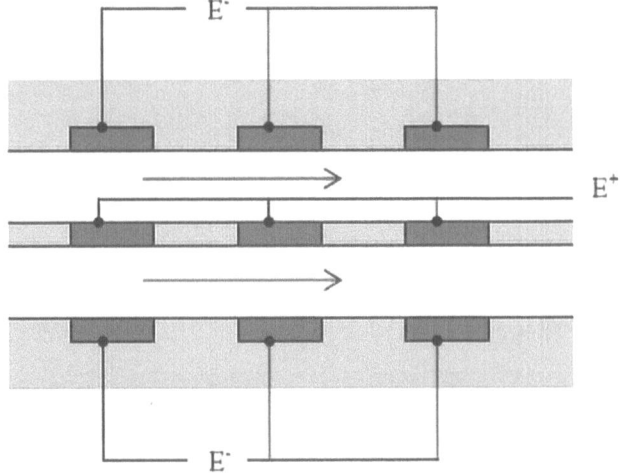

FIGURE 8.10 Electrostatic forces with wall-integrated probes for a microchannel.

provided by a pulsating flow. Hessami et al. (2003) studied the effect of flow pulsation on a two-phase flow in a 25 mm pipe. They determined that the enhancement could be as much as 15% depending upon the frequency. This technique could be applied in a microchannel. The requirement of delivering constant mass flow rates to a cooling device could be eliminated.

8.1.11 VARIABLE ROUGHNESS STRUCTURES

Another possibility exists for a variable roughness structure in a microchannel flow. With such a structure, the heat transfer enhancement could become variable as well. Piezoelectric actuators could be used to control the local surface roughness along the wall. Therefore, the heat transfer enhancement could be customized. Figure 8.11 shows a possible variable roughness structure using actuators of some type.

8.1.12 ACOUSTIC ENHANCEMENT

Decreasing heat losses and/or increasing the efficiency of the energy transfer process are also of major importance from both economical and ecological viewpoints. The enhancement of pool boiling heat transfer and natural convection heat transfer has been studied for many years. Various investigators (e.g., Yamashiro et al., 1998) have shown the enhancement of pool boiling and natural convection heat transfer by imposing acoustic waves into a liquid medium that is also externally heated by means of metallic wires or other kinds of heat sources such as immersion heaters, etc.

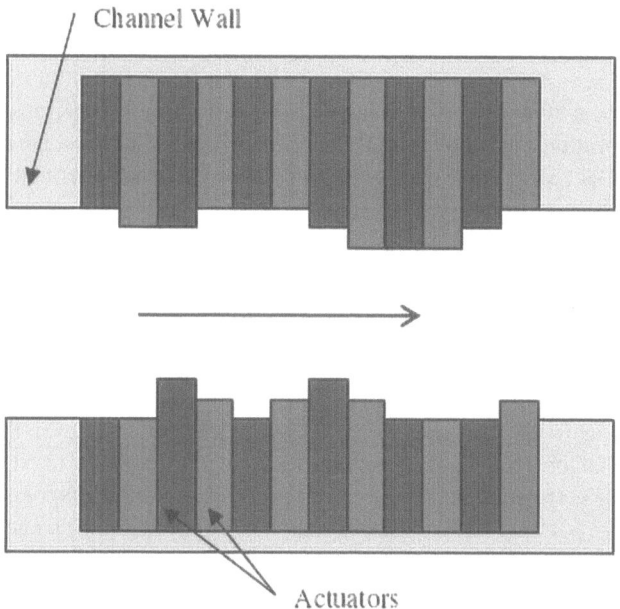

FIGURE 8.11   Variable roughness structure using actuators.

Bergles et al. (1978) reviewed heat-transfer enhancement equipment and techniques to reduce the size of the processing equipment or to reduce the approach temperature difference for the process streams. The effective enhancement of these techniques depends on the mode of heat transfer to dispersed flow film boiling. The techniques can be either passive or active. Some of the passive techniques described are treated surfaces, rough surfaces, extended surfaces, displaced enhancement devices which are inserted into the flow channel, swirl flow devices, surface tension devices, and some additives for gases and liquids. Some of the active techniques include mechanical aids such as mechanical stirrers, scrapers, etc., surface vibration using low or high frequency, fluid vibration ranging from 1 Hz to ultrasound, electrostatic fields applying either direct or alternating current, injection of gases into stream or degassing, suction involving vapor removal in nucleate or film boiling, and so on. Two or more of the above techniques can be combined together to get a compound enhancement effect. Among the possible new technologies that could be developed and optimized in order to improve heat-transfer processes, the use of ultrasonic waves appears to be a particularly attractive technical solution.

Kim et al. (2004) showed that the highest heat transfer enhancement ratio is obtained in natural convection regime where the effect of ultrasonic vibration is manifested through violent motion of cavitation bubbles. Also in natural convection and sub-cooled boiling regimes, behavior of cavitation bubbles strongly affects the degree of heat-transfer enhancement. Hyun et al. (2005) performed CFD analyses and experiments to characterize the effect of acoustic streaming on heat-transfer enhancement. Other researchers such as Gould (1996) and Nyborg (1958) have investigated this phenomenon as well.

Oh et al. (2002) studied the melting process of a phase change material, PCM, with ultrasound, and compared the results to the case without ultrasound. They observed that melting time with ultrasound was less compared to natural melting. They also observed that acoustic streaming enhances both thermal convective streaming and mass transport, where convective transport is superimposed on diffusion transport. Ultrasonic vibrations accelerate the melting process 2.5 times compared to natural melting, an effect associated with cavitation, streaming, and thermally oscillating flow initiated by ultrasound.

Gondrexon et al. (2010) investigated the intensification of heat transfer in a shell and tube heat exchanger using low-frequency ultrasound. They used a correlation for concentric annular ducts for calculating the Nusselt number with and without ultrasound, assuming all other terms of the correlation to be constant, thereby estimating corresponding values of the heat transfer coefficient from experimental values. They concluded that ultrasound is an efficient way to enhance heat-transfer performance of liquids in the laminar regime. They also observed that even for the lowest enhancement-factor conditions, the application of ultrasound resulted in a virtual liquid flow rate at the shell side higher by 10X compared to the real one.

Legay et al. (2011) investigated the overall heat-transfer rate with and without the influence of ultrasound on cold and hot water flow in shell and tube heat exchangers. They also found that the ultrasound effect is palpable only at low flow rate of cold water in the absence of agitation and turbulence, and that it becomes insignificant at higher flow rate.

Most of these studies have involved the use of a sonicator or horn-type ultra-sonic vibrator. Furthermore, every study has involved the use of either an immersion heater or a parallel arrangement of heater and vibrating plates. Many have used a single-frequency sonicator, or have used two different equipment that vibrate with different frequencies. In such cases, a generalization of the results is difficult as it is likely that different sonicators have different energy efficiencies, and that similar conditions may not be achieved for comparing two different ultrasonic vibrators considering the fact that any ultrasonic process is highly randomized and erratic.

Vetrimurugan (2007) has employed single ultrasonic equipment that can be operated at different frequencies, in which similar conditions of operation prevail, thus enabling a detailed study to generate heat transfer enhancement data and relevant dimensionless parameters. This was the first time that an orthogonal arrangement of the heater and vibrating plates has been attempted; moreover, a range of frequencies and power densities were used to get a detailed picture of heat transfer enhancement in a typical ultrasonic process.

Experiments were conducted using a digital thermometer (with data acquisition card), cavitation energy intensity meter, and 10 copper-constantan thermocouples placed at strategic positions in the ultrasonic tank. The experimental setup is shown in Figure 8.12. The experimental procedure involves continuous and simultaneous monitoring of temperature at the heater walls and various strategic positions in the tank using thermocouples. The temperature profiles are then used in calculating the average heat transfer coefficient which is both time-averaged as well as spatially averaged.

The heat transfer coefficient is obtained by macroscopic heat balance as:

$$h = \frac{\Phi_T - \Phi_V - \Phi_L}{A\left(\theta_W - \theta_B\right)} \tag{8.1}$$

where Φ_T, Φ_V, Φ_L, represent total heat flux supplied to the liquid medium in the ultrasonic tank, calorific power due to viscous dissipation, and other heat flux losses, respectively, A is the area of the embedded heater plate in the vertical wall, θ_W the

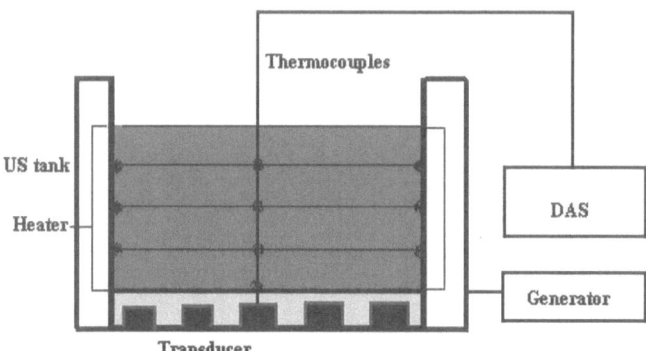

FIGURE 8.12 Schematic of experimental setup to measure heat-transfer augmentation.

wall temperature, and θ_B the bulk fluid temperature (spatial average of the temperatures at any instant). Given that:

$$\Phi_T = mC_p \frac{\partial \theta_B}{\partial t} \tag{8.2}$$

(where m is the mass of the agitated fluid (kg), and C_p is the specific heat capacity kJ/kg C)

This reduces to:

$$h = \frac{mC_p \frac{\partial \theta_B}{\partial t}}{A\left(\theta_W - \theta_B\right)} \tag{8.3}$$

where $\partial \theta_B / \partial t$ is the time-derivative of bulk fluid temperature. Here, heat losses and viscous losses have been neglected. The heating due to ultrasonic operation (about 3°C rise on the continuous mode of operation for 15 minutes) is neglected as well. All the experiments are conducted in pulse mode which results in negligible heating of the medium.

Enhancement ratio is defined here as:

$$\zeta_{Enhancement} = \frac{h_{withUS}}{h_{withoutUS}} \tag{8.4}$$

The heat transfer experiments were carried out with frequencies of 58, 132, 172, 192 kHz, and varying power densities. An increase in heat transfer coefficient or the enhancement ratio is observed with increase in power level for a given frequency of operation. The enhancement ratio increases from 1.74 to 4.76 with increase in power density from 6.45 kW/m³ to 16.13 kW/m³ when operated with 58 kHz frequency. The effect of power is more pronounced for higher frequencies such as 172 and 192 kHz. For 132 kHz frequency, increase in power level seems to have a marginal effect on the enhancement ratio. Higher enhancement at higher power densities can be directly attributed to higher cavitation intensities and higher acoustic streaming velocities at higher power densities. The number density or the cavitational activity of the cavitating bubbles is directly related to the power density. As the number density increases, the intensity of turbulence, which is responsible for local heat transfer enhancement, increases. Furthermore, the increase in acoustic streaming velocities results in convection patterns favorable to enhanced heat transfer.

Higher heat transfer coefficients are obtained for higher frequencies for a given power level. For 60, 80, and 100% power levels, the enhancement ratio first shows a marginal increase with increase in frequency from 58 to 132 kHz, and then it increases more rapidly with increase in frequency from 132 to 172 and 192 kHz. However, for a 40% power level there is a constant rate of increase in the enhancement ratio with an increase in frequency. From Figure 8.13, it can be observed that for single-frequency operation, higher enhancement ratios are observed for higher frequencies at similar power densities. This can be attributed to higher acoustic streaming velocities at higher frequencies at constant input power densities. Higher acoustic streaming velocities result in efficient transfer of cold fluid from the bulk to

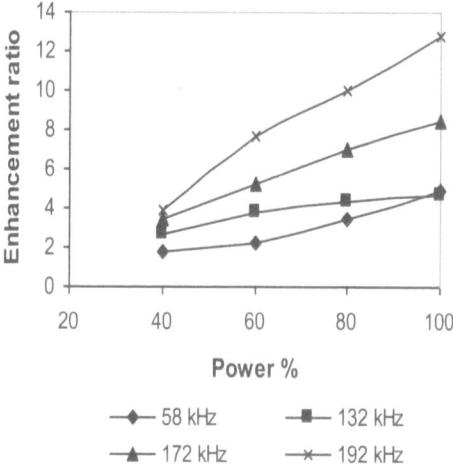

FIGURE 8.13 Effect of input power.

the heater surface and help in establishing convective patterns conducive to enhanced heat transfer. Figure 8.14 shows the effect of ultrasonic frequency at various input power levels. Higher frequencies result in greater enhancement, regardless of input power.

The enhancement ratio shows a decrease with increase in cavitation intensity. Figure 8.15 shows the observed trend. There is a sudden drop in enhancement ratio with cavitation intensity and then there is a marginal decrease with further increase in cavitation intensity. From Figure 8.15, a limiting effect of cavitation intensity can be established. For cavitation intensities greater than the limiting cavitation intensity, there is no effect of cavitation intensity on heat transfer enhancement ratio which

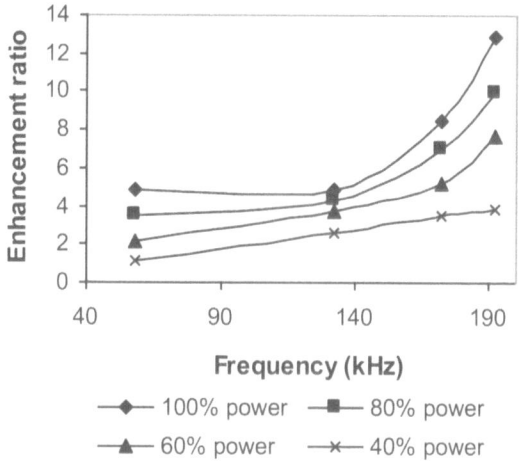

FIGURE 8.14 Effect of ultrasonic frequency.

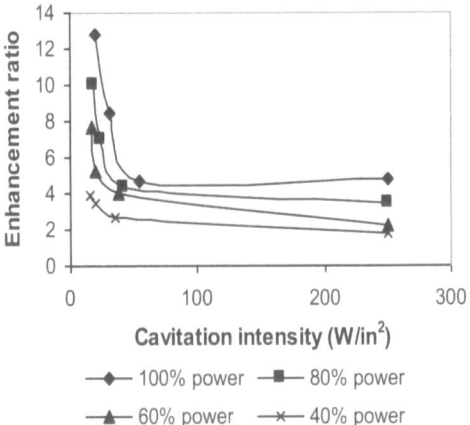

FIGURE 8.15 Effect of cavitation intensity.

stabilizes to a constant value. The experiments show an increase in limiting cavitation intensity with increase in power density. This can be observed in Figure 8.16.

Though high cavitation intensity increases in local turbulence intensity, which, in turn, results in higher heat transfer in localized regions in vicinity of the heater plate for macroscopic heat transfer, the movement of the bulk fluid is the deciding factor which occurs more efficiently at higher frequencies. At higher frequencies, cavitation intensities are lower, as cavitation intensity scales as $1/f^3$ (where f is frequency). Hence, a lower heat transfer enhancement ratio is obtained at higher cavitation intensity up to limiting cavitation intensity.

The heat transfer enhancement ratio also shows an increase with increase in the non-dimensionalized acoustic streaming velocity. Figure 8.17 shows the trend observed experimentally.

When an ultrasonic wave is imposed on a liquid medium, the phenomena of cavitation and acoustic streaming come into play and all the observed effects are due

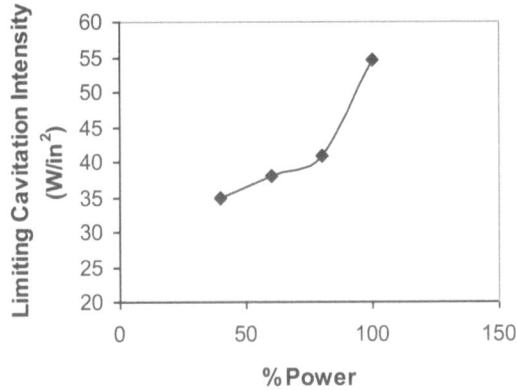

FIGURE 8.16 Variation of limiting cavitation intensity with % power.

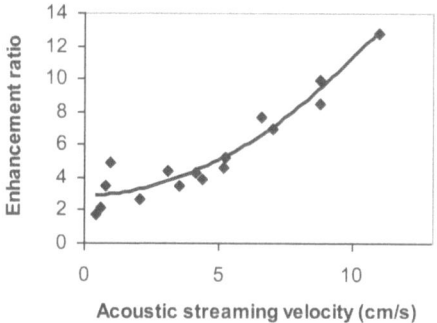

FIGURE 8.17 Variation of enhancement ratio with increase in acoustic streaming velocity.

to the interplay of these two phenomena. The input ultrasonic power is distributed among the various effects as follows,

$$P_{INPUT} = P_{CAVITATION} + P_{ACOUSTICSTREAMING} + P_{BULKHEATING} + P_{LOSSES} \qquad (8.5)$$

The manner in which the constant input ultrasonic power is distributed among the effects depends on the frequency of the operation.

At high frequencies	$P_{ACOUSTIC\ STREAMING} > P_{CAVITATION}$
At low frequencies	$P_{CAVITATION} > P_{ACOUSTIC\ STREAMING}$

For heat transfer enhancement, acoustic streaming is the most desired, and higher acoustic streaming velocities are achieved when $P_{ACOUSTIC\ STREAMING}$ is higher. This is ensured by high-frequency operation.

For processes that necessitate simultaneous heat and mass transfer, a trade-off is required. This can be achieved by dual frequency operation comprising of an optimized mix of low frequency (that increases local cavitation intensity for efficient mass transfer), and high frequency (that increases bulk fluid motion for efficient heat transfer).

PIF/CIF ANALYSIS

Intensification of heat transfer is certainly possible by the various techniques outlined in this section, and many are cost-effective in view of the improvement in heat-transfer efficiency achieved. Pan et al. (2016) have shown that the overall retrofit profit can be maximized based on the best trade-off among energy savings, intensification implementation costs, exchanger cleaning costs, and pump power costs. To solve such a complex optimization problem, they have developed a new mixed-integer linear programming (MILP) model. Kapustenko et al. (2015) point out that combining several enhancement techniques can achieve higher energy/cost savings when compared to implementing a single technique. Active and passive techniques need to be optimally combined to obtain a high PIF/CIF ratio. This requires a high level of creativity in design and an ability to think out of the box.

8.2   MASS TRANSFER (WITHOUT CHEMICAL REACTION)

Mass transfer is a core chemical engineering operation that is frequently rate-limiting from a system viewpoint. Hence, its intensification is always a holy grail for chemical engineers. Dyeing of leather and textiles, and removal of dyes from the effluent stream are examples of such challenges encountered in industry.

Burkinshaw et al. (2012) investigated the dyeing of non-woven fabrics using direct dyes (C.I direct Black 22 and C.I direct Red 89) in an infrared dyeing machine. The rate of uptake of direct dyes on different cotton fabrics as well as the rate of uptake of dispersed dyes onto polyethylene terephthalate was compared. They found that dyeing depends on sorption as well as dye molecule diffusion onto the surface of fibers. The transportation of dye to the proximity of fiber surface in solution and mass transfer processes in the fluid phase are influenced by forced flow of liquid and diffusion in liquid phase, i.e., by convective diffusion. The mass transfer rate of dye, therefore, depends on both yarn and fabric construction, i.e., their porosity. They concluded that the uptake of both dyes was independent of fiber and dye, but dependent on fabric construction.

Gupta et al. (2011) investigated the removal of acid blue dye 113 from wastewater using mesoporous activated carbon prepared from waste rubber tire. This dye has a significant presence of organic materials in it and is more stable, and hence, conventional physicochemical and biological methods of removing appear to be ineffective. They concluded that the adsorption of dye depends on surface properties as well as on porosity.

Moholkar et al. (2003) studied the intensification of mass transfer in wet textile processing by the application of ultrasound. They studied the increase in intra-yarn mass transfer. The resistance offered by small intra-yarn pores is much higher than large inter-yarn pores. Mass transfer through yarns is driven by diffusion, whereas between yarns, it is by convection. Diffusion is much slower than convection, and ultrasound micro-jet was effective in reducing diffusion resistance.

Haydock and Yeomans (2003) simulated the effects of attenuation-driven acoustic streaming on diffusion within a porous material and compared to a similar flow induced by more conventional means such as a fluid jet. Their simulations have shown both methods to have a significant effect on diffusion. In addition, the simulated results showed that in acoustic streaming, the force inducing the flow acts through the material leading to a higher average internal flow rate, resulting in a higher diffusion rate.

Sivakumar et al. (2008) studied the enhancement of the diffusion process in leather dyeing. Leather and powdered leather were compared to study the influence of ultrasound, and a diffusion model was proposed for the same. The results indicated that in powdered leather, diffusion rate is higher than in leather as its surface area is higher and diffusion resistance is lower. SEM analysis also showed that the morphology of leather is unaltered.

Sun et al. (2010) investigated the effect of ultrasound on dye exhaustion and dye fixing, both in continuous and intermittent mode of operation. They also showed that pre-treatment of fabrics with ultrasound results in a slight improvement in dye exhaustion and fixing, but not in any significant improvement in fastness properties.

Comparing the efficiencies of continuous and intermittent processes, the latter proved to be more beneficial.

Ferrero and Periolatta (2012) studied wool dyeing at lower temperature using ultrasound and compared it with conventional dyeing at higher temperature. The latter method caused damage to the fiber on prolonged exposure to higher temperature. They obtained better dyeing kinetics by coupling mechanical stirring and ultrasound. Also with ultrasound, 50% increase in absorption rate was obtained at every temperature. SEM results showed that the quality of wool fiber was unaffected by ultrasound. Thus, they concluded that ultrasound acts on dye dispersion, mass transfer, and diffusion without affecting the fiber structure.

Kamel et al. (2003) investigated nylon-6 fiber dyeing using ultrasound of 38.5 kHz, 350 W with reactive red 55 dye. They showed that ultrasonic dyeing is better at both neutral and acidic pH compared to conventional heating. The color strength values too proved to be superior with ultrasound. They indicated that the overall enhancement effect is mainly due to dye de-aggregation and dye-fiber covalent bond fixation using ultrasound.

Sivakumar et al. (2009) studied the natural dyeing of leather from beetroot using ultrasound for both extraction of dye as well as dyeing of leather. Ultrasound-assisted leaching and dyeing of leather showed significant improvement in extraction efficiency and dye exhaustion compared to magnetic stirrer. They used 20 kHz probe-type sonicator with maximum output of 400 W.

Kannan and Pathan (2004) investigated the enhancement of dissolution rate of sparingly soluble benzoic acid using three means, namely rotating cylinders, ultrasound, and chemical reaction. They found that among the three methods, ultrasonic enhancement of mass transfer coefficient was higher than the other two. This was attributed to enhanced turbulence and consequent reduction of mass transfer resistance by solvent impingement on a solid surface.

Sivakumar and Gangadhar Rao (2003) studied diffusion rate enhancement using ultrasound in leather dyeing of acid red dye. They performed the experiments using 20 kHz, 400 W ultrasonic probe both in pulsed and continuous mode and compared the results with static control and conventional drumming process. They found that the use of surfactant increased the percentage exhaustion of dye. They concluded that the ultrasonic method was more economical and cost-effective for dyeing of leather with improved color values.

Javier et al. (1999) investigated mass transfer and cavitation effect of 20 kHz ultrasound on low-temperature sono-electrochemical processes. They demonstrated the enhancement of mass transfer, mixing, and dissolution kinetics using ultrasound at low temperature. They also found a linear increase in mass transfer coefficient with increase in ultrasonic intensity, which indicates that the diffusion layer thickness decreased with increase in ultrasonic intensity, thereby increasing mass transfer rate.

Sivakumar et al. (2010) studied the application of ultrasound on various unit operations in leather processing and concluded that it would be a viable option for cleaner production with a good possibility of scale-up. They analyzed the effect of ultrasound on particle sizes of chemicals employed in leather industry and found that sono-fragmentation helps in better penetration of dye, or enhanced diffusion through leather matrix. They have also studied the influence of ultrasound on enzyme-based unhairing process and in chrome tanning process. The major steps in leather dyeing are: (1) mass transfer from liquid boundary between leather and bulk liquid to inter-fiber pores, and (2) mass transfer from inter- to intra-pores of fiber matrix. The process of mass transfer through the matrix will therefore be driven by diffusion, while mass transfer between matrices will be driven by convection. Since the diffusion process is much slower than convection, the rate-determining step will be diffusion from inter- to intra-pores. Hence, it is considered to be extremely beneficial to employ power ultrasound to assist leather dyeing process at room temperature. Enhanced diffusion is due to reversible opening and contraction of pores in the leather matrix under the influence of ultrasound propagation. Ultrasound achieves near-complete and uniform penetration of dye throughout the cross-section of leather, thereby improving its quality.

PIF/CIF ANALYSIS

Intensifying mass transfer, in comparison to heat transfer, is typically a greater challenge as the barriers to mass transfer are generally more difficult to overcome. Radiation, for example, is a heat-transfer mechanism that simply does not have a counterpart in mass transfer. In the absence of chemical reactions, mass transfer to surfaces frequently becomes diffusion-limited, even in systems with convective flow. As a process, diffusion is an especially difficult one to intensify through cost-effective measures. External fields, such as acoustic, can play a key role here. The "boundary layer" is the principal obstacle to overcome, and any mechanism that can reduce its thickness will greatly augment diffusive mass transfer.

8.3 CHEMICAL REACTIONS

Cavitational phenomena are widely used in wastewater treatment for the destruction of contaminants because of the localized high concentrations of oxidizing species such as hydroxyl radicals and hydrogen peroxide, and locally higher temperatures. Sometimes, addition of ferrous sulfate or iron will cause enhanced generation of free radicals in sonication system, helping in further reduction of chemicals in effluent treatment process. The cavitational effect is also used widely in preparation of polymers with nanomaterials, e.g., in applications for initiation of polymerization reactions for destruction of complex polymers, and in crystallization operations for controlling the crystal size distribution.

Psillakis et al. (2004) investigated the effect of horn-type sonicators (24 and 80 kHz) in degrading PAH (polyaromatic hydrocarbons) in an aqueous solution. They studied the conversion rate as a function of initial concentration, temperature, frequency, power, and dissolved salts. They highlighted the fact that oxidation plays a major role in degradation. They also observed that a low concentration of ferrous ion enhances PAH degradation through a Fenton-like reaction.

Suslick (1989) proposed that liquid sonochemistry caused by cavitation depends on physical effects such as heating and cooling. Cavitational heating decomposes water, forming extremely reactive radicals and atoms, whereas cooling leads to recombination of the above to hydrogen peroxide and molecular hydrogen. In combination with other compounds, it may result in a wide range of secondary reactions, which in turn help in degradation of organic compounds and oxidation and reduction of inorganic compounds.

Tantak and Chaudhari (2006) investigated azo dye degradation using Fenton process to enhance decolorization and demineralization of Reactive black 5, Reactive blue 13, and acid orange 7. Bouasla et al. (2010) investigated the degradation of methyl violet using Fenton process alone and found the optimum conditions for the reaction in terms of pH, concentration of reagents, and initial concentration of dye at 30°C. Inoue et al. (2008) investigated the degradation of bisphenol-A using 404 kHz, and found that high intensities resulted in the best TOC reduction and faster reactions. The addition of Fenton-like reagent to the above process slightly enhanced the degradation reaction but enhanced faster mineralization, thereby resulting in reduced TOC values.

Gogate et al. (2003a) studied the degradation of formic acid using different types of sonicators such as ultrasonic horn, bath, a dual-frequency, and triple-frequency flow cells. They observed that the bath-horn combination enhanced the results more than individual effects, as the OH–radical generation was higher. This radical generation is the overall rate-controlling step, and the combined system gives higher energy efficiency.

Koda et al. (2003) investigated the Fricke reaction for KI using a wide range of frequencies from 19.5 kHz to 1.2 MHz. They proposed a standard method for calibrating sonochemical efficiency, which was found to depend on the ultrasonic frequency and to be independent of experimental conditions.

Sohmiya et al. (2004) studied the sonochemical oxidation of alkyl and thiols in a heterogeneous solvent system using 20 kHz horn-type sonicators. They showed that the reaction is affected mainly by polarities, surface activities, vapor pressure, and hydrophobicity.

Zhao et al. (2002) studied the effect of a bi-frequency system, 28 kHz and 1.7 MHz, and compared results with individual systems. They found that terephthalate ion (TA) is non-fluorescent but emits fluorescence when it reacts with hydroxyl radicals to produce the hydroxyl-terephthalate ion. Thus, the radicals generated during sonication increase the fluorescence intensity of TA.

Okitsu et al. (2005) proposed a heterogeneous kinetic model considering a local reaction site at the cavitation bubble interface, where OH radical concentration is higher and hence helps in decomposition of azo dyes. They used an ultrasound of 200 kHz for decolorizing Reactive red 22 and methyl orange to confirm their proposed model.

Wang et al. (2003) studied the sonochemical degradation of methyl violet, a basic dye, using 20 kHz. They observed that degradation rate decreases with increase in initial concentration, and that the reaction rate was invariant for a change in temperature from 20°C to 40°C; beyond 80°C; however, the decrease in rate was remarkable.

Chang-qi et al. (2008) investigated azo dye degradation (methyl orange) using ultrasonics with Fenton reagent. They found that the combination enhanced the degradation rate drastically due to the augmentation of active species, OH radicals generated during the reaction between hydrogen peroxide, and ferrous and ferric ions.

Gogate et al. (2003b) analyzed the degradation of formic acid using US. They observed that high-frequency irradiation gives more beneficial effects than low frequencies, which was also reported in their earlier work in 2001.

Little et al. (2002) investigated the sono-degradation of phenanthrene (PAH) using 30 kHz probe-type sonicator, achieved around 88% reduction in initial concentration and found no traces of higher order PAH, considered hazardous.

PIF/CIF ANALYSIS

Given the extremely high pressures and temperatures generated during cavitation-induced implosion events, chemical reactions are greatly intensified by ultrasonic fields. Megasonic fields also augment reactions by providing reactive radicals and increasing the interfacial area of the bubbles. For a given cost, the biggest bang for the buck for acoustic intensification is likely to be in accelerating chemical reactions. For this reason, sonochemical reactors, employing ultrasonic as well as hydrodynamic cavitation, are in widespread commercial use. The batch nature of the process is a limitation that can be overcome in several creative ways—such as multiple sono-probes along a flow path—but may involve an additional cost.

8.4   MASS TRANSFER (WITH CHEMICAL REACTION)

In many chemical processes, transport phenomena are coupled to chemical reactions. Intensification efforts are then directed to both, in particular the rate-determining mechanism.

Mukherjee and Borthakur (2003) investigated the effect of leaching of coal using KOH at 95°C and 150°C followed by mild acid (10% HCl) demineralization and desulfurization. They found that both temperature and KOH concentration are kinetic factors impacting demineralization and desulfurization. Demineralization of about 28–45% to 39–68% and desulfurization of about 22–35% and 34–53% at 95°C and 150°C, respectively, was achieved. This method almost completely removes inorganic sulfur and up to 37% of the organic sulfur.

Li and Guo (1996) studied supercritical desulfurization of coal with alcohol/water and alcohol/KOH mixtures. They found that supercritical desulfurization in a semi-continuous reactor took place within 1 hour. The pretreatment using KOH (less than 5%) showed positive results. Ethanol/KOH solution improved desulfurization significantly, with inorganic sulfur being removed preferentially. Sulfur removal from a coal containing organic sulfur reached a maximum when ethanol concentration was 95 vol% in ethanol-water mixtures.

Ali et al. (1992) investigated the chemical desulfurization of high-sulfur coal using various inorganic reagents. They found that among them, H_2O_2 was most suitable in removing 50–90 wt % of sulfur at room temperature within 30 minutes and with minimum destruction of coal structure; however, organic sulfur was found to be difficult to remove.

Liu et al. (2008) employed a different method combining atmospheric oxidation and chemical cleaning with an alkali solution to desulfurize coal at a low temperature of around 90°C. The result showed 66% organic sulfur removal, 44% sulfide sulfur removal, and 15% pyrite sulfur removal when coal is treated in 0.25 M NaOH at 90°C, and aerated at a flow rate of 0.136 m^3 h$^-$. On further treatment with HCl at pH 1 for an hour, the rate further increased to 73% organic, 83% sulfide, and 84% pyrite sulfur removal.

Rodriguez et al. (1996) studied the influence of process parameters such as temperature, acid concentration, particle size, and physical pretreatment on coal desulfurization by nitric acid leaching. They found it to be a good method for intermediate-rank coal, especially with reference to inorganic sulfur, and the process to be greatly affected by temperature and to a lesser extent by acid concentration and time. They found moderate reduction of organic sulfur 19 (~15%) and much higher reduction of inorganic sulfur (~90%) even in the presence of very fine sulfur inclusions.

Sonmez and Giray (2001) investigated the influence of process parameters on desulfurization of two lignites using a mixture of H_2O_2 in acetic acid. They measured sulfur removal with respect to particle size, reaction time, reaction temperature, and ultrasonic irradiation. Under mild conditions, much of the inorganic and some organic sulfur were removed. Reaction time and temperature slightly changed the level of sulfur removal from different coals. The level of desulfurization was largely independent of the particle size.

Mukherjee and Borthakur (2001) studied the chemical desulfurization of high-sulfur coal using NaOH followed by HCl acid. Compared to the individual effects of alkali and acid, the combined effect resulted in the significant removal of minerals and sulfur from coal. They removed 43–50% of the ash, all of the inorganic sulfur, and around 10% of organic sulfur from coal using 16% NaOH followed by 10% HCl.

Prasassarakich and Thaweesri (1996) investigated the kinetics of coal desulfurization with the organometallic compound, sodium benzoxide. They showed that ˜68% of pyritic sulfur and ~33% of organic sulfur were leached. The kinetics of pyritic sulfur removal were investigated and the data found to be well-described by the shrinking core model, indicating the pyritic sulfur reaction to be predominantly diffusion-controlled. The organic sulfur reaction was zero order with respect to the organic sulfur.

Alam et al. (2009) studied the desulfurization of coal using froth floatation followed by leaching using two reagents—HNO_3 and HCl. The result showed that HNO_3 is more effective than HCl. They used a Taguchi L9 orthogonal array for experiments and found the optimum conditions using Analysis of Means (ANOM) confirmed via Analysis of Variance (ANOVA). The significance of process parameters followed the order: Acid concentration > temperature > stirring speed > time.

Zaidi (1993) investigated the application of ultrasound for de-sulfurization of low-rank coals using dilute solutions of sodium hydroxide (0.025–0.2 M) at 30°C and 70°C. With low-temperature sonication, sulfur removal was found to be higher, possibly due to shear forces produced by the ultrasound energy that are responsible for exposing the finely disseminated sulfur sites in coal to alkali attack. However, the mechanism involved in the interaction between sonication and dilute sodium hydroxide is not yet fully explained.

Ze et al. (2007) investigated the enhancement of de-sulfurization and de-ashing of coal. 100 grams of Zibo coal and 300 ml of water mixture were sonicated for 10 minutes using 20 kHz frequency and 200 W power. Then, the sample was wet screened. The same procedure was followed without sonication. Yield, sulfur and ash analysis were performed, and results revealed that ultrasonic conditioning can drive physical separation of pyrite and refuse from coal. On the other hand, ultrasonic conditioning can change the surface of the coal and pyrite particles, and increase the hydrophobicity of slime and the hydrophilicity of pyrite and refuse. It may be concluded that ultrasonic conditioning can, in general, enhance the performance of coal flotation methods used for desulfurization and de-ashing.

Mello et al. (2009) investigated an ultrasound-assisted oxidative process for sulfur removal from petroleum product feedstock. Dibenzothiophene was used as a model sulfur compound. The effect of sonication time, volume of oxidizing reagents, type of solvent for the extraction step, and type of organic acid were investigated. Higher efficiency of sulfur removal was achieved using sonication in comparison to experiments performed without its application, under the same reaction conditions.

Wang and Yang (2007) used several carbon-based sorbents for de-sulfurization of a model jet fuel. The results showed that the selective adsorption ability of $PdCl_2$ was higher than those of CuCl and metallic Pd. The results of desorption experiments showed that ultrasound-assisted regeneration was an effective method for $PdCl_2$/AC that was saturated with benzothiophene and substituted compounds. The amount of sulfur desorbed was higher with ultrasound, i.e., 65 wt % desorption vs. 45 wt % without ultrasound.

Grobas et al. (2007) investigated hydrogenation of cyclohexene, biphenyl, and quinoline, as well as hydro-desulfurization of benzothiophene in the presence of formic acid (a hydrogen precursor), and a Pd/C catalyst; ultrasound irradiation was investigated too. It was found that the use of formic acid in the presence of ultrasonic irradiation was effective in promoting hydrogenation and desulfurization under very mild conditions (i.e., ambient temperature and pressure).

In the next chapter, more specific examples of acoustic intensification of various transport phenomena and coupled chemically reactive processes will be presented as case studies to further illustrate the principles of the two factors—PIF and CIF.

PIF/CIF ANALYSIS

Many limitations to mass transfer intensification, listed in the earlier section, may be overcome when a reactive mass transfer is enabled. The rate-limiting step is still a key consideration. If the mass transfer is diffusion-limited, many of the limitations will remain. However, if the regime can be changed to one that is reaction-controlled, cost-effective intensification becomes a real possibility.

EXERCISE QUESTIONS

1. Ligrani et al. (2012) have reviewed various heat-transfer enhancement techniques. The technique of choice is often dictated by practical considerations such as accessibility to heat transfer surfaces, exposed surface area, cost and feasibility of retrofitting, etc. For the techniques listed in this chapter, enumerate factors that need to be taken into account in making the decision. To the extent possible, use quantitative data from operational reports. There are many hidden costs that only surface during actual production experience.

2. Mass transfer in quiescent environments is dominated by diffusion. Effective diffusivity under "intensified" process conditions is then a good metric to assess potential enhancement. Given that diffusion is a size-dependent mechanism, and power ultrasound has the ability to fragment particles, estimate the net effect of a high-energy high-frequency ultrasonic field on augmentation of particulate diffusion in a high-temperature-gradient environment where thermophoresis effects can become important as well.

3. Sonochemical reactors are now commonplace in the industry. The design features of sonochemical reactors are defined by different types, number, and position of ultrasonic transducers, whereas their operational parameters are determined by ultrasonic frequency and intensity. As in the case of stirred vessels, sonochemical reactors can also be characterized for their performance based on mass transfer, mixing time, and flow pattern. Sonochemical reactors have potential to be more energy efficient compared to the stirred vessel if designed and operated appropriately (Asgharzadehahmadi et al., 2016). There is immense potential for sonochemical reactors for process intensification leading to greener processing and economic benefits. Identify some products manufactured in sonochemical reactors, and assess how their production has benefitted from sonochemistry.

4. A Damköhler number (Da) is a useful ratio for determining whether diffusion rates or reaction rates are more important for defining a steady-state chemical distribution over the length and time scales of interest. For $Da \gg 1$, the reaction rate is much greater than the diffusion rate, and transport is said to be diffusion-limited; for $Da \ll 1$, diffusion occurs much faster than the reaction, and thus diffusion reaches an "equilibrium" well before the reaction does. Ultrasonic fields intensify reactions and diffusive processes to different extents based primarily on frequency. Discuss the effect of ultrasonic frequency on the Damköhler number in the cavitational and streaming limits.

REFERENCES

Alam, H.G., A.Z. Moghaddam, and M.R. Omidkhah, "The Influence of Process Parameters on Desulfurization of Mezino Coal by HNO₃/HCl Leaching", *Fuel Process. Technol.*, vol. 90, 2009, pp. 1–7.

Ali, A., S.K. Srivastava, and R. Haque, "Chemical Desulphurization of High Sulphur Coals", *Fuel*, vol. 71, 1992, pp. 835–840.

Allen, P.H.G. and T.G. Karayiannis, "Electrohydrodynamic Enhancement of Heat Transfer and Fluid Flow", *Heat Recov. Syst. CHP.*, vol. 15, no. 5, 1995, pp. 389–423.

Asgharzadehahmadi, S., A.A.A. Raman, R. Parthasarathy, and B. Sajjadi, "Sonochemical Reactors: Review on Features, Advantages and Limitations", *Renew. Sustain. Energy Rev.*, vol. 63, September 2016, pp. 302–314.

Bergles, A.E., R.L. Webb, and G.H. Junkan, "Energy Conservation via Heat Transfer Enhancement", *Energy*, vol. 4, 1978, pp. 193–200.

Bondar, F. and F. Battaglia, "A Computational Study on Mixing of Two-Phase Flow in Microchannels", *Proc. ASME International Mechanical Engineering Congress*, Washington, D.C., November 15–21, 2003, Paper # IMECE03-43957, ASME Publications.

Bouasla, C., S.M. El-Hadi, and F. Ismail, "Degradation of Methyl Violet 6B Dye by the Fenton Process", *Desalination*, vol. 254, 2010, pp. 35–41.

Burkinshaw, S.M., A.D. Hewitt, R.S. Blackburn, and S.J. Russell, "The Dyeing of Nonwoven Fabrics Part 1: Initial Studies", *Dyes Pigments*, vol. 94, 2012, pp. 592–598.

Chang-qi, C., C. Jian-ping, and T. Zhiguo, "Degradation of Azo Dyes by Hybrid Ultrasound-Fenton Reagent", *IEEE Trans.*, 2008, pp. 3600–3603.

Ferrero, F. and M. Periolatta, "Ultrasound for Low Temperature Dyeing of Wool with Acid Dye", *Ultrason. Sonochem.*, vol. 19, 2012, pp. 601–606.

Go, J.S., "Design of a Microfin Array Heat Sink Using Flow-Induced Vibration to Enhance the Heat Transfer in the Laminar Flow Regime", *Sensors Actuators A*, vol. 105, 2003, pp. 201–210.

Gogate, P.R., A.M. Wilhelm, and A.B. Pandit, "Some Aspects of the Design of Sonochemical Reactors", *Ultrason. Sonochem.*, vol. 10, 2003a, pp. 325–330.

Gogate, P.R., S. Mujumdar, and A.B. Pandit, "Sonochemical Reactors for Waste Water Treatment: Comparison Using Formic Acid Degradation as a Model Reaction", *Adv. Environ. Res.*, vol. 7, 2003, pp. 283–299.

Gondrexon, N., Y. Rousselet, M. Legay, P. Boldo, S.L. Person, and A. Bontemps, "Intensification of Heat Transfer Process: Improvement of Shell- and Tube Heat 132 Exchanger Performances by Means of Ultrasound", *Chem. Eng. Process.: Process Intensif.*, vol. 49, 2010, pp. 936–942.

Gould, R.K., "Heat Transfer across a Solid-Liquid Interface in the Presence of Acoustical Streaming", *J. Acoust. Soc. Am.*, vol. 40, 1996, pp. 219–225.

Grobas, J., C. Bolivar, and C.E. Scott, "Hydro-Desulfurization of Benzothiophene and Hydrogenation of Cyclohexene, Biphenyl, and Quinoline, Assisted by Ultrasound, Using Formic Acid as Hydrogen Precursor", *Energy Fuels*, vol. 21, 2007, pp. 19–22.

Gui, F. and R.P. Scaringe, "Enhanced Heat Transfer in the Entrance Region of Microchannels", *Proc. 1995 30th Intersociety Energy Conversion Engineering Conference, IECEC*, July 30–August 4, 1995, Orlando, FL, ASME Publications, 2, 1995, pp. 289–294.

Gupta, V.K., B. Gupta, A. Rastogi, S. Agarwal, and A. Nayak, "A Comparative Investigation on Adsorption Performances of Mesoporous Activated Carbon Prepared from Waste Rubber Tire and Activated Carbon for a Hazardous Azo Dye- Acid Blue 113", *J. Hazard. Mater.*, vol. 186, 2011, pp. 891–901.

Haydock, D. and J.M. Yeomans, "Acoustic Enhancement of Diffusion in a Porous Material", *Ultrasonics*, vol. 41, 2003, pp. 531–538.

Hessami, M.-A., A. Berryman, and P. Bandopdhayay, "Heat Transfer Enhancement in an Electrically Heated Horizontal Pipe Due to Flow Pulsation", *Proc. ASME International Mechanical Engineering Congress*, Washington, D.C., November 15–21, 2003, Paper # IMECE03-55146, ASME Publications.

Hu, X. and Y. Zhang, "Novel Insight and Numerical Analysis of Convective Heat Transfer Enhancement With Microencapsulated Phase Change Material Slurries: Laminar Flow in a Circular Tube With Constant Heat Flux", *Int. J. Heat and Mass Transf.*, vol. 45, no. 15, 2002, pp. 3163–3172.

Hyun, S., D.R. Lee, and B.G. Loh, "Investigation of Convective Heat Transfer Augmentation Using Acoustic Streaming Generated by Ultrasonic Vibrations", *Int. J. Heat and Mass Transfer*, vol. 48, 2005, pp. 703–718.

Inoue, M., Y. Masuda, F. Okada, A. Sakurai, T. Takahashi, and M. Sakakibara, "Degradation of Bisphenol A Using Sonochemical Reactions", *Water Research*, vol. 42, 2008, pp. 1379–1386.

Javier, D.C., A. Neudeck, C. Richard, and F. Marken, "Low-temperature Sonoelectrochemical Processes Part 1. Mass Transport and Cavitation Effects of 20 kHz Ultrasound in Liquid Ammonia", *J. Electroanal. Chem.*, 477, 1999, pp. 71–78.

Kamel, M., M. El-Shishtawy, L. Hanna, and E. Ahmed, "Ultrasonic-assisted Dyeing: 1. Nylon Dyeability with Reactive Dyes", *Polym Int.*, 52, 2003, pp. 373–380.

Kandlikar, S.G., "Fundamental Issues Related to Flow Boiling in Minichannels and Microchannels", *Exp. Therm. Fluid Sci.*, vol. 26, no. 2–4, 2002, pp. 389–407.

Kandlikar, S.G., S. Joshi, and S. Tian, "Effect of Surface Roughness on Heat Transfer and Fluid Flow Characteristics at Low Reynolds Numbers in Small Diameter Tubes", *Heat Transf. Eng.*, vol. 24, no. 3, 2003, pp. 4–16.

Kannan, A. and S.K. Pathan, "Enhancement of Solid Dissolution Process", *Chem. Eng. J.*, vol. 102, 2004, pp. 45–49.

Kapustenko, P.O., D.J. Kukulka, and O.P. Arsenyeva, "Intensification of Heat Transfer Processes", *Chem. Eng. Trans.*, 45, 2015, pp. 1729–1734. DOI:10.3303/CET1545289.

Kim, H.Y., Y.G. Kim, and B.H. Kang, "Enhancement of Natural Convection and Pool Boiling Heat Transfer via Ultrasonic Vibration", *Int. J. Heat Mass Transf.*, vol. 47, 2004, pp. 2831–2840.

Koda, S., T. Kimura, T. Kondo, and H. Mitome, "A Standard Method to Calibrate Sonochemical Efficiency of an Individual Reaction System", *Ultrason. Sonochem.*, vol. 10, 2003, pp. 149–156.

Legay, M., N. Gondrexon, S. Le Person, P. Boldo, and A. Bontemps, "Enhancement of Heat Transfer by Ultrasound: Review and Recent Advances", *Int. J. Chem. Eng.*, 2011, pp. 1–17.

Li, W. and S. Guo, "Supercritical Desulfurization of High Rank Coal with Alcohol/Water and Alcohol/KOH", *Fuel Process. Technol.*, vol. 46, 1996, pp. 143–155.

Ligrani, P., M. Goodro, M. Fox, and H.-K. Moon, "Full-Coverage Film Cooling: Film Effectiveness and Heat Transfer Coefficients for Dense and Sparse Hole Arrays at Different Blowing Ratios", *J. Turbomach.*, vol. 134, no. 6, November 2012, 061039; https://doi.org/10.1115/1.4006304.

Little, C., M.J. Hepher, and M. El-Sharif, "The Sono-Degradation of Phenanthrene in an Aqueous Environment", *Ultrasonics*, vol. 40, 2002, pp. 667–674.

Liu, K., J. Yang, J. Jia, and Y. Wang, "Desulfurization of Coal via Low Temperature Atmospheric Alkaline Oxidation", *Chemosphere*, vol. 71, 2008, pp. 183–188.

Mello, P.A., F.A. Duarte, M.A.G. Nunes, M.S. Alencar, E.M. Moreira, M. Korn, V.L. Dressler, and E.M.M. Flores, "Ultrasound-Assisted Oxidative Process for Sulfur Removal from Petroleum Product Feedstock", *Ultrasonics Sonochem.*, vol. 16, 2009, pp. 732–736.

Moholkar, V.S., V.A. Nierstrasz, and M.M.C.G. Warmoeskerken, "Intensification of Mass Transfer in Wet Textile Processes by Power Ultrasound", *AUTEX Res. J.*, vol. 3, 2003, pp. 129–138.

Mukherjee, S. and P.C. Borthakur, "Chemical Demineralization/Desulphurization of High Sulphur Coal Using Sodium Hydroxide and Acid Solutions", *Fuel*, vol. 80, 2001, pp. 2037–2040.

Mukherjee, S. and P.C. Borthakur, "Effect of Leaching High Sulphur Subbituminous Coal by Potassium Hydroxide and Acid on Removal of Mineral Matter and Sulphur", *Fuel*, vol. 82, 2003, pp. 783–788.

Nyborg, W.L., "Acoustic Streaming Near a Boundary", *J. Acoust. Soc. Am.*, vol. 30, 1958, pp. 329–339.

Oh, Y.K., S.H. Park, and Y.I. Cho, "A Study of the Effect of Ultrasonic Vibrations on Phase-Change Heat Transfer", *Int. J. Heat and Mass Transf.*, vol. 45, 2002, pp. 4631–4641.

Okitsu, K., K. Iwasaki, Y. Yobiko, H. Bandow, R. Nishimura, and Y. Maeda, "Sonochemical Degradation of Azo Dyes in Aqueous Solution: A New Heterogeneous Kinetics Model Taking into Account the Local Concentration of OH Radicals and Azo Dyes", *Ultrasonics Sonochem.*, vol. 12, 2005, pp. 255–262.

Pan, M., I. Bulatov, and R. Smith, "Improving Heat Recovery in Retrofitting Heat Exchanger Networks With Heat Transfer Intensification, Pressure Drop Constraint and Fouling Mitigation", *Appl. Energy*, vol. 161, January 2016, pp. 611–626.

Peng, X.F. and G.P. Peterson, "Forced Convection Heat Transfer of Single-Phase Binary Mixtures Through Microchannels", *Exp. Therm. Fluid Sci.*, vol. 12, no. 1, 1996, pp. 98–104.

Prasassarakich, P. and T. Thaweesri, "Kinetics of Coal Desulfurization With Sodium Benzoxide", *Fuel*, vol. 75, 1996, pp. 816–820.

Psillakis, E., G. Goula, N. Kalogerakis, and D. Mantzavinos, "Degradation of Polycyclic Aromatic Hydrocarbons in Aqueous Solution by Ultrasonic Irradiation", *J. Hazardous Materials*, vol. B108, 2004, pp. 95–102.

Rodriguez, R.A., C.C. Jul, and D. Gomez-Limon, "The Influence of Process Parameters on Coal Desulfurization by Nitric Leaching", *Fuel*, vol. 75, 1996, pp. 606–612.

Sivakumar, V. and P. Gangadhar Rao, "Diffusion Rate Enhancement in Leather Dyeing with Power Ultrasound", *J. Am. Leather Chem. Assoc.*, vol. 98, 2003, 230–237.

Sivakumar, V., J. Vijayeeswarri, and G. Swaminathan, "Ultrasound Assisted Enhancement in Natural Dye Extraction from Beetroot for Industrial Applications and Natural Dyeing of Leather", *Ultrasonics Sonochem.*, vol. 16, 2009, pp. 782–789.

Sivakumar, V., G. Swaminathan, P.G. Rao, and T. Ramasami, "Influence of Ultrasound on Diffusion Through Skin/Leather Matrix", *Chem. Eng. Process.*, vol. 47, 2008, pp. 2076–2083.

Sivakumar, V., G. Swaminathan, P.G. Rao, C. Muralidharan, A.B. Mandal, and T. Ramasami, "Use of Ultrasound in Leather Processing Industry: Effect of Sonication on Substrate and Substances - New Insights", *Ultrason. Sonochem.*, vol. 17, 2010, pp. 1054–1059.

Sohmiya, H., T. Kimura, M. Fujita, and T. Ando, "The Effect of Heterogeneous Solvent Systems on Sonochemical Reactions: Accelerated Degradation of Alkyl Thiols in Emulsions", *Ultrason. Sonochem.*, vol. 11, 2004, pp. 435–439.

Sonmez, O. and E.S. Giray, "The Influence of Process Parameters on Desulfurization of Two Turkish Lignites by Selective Oxidation", *Fuel Process. Technol.*, vol. 70, 2001, pp. 159–169.

Sturgis, J.C. and I. Mudawar, "Single-Phase Heat Transfer Enhancement in a Curved, Rectangular Channel Subjected to Concave Heating", *Int. J. Heat and Mass Transf.*, vol. 42, no. 7, 1999, pp. 1255–1272.

Sun, D., Q. Guo, and X. Liu, "Investigation into Dyeing Acceleration Efficiency of Ultrasound Energy", *Ultrasonics*, vol. 50, 2010, pp. 441–446.

Suslick, K.S., "The Chemical Effects of Ultrasound", *Sci. Am.*, vol. 2, 1989, pp. 80–86.

Tantak, N.P. and S. Chaudhari, "Degradation of Azo Dyes by Sequential Fenton's Oxidation and Aerobic Biological Treatment", *J. Hazard. Mater.*, vol. B136, 2006, pp. 698–705.

Vetrimurugan, R., "Investigation of High-Frequency, High-Intensity Ultrasound for De-stratification of Liquids Stored in Insulated Containers", Ph.D. Thesis, Indian Institute of Technology Madras, 2007.

Vinay, R., "Investigation of High-Frequency, High-Intensity Ultrasonics for Enhancement of Heat Transfer and Gas-Liquid Mass Transfer", B.Tech. Project Report, May 2007.

Wang, X.K., G.H. Chen, and W.L. Guo, "Sonochemical Degradation Kinetics of Methyl Violet in Aqueous Solutions", *Molecules*, vol. 8, 2003, pp. 40–44.

Wang, Y. and R.T. Yang, "Desulfurization of Liquid Fuels by Adsorption on Carbon-Based Sorbents and Ultrasound-Assisted Sorbent Regeneration", *Langmuir*, vol. 23, 2007, pp. 3825–3831.

Yamashiro, H., H. Kakamatsu, and H. Honda, "Effect of Ultrasonic Vibration on Transient Boiling Heat Transfer During Rapid Quenching of a Thin Wire in Water", *J. Heat Transf.*, vol. 120, 1998, pp. 282–286.

Zaidi, S.A.H., "Ultrasonically Enhanced Coal Desulphurization", *Fuel Process. Technol.*, vol. 33, 1993, pp. 95–100.

Ze, K.W., X.H. Xin, and C.J. Tao, "Study of Enhanced Fine Coal Desulfurization and De-Ashing by Ultrasonic Floatation", *J. China Univ. Min. Technol.*, vol. 17, 2007, pp. 358–362.

Zhao, Y., C. Zhu, R. Feng, J. Xu, and Y. Wang, "Fluorescence Enhancement of the Aqueous Solution of Terephthalate Ion after Bi-Frequency Sonication", *Ultrason. Sonochem.*, vol. 9, 2002, pp. 241–243.

9 Intensification of Heat Transfer

Case Studies

9.1 HEAT TRANSFER IN FURNACE TUBES

Enhancement of heat transfer from a heat source to a flowing fluid within a tube is a challenging problem with many practical applications. In general, techniques used for heat transfer enhancement can be classified into two types: (1) active and (2) passive. Passive techniques use specific surface geometries for augmentation, whilst active techniques supply external power to the heat-transfer surface, such as surface vibration, and acoustic or electric fields. The effectiveness of both types of techniques is strongly dependent on the mode of heat transfer (viz., single-phase, boiling, or condensing; natural or forced flow). Two or more of the above techniques may be utilized simultaneously to produce an enhancement that is larger than that achievable by the individual technique applied separately, i.e., compound enhancement. This so-called "third generation" heat transfer technology dramatically raises surface area per unit volume of heat exchangers, while avoiding potentially significant fouling and pressure-drop penalties. The majority of commercially viable enhancement techniques are, however, currently limited to passive techniques.

A number of active techniques have been developed for in-tube enhancement during two-phase and single-phase flows. One promising technology involves the use of high-intensity acoustic (ultrasonic) fields. Ultrasonic vibration induces an acoustic streaming velocity, as well as a cavitational velocity produced by the violent implosion of cavitation bubbles. These mechanisms of microscopic, even nano-dimensional, flow make ultrasonically induced acoustic streaming and cavitation an interesting alternative to enhance conventional heat transfer rates in the water tube of a boiler.

Experimental investigation of a low-frequency (20–33 kHz), high-intensity (500–1000 W) ultrasonic field as a potential heat-transfer process intensifier was undertaken by Dhanalakshmi et al. (2012). Heat-transfer enhancement data collected in a miniaturized furnace tube over a range of flow conditions and ultrasonic process parameters indicate that sonication provides significant augmentation only under near-static (e.g., stagnant) and low-Reynolds number flow conditions. With increasing flow velocity, cavitational and acoustic-streaming fields associated with ultrasound are rapidly diminished in importance, and hence, play no role in bulk fluid heat transfer (unless input power levels or frequencies are suitably increased). However, the relevance to some locations, such as those under porous deposits in water-wall tubes of boilers near the goose-neck portion, can spur further study to exploit the impact of ultrasonic heat-transfer enhancement. The critical parameter

DOI: 10.1201/9781003283423-9

FIGURE 9.1   Experimental setup.

that determines the efficacy of ultrasonic enhancement of heat transfer appears to be the ratio of the characteristic ultrasonic field velocity (sum of cavitational and acoustic streaming velocities) to the prevailing flow velocity.

In this study (Dhanalakshmi, 2013), a furnace tube miniature model was designed as shown in Figure 9.1 to have better control over the set temperature, flow, precision of measurement of fluid temperature at various locations, and influence of ultrasonics in the direction of flow. In order to achieve this, the model was designed in such a way that the bottom half of the pipe is heated, and the top half was designed to have thermocouples at different locations and to provide three different locations for fixing the ultrasonic probe. This was planned to facilitate a better understanding of the influence of ultrasonics. Thermocouples were inserted at locations A, and the ultrasonic probes were placed in locations B as shown in Figure 9.2.

The heater setup was connected to a stainless steel pipe of inner diameter 50 mm and 6 mm thickness. The heater was connected to the bottom half of the pipe and in the top half, and provisions were made to fix ten 4-wire RTDs. A thermostat was provided in the control panel to maintain the set heater temperature. A 'K' type thermocouple, used to measure the heater temperature, was placed between the heater coils, near the bottom surface of the pipe. The thermocouple measurement was used by the thermostat to maintain the temperature. Heater coils would be energized for an appropriate period if the heater temperature fell below the set value. Once the set temperature is achieved, the supply to the coils would be cut off. Thus, the coils would be energized and be kept idle by the thermostat alternatively to maintain the set temperature. Even though the coils were energized only for a short period, this resulted in oscillations in the heater temperature, and the effect was felt on exit fluid temperature. Maintaining constant exit fluid temperature was of primary importance since the effect of ultrasonic heat transfer enhancement would be calculated on the

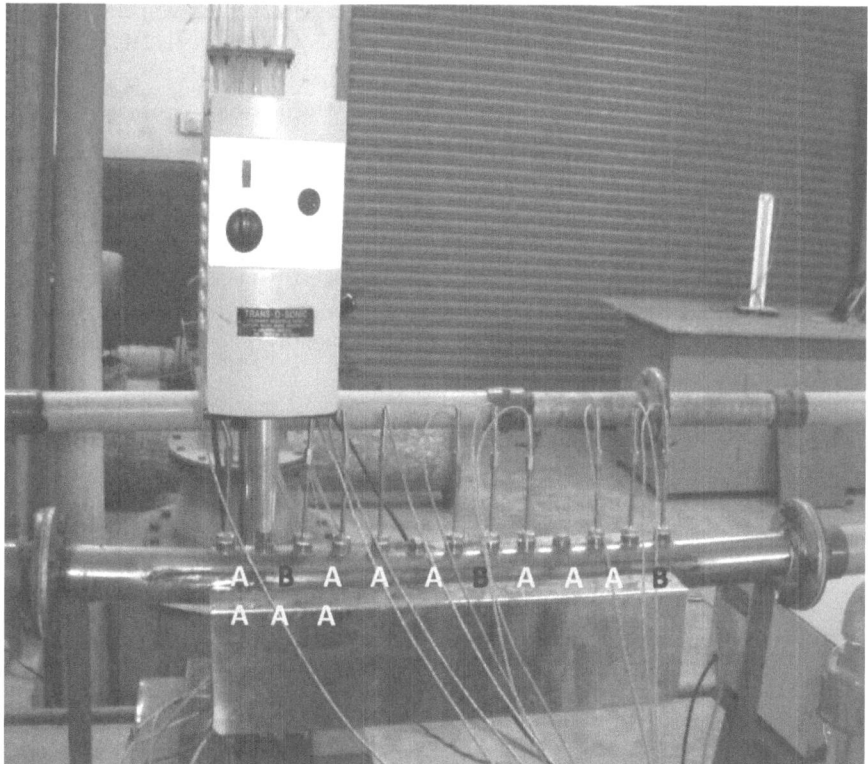

FIGURE 9.2   Placement of thermocouples and probes.

basis of bulk fluid temperature rise. Hence, a 7.5 kW variable rheostat was connected to the center heater element, while the other two heaters were connected through the thermostat. Since the input was being controlled with the rheostat to maintain a constant power being absorbed by the heater elements, the oscillations in the exit fluid temperature were eliminated.

The desired mass flow rate inside the pipe was set initially with the help of a flow-controlling valve. All the 4-wire RTDs were connected from the pipe to a data acquisition unit, which was set to read temperatures and was connected to a computer. Power was applied to the heaters, and all three heater ampere-meters were set to "on" (total power – 9 kW). By adjusting the rheostat, the heater temperature was set to the desired value. Once the set temperature was reached, the temperature was sustained at the set value by rheostat adjustment. The temperature profile in the data acquisition unit for various RTDs was monitored to ensure that the system had attained a steady state. When steady-state was reached, the ultrasonic field was switched ON for a certain period, say 5 minutes, and hence the pulsed mode of operation of ultrasonic was achieved, e.g., 5 minutes ON and 5 minutes OFF. At each time step (2 seconds), data from all the RTDs were recorded on the computer. Heat-transfer enhancement was calculated by analyzing the recorded data. The same procedure was repeated for various mass flow rates and heater temperatures. The heat transfer enhancement ratio (PIF for this case study) was calculated as:

$$HTER = \left(\text{Heat Transfer Coefficient with ultrasonics}\right) \Big/ \left(\text{Heat Transfer Coefficient w/o ultrasonics}\right) \tag{9.1}$$

The actual heat-transfer enhancement due to the introduction of ultrasonics was noticeable once the system attained a steady state, i.e., there were no significant oscillations in the exit fluid temperature. To ensure steady state attainment of the system and eliminate oscillation in the exit fluid temperature, a typical mass flow rate of 0.029 kg/s (Re = 868) was selected and the experiment was conducted continuously for a certain period of time.

Figure 9.3 shows the dependence of heat-transfer enhancement on the Reynolds number in the case of applying a 20 kHz probe, for different heater temperatures. A decrease in enhancement was observed with increase in Re for all the heater temperatures. This confirms that ultrasonic enhancement of heat transfer to a flowing fluid will very much depend on flow conditions—laminar versus turbulent—as determined by the Reynolds number.

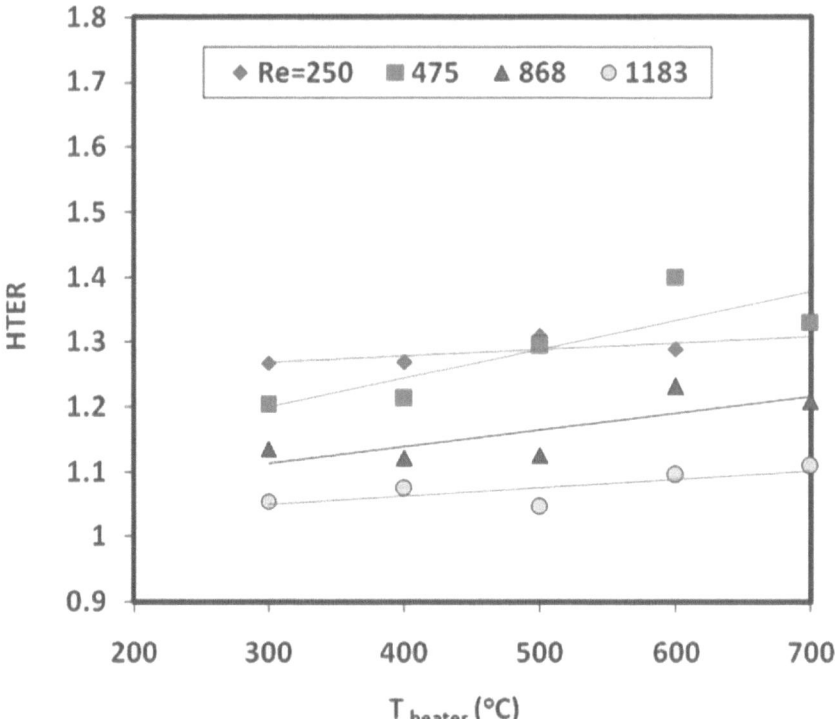

FIGURE 9.3 Heat transfer enhancement ratio as a function of heater temperature for various Reynolds numbers.

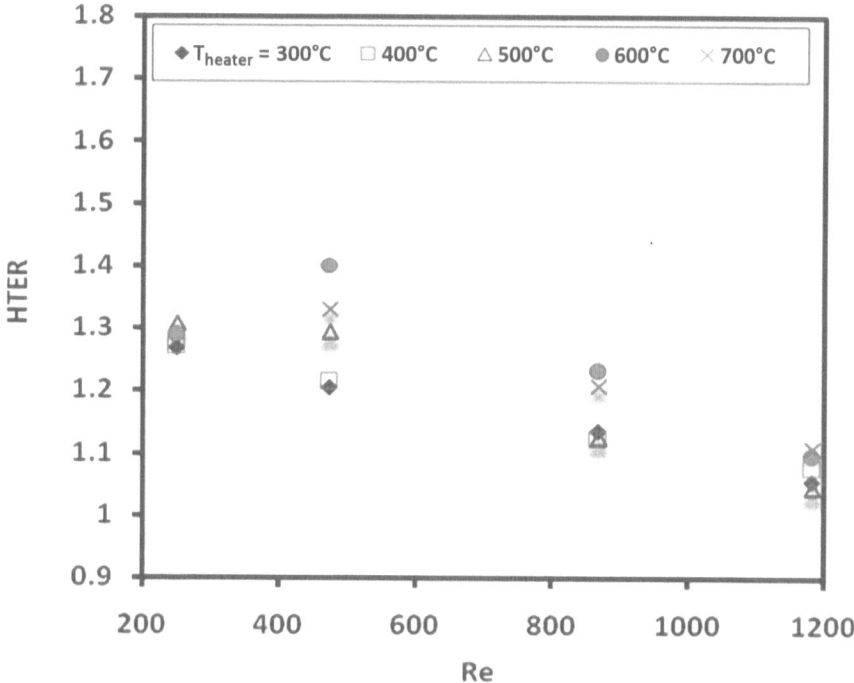

FIGURE 9.4 Heat transfer enhancement ratio as a function of Reynolds number for various heater temperatures.

From Figure 9.4, it is clear that the enhancement ratio of heat transfer (at 20 kHz) increases until Re reaches 500, which can be viewed as a "threshold" value. Beyond this threshold Reynolds number, the enhancement ratio gradually decreases till Re = 1200, and thereafter, no significant changes occur.

Figure 9.5 shows the effectiveness of ultrasonics for 33, 20, and (20 + 33) kHz probes (in sequence) for Reynolds number close to 500 at different heater temperatures. 20 kHz produces more enhancement than 33 kHz, and when 20 and 33 kHz are combined, a minimal incremental effect is observed when compared with a single 20 kHz. This is attributable to the more-intense cavitation of 20 kHz compared to 33 kHz; the net effect is due to turbulence generated by cavitation implosions.

Experiments carried out in a miniature furnace-tube model can serve as a bridge to integrate laboratory-scale results with actual furnace-tube operation. The ratio of external flow velocity to ultrasonic velocity is the linking factor.

Cavitational velocity is associated with bubble implosion and is related to volume of the bubble before collapse. Since the size of a spherical bubble scales inversely with frequency, bubble volume scales inversely with the cube of frequency. Streaming velocity, on the other hand, is associated with convective, uni-directional flow induced by the high-frequency field, and scales as square of frequency. Total ultrasonic velocity at any frequency may then be expressed as the sum of the two.

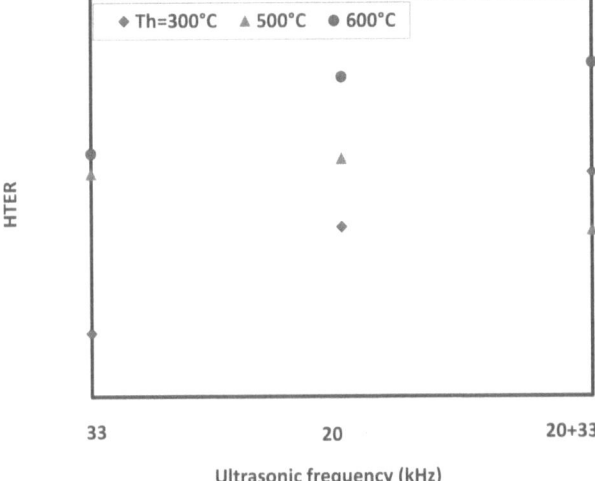

FIGURE 9.5 Heat transfer enhancement ratio as a function of ultrasonic frequency for various heater temperatures.

Measured cavitational (at 20 kHz frequency) and streaming velocities (at 850 kHz) in water at room temperature have been reported by Awad (1996), Busnaina et al. (1994), and Elsawy and Busnaina (1998) to be: $V_{cavitation} = 100$ m/s, and $V_{streaming} = 4$ m/s. These values have been scaled here for frequencies used in the study and used to generate the data reported in Figure 9.6.

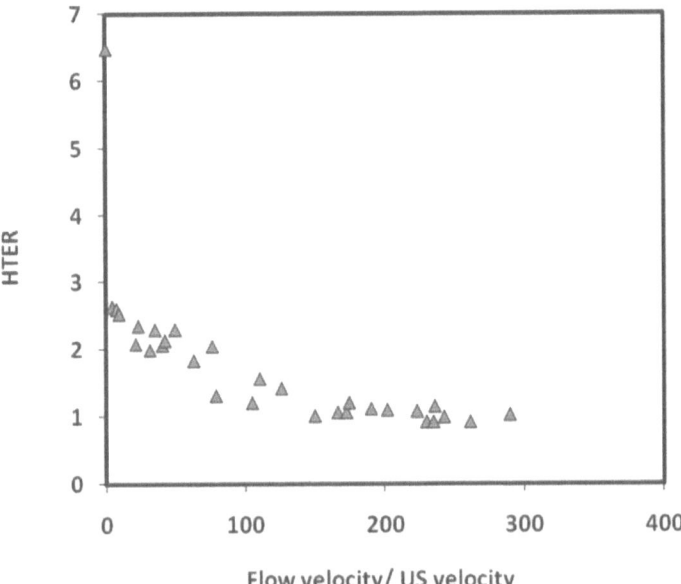

FIGURE 9.6 Effect of ratio of flow velocity to net ultrasonic velocity of heat transfer enhancement coefficient.

From Figure 9.6, it is evident that external flow conditions have a negative effect on ultrasonic heat-transfer enhancement rate. This makes high-intensity ultrasound ineffective for high flowrate applications; however, for controlled laminar flow cases, significant enhancement in heat transfer rates is observed with the use of high-intensity ultrasonic fields. Alternatively, cavitational or streaming velocity must be substantially enhanced (by increasing field amplitude or frequency) to overcome the external flow effect. In particular, acoustic frequencies in the 1 MHz range ("magasonics") are worthy of further study, incorporating, as they do, significant acoustic-streaming effects.

A key learning from the study was that ultrasonic enhancement of heat transfer while approaching an order-of-magnitude under static and near-static conditions, was limited to a maximum of 30–40% (still quite significant) under flow conditions characterized by Reynolds numbers approaching 500. The crucial scaling factor appears to be a ratio of ultrasonically induced velocity to the prevailing flow velocity; as this ratio increases, for a given mean velocity of flow, so does the enhancement effect. Hence, high flow rate conditions are not conducive to sono-augmentation of heat-transfer. If the effort is taken to identify locations in process equipment where stagnation or recirculation zones are likely to be present, localized placement of ultrasonic probes at such locations may be expected to provide significant heat transfer benefits.

It is crucial to note that an ultrasonic field, in the context of heat transfer enhancement, may be essentially described by two characteristic velocities—cavitational, predominant at lower frequencies (40 kHz), and streaming, dominant in the megasonic range. In the intermediate frequency zone, 100–200 kHz (also known as the trans-sonic regime), both velocities are present at comparable magnitudes. When an ultrasonic field is coupled to a flow field, the effect is not merely to induce turbulence, although that certainly does occur. Since the ultrasonic field is a source of linear momentum, the effective mean flow velocity is increased as well, resulting in higher "effective" Reynolds numbers, particularly in the laminar regime. The magnitude of this velocity augmentation may or may not be significant compared to the prevailing mean flow velocity; similarly, the induced turbulence may or may not be of a magnitude comparable to the prevailing flow turbulence. In order for ultrasonic augmentation of heat transfer to a flowing liquid to be appreciable, either condition (or both) must be met.

PIF/CIF ANALYSIS

The finding that acoustic enhancement of heat transfer in boiler tubes is significant only in locations of reduced flow velocity limits its practical application potential. However, it is also a fact that heat transfer is most constrained at such locations of virtual stagnation. It may thus be possible to apply acoustic fields and other intensification mechanisms in isolated areas where they would do the most good. Mapping out such quiescent regions in flow systems would be a useful precursor to implementing intensification strategies. It would minimize expenditure while maximizing impact.

9.2   HEAT TRANSFER IN TANK WITH SIDE HEATERS

A continuous flow sonicated system was set up for this study (Vinay, 2007) employing control valves both at the inlet and the outlet ports to regulate the flow. The system, and partial results obtained under quiescent conditions, were presented earlier in Chapter 8, and are reproduced here to provide a baseline for comparison with results obtained under flow conditions. The inlet port was directly connected to the tap to facilitate a continuous supply of water to the ultrasonic tank. The heat transfer experiments were conducted using a digital thermometer (with a data acquisition card), cavitation energy intensity meter (ppb500™), and 10 copper-constantan thermocouples placed in the tank. The experimental procedure involved the induction of constant heat flux from an embedded plate heater at the walls of the ultrasonic tank that are orthogonal to the piezo-ceramic transducers placed at the bottom of the tank, and subsequent temporal monitoring of temperature at the heater walls and various positions in the tank using thermocouples. The temperature profiles were then used in calculating the average heat transfer coefficient, which is time-averaged as well as spatially averaged, using the conventional heat balance technique. The ultrasonic operation was done in pulse mode in order to avoid ultrasonic heating which may distort the experimental data.

The calculation procedure involved some basic assumptions and simplifications:

- The specific heat capacity of water was assumed to be constant, even though the collapse of cavitating bubbles results in local spots character- ized by high temperatures of the order of 1000 K. The exact nature of the variation in temperature with time, which is stochastic in nature, has not been established so far and may be irrelevant to a macroscopic view of heat transfer enhancement using power ultrasound.
- The bulk temperatures were obtained by taking an average of temperatures measured at various positions within the tank. This might result in an inher- ent error if the sample points do not correspond to that of a well-mixed region. Such regions occur at higher frequencies when the global acoustic streaming velocities are higher (see following sections), but for lower fre- quencies, the mixed region is not well established and there can be erratic variation in the temperature within the tank. This problem is averted by increasing the number of sample points at which temperatures are mea- sured. Though there are physical limitations in taking a large number of sample points, about 8 points within the tank serve the purpose.
- There are limitations to the response of the digital thermometer to the local temperature variations. Nevertheless, only an average heat transfer coefficient is sought, for which temperatures measured using the setup proved to be satisfactory.

The heat transfer coefficient is obtained by macroscopic heat balance as (see Chapter 8),

$$h = \frac{\Phi_T - \Phi_V - \Phi_L}{A(\theta_W - \theta_B)} \tag{9.2}$$

where Φ_T, Φ_V, Φ_L, are total heat flux supplied to the liquid medium in the ultra- sonic tank, calorific power due to viscous dissipation, and other heat flux losses, respectively. A is the area of the embedded heater plate in the vertical wall,

θ_W = wall temperature, and θ_B = bulk fluid temperature (spatial average of the temperatures at any instant)

$$\Phi_T = mC_P \frac{\partial \theta_B}{\partial t} \tag{9.3}$$

(m is the mass of the agitated fluid (kg), and C_P is the specific heat capacity kJ/kg C)
Equation (9.2) then reduces to:

$$h = \frac{mC_P \dfrac{\partial \theta_B}{\partial t}}{A(\theta_W - \theta_B)} \tag{9.4}$$

where $\partial\theta_B/\partial t$ is the time-derivative of bulk fluid temperature; here, the heat losses and viscous losses have been neglected.

The heating due to ultrasonic operation (about 3°C rise on the continuous mode of operation for 15 minutes) is neglected. All the experiments were conducted in pulse mode, which results in negligible heating of the medium. The errors in the calculation of the heat transfer coefficients arise due to the heat losses and viscous dissipation losses; however, these losses are difficult to estimate due to the complex flow pattern involved in an ultrasonic operation. Several investigators (e.g., Thompson and Doraiswamy, 1999) have reported an error of about 25–60% in the determination of heat transfer coefficients using this method. An alternative to this conventional procedure would be the use of heat flux sensors, which gives an idea of instantaneous heat transfer coefficient. Thompson and Doraiswamy (1999) have shown the accuracy and usefulness of a heat flux sensor in determining instantaneous process-side heat transfer coefficients. This is difficult to achieve with the conventional thermocouple technique, no matter how good the response of the data acquisition system is to the temperature fluctuations.

The schematic of flow conditions is shown in Figure 9.7.

The liquid enters with a temperature T_i and exits with a temperature T_f. The temperature of the plate heater surface is T_S (constant as verified by measurement). L1,

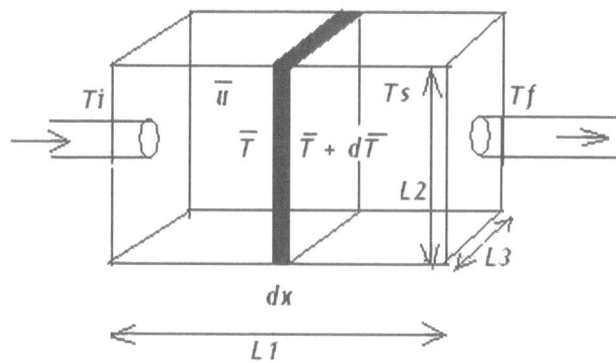

FIGURE 9.7 Determination of heat transfer coefficient under flow conditions.

L2, and L3 are the dimensions of the tank. For an infinitesimal element as shown in Figure 9.7, heat balance yields the following

$$\rho C_P \, \bar{u} \, L_2 L_3 \, \partial \bar{T} = h_{avg} (T_s - \bar{T}) L_2 \, \partial x \qquad (9.5)$$

Upon simplification and integrating between the limits T_i to T_f for temperature, and 0 to L_1 for x dimension, the expression for average heat transfer coefficient is obtained as:

$$h_{avg} = \frac{\rho C_P \, \bar{u} \, L_3}{L_1} \ln\left(\frac{T_s - T_i}{T_s - T_f}\right) \qquad (9.6)$$

where the average velocity is approximated as:

$$\bar{u} = \frac{Q}{A_{Pipe}} \qquad (9.7)$$

The heat transfer coefficient enhancement ratio, or the PIF in this case, is defined here as:

$$\zeta_{Enhancement} = \frac{h_{withUS}}{h_{withoutUS}} \qquad (9.8)$$

Parametric studies may then be conducted taking this as the dependent variable.

9.2.1 QUIESCENT CONDITIONS

The heat transfer experiments were carried out under no-external flow conditions (quiescent conditions) using frequencies of 58, 132, 172, 192 kHz and varying power densities or power levels. The heat transfer coefficient obtained for tank I (volume = 31 liters) without ultrasound is 317.83 W/m²K. The heat transfer coefficient obtained for tank II (volume = 20 liters), which operates at 132 kHz without ultrasound is 167.94 W/m²K.

An increase in heat transfer coefficient, or the enhancement ratio/PIF, is observed with increase in power level for a given frequency of operation and operating under the same conditions. The enhancement ratio increases from 1.74 to 4.76 with increase in power density (6.45 kW/m³–16.13 kW/m³), when operated with 58 kHz frequency. The effect of power is more pronounced for higher frequencies, such as 172 and 192 kHz, than for lower frequencies. For 132 kHz frequency, increase in power level seems to have marginal effect on the enhancement ratio. This can be observed in Figure 9.8. Higher enhancement at higher power densities can be directly attributed to prevailing higher cavitation intensities and higher acoustic streaming velocities. The number density, or the activity of the cavitating bubbles, is directly related to the power density. As the number density increases, the intensity of turbulence, which is responsible for local heat transfer enhancement, increases. Furthermore, the increase in acoustic streaming velocities results in convection patterns favorable to enhanced heat transfer.

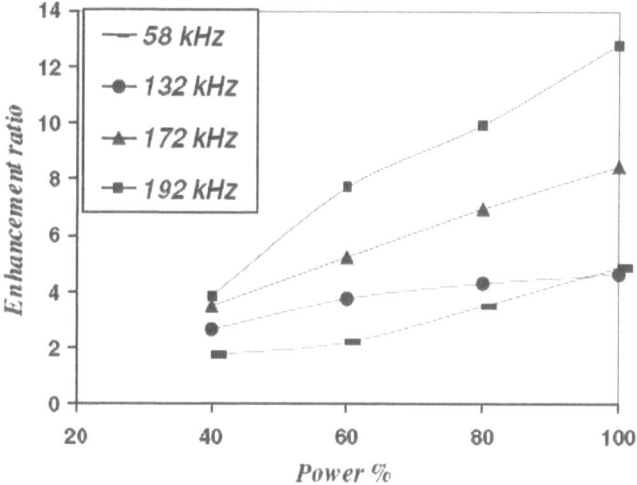

FIGURE 9.8 Variation of $\zeta_{Enhancement}$ with power.

Higher heat transfer coefficients are obtained for higher frequencies at a given power level. This trend is seen in Figure 9.9.

For 60, 80, and 100% power levels, the enhancement ratio first shows a marginal increase with increase in frequency from 58 to 132 kHz, and then increases more rapidly with increase in frequency from 132 to 172 and 192 kHz. However, for 40% power level, there is a constant rate of increase in the enhancement ratio with

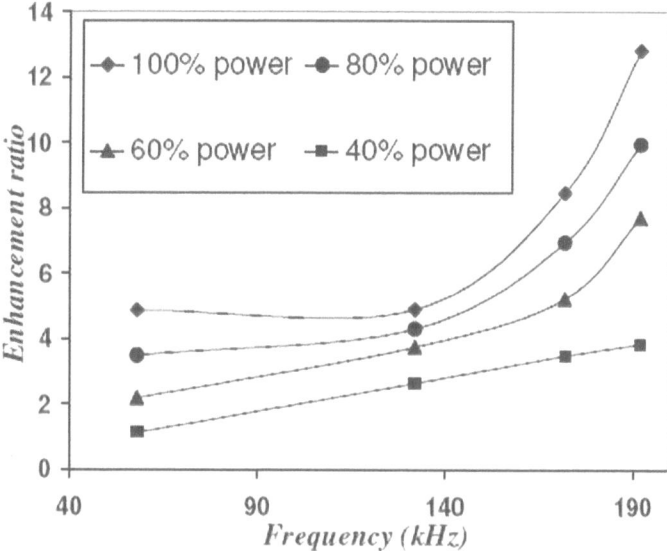

FIGURE 9.9 Variation of $\zeta_{Enhancement}$ with frequency.

increase in frequency. From Figure 9.8, it can be observed that for single-frequency operation, higher enhancement ratios are observed for higher frequencies at similar power densities. This can be attributed to higher acoustic streaming velocities at higher frequencies and constant input power densities. Higher acoustic streaming velocities result in efficient transfer of cold fluid from the bulk to the heater surface and help in establishing convective patterns conducive to enhanced heat transfer. These results also illustrate the inter-dependent nature of acoustic frequency and amplitude effects on heat transfer.

The enhancement ratio shows a decrease with increase in cavitation intensity, for a given power input. Figure 9.10 shows the observed trend.

There is a sudden drop in enhancement ratio with cavitation intensity, followed by a marginal decrease with further increase in cavitation intensity. In Figure 9.10, a "threshold" effect of cavitation intensity can be established. For cavitation intensities greater than the threshold value, there is no effect of cavitation intensity on heat transfer enhancement ratio. The experiments show an increase in the threshold cavitation intensity with increase in power density. This can be observed in Figure 9.11.

This counter-intuitive effect may be rationalized as follows: High cavitation intensity results in increase in local turbulence intensity, which, in turn, results in higher heat transfer in localized regions in the vicinity of the heater plate. However, for macroscopic heat transfer, the movement of the bulk fluid is the deciding factor, and this occurs more efficiently at higher frequencies. Fields with higher cavitation

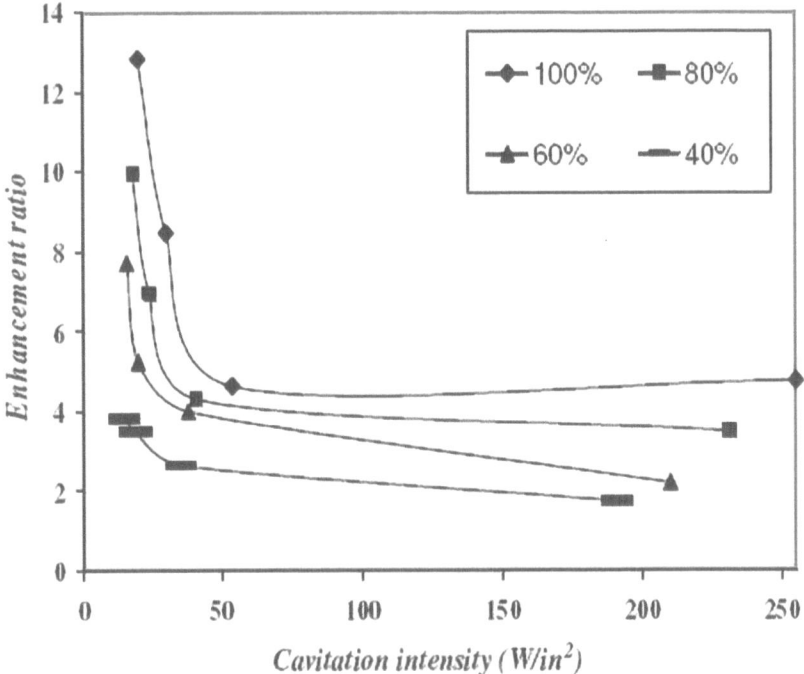

FIGURE 9.10   Variation of $\zeta_{Enhancement}$ with cavitation intensity.

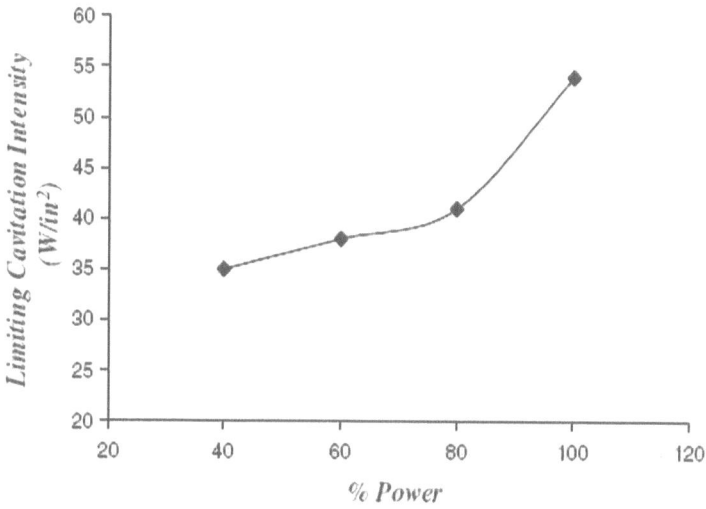

FIGURE 9.11   Variation of threshold cavitation intensity with % power.

intensities are characterized by lower acoustic streaming velocities since cavitation intensity scales as $1/f^3$ (where f is frequency). Hence, a lower heat transfer enhancement ratio is obtained at higher cavitation intensity up to the threshold value.

For heat transfer enhancement in a static tank with quiescent conditions prevailing, acoustic streaming is more impactful, and higher acoustic streaming velocities are achieved when $P_{ACOUSTIC\ STREAMING}$ is higher. This is ensured by high-frequency operation. However, when an external flow is imposed on the liquid, this conclusion is no longer valid, as will be shown in the next section.

9.2.2   Flow Conditions

The heat transfer experiments were repeated in the presence of external flow conditions with varying flow velocities, acoustic frequencies, and power levels.

Figure 9.12 shows the effect of flow conditions on heat transfer enhancement for single-frequency operation at 80% input power (total input power = 400 W).

Experimental data indicate a marked decrease in observed enhancement with increase in external flow rate. This is true for all frequencies of operation. For 172 and 192 kHz, hardly any enhancement is observed above a flow rate of 0.05 liter/s. Lower frequency operation (40 kHz) seems to yield higher enhancement in heat transfer rate than higher frequency operations (172 and 192 kHz)—which is in sharp contrast to quiescent conditions for which higher frequencies yield higher enhancements in heat transfer rates. This is indicative of the destruction of acoustic streaming pattern due to the imposition of external flow.

Acoustic streaming pattern is responsible for high-frequency enhancement in heat transfer rates, and hence, with the destruction of this pattern, there is a marked decrease in enhancement. 40 and 58 kHz frequency operations seem to yield higher heat transfer enhancement due to turbulence generated by cavitation implosions.

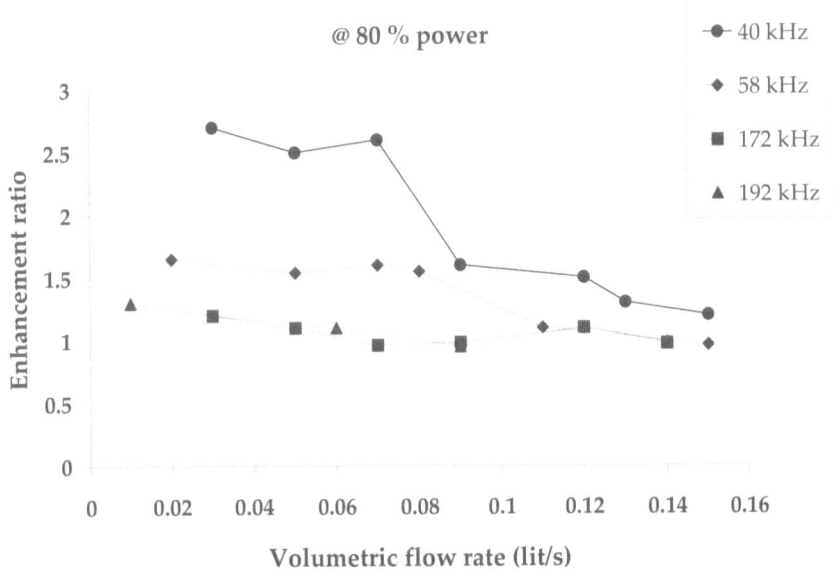

FIGURE 9.12 Heat transfer enhancement under flow conditions at different frequencies of operation with input ultrasonic power kept constant at 100%.

Figure 9.13 shows the effect of flow conditions on heat transfer enhancement for dual frequency operation at 100% power (in this case total input power = 1 kW).

In Figure 9.13, an almost linear decrease is seen in enhancement ratio with increase in external flow rate. This is again attributable to the destruction in acoustic streaming patterns with the imposition of external flow. A combination of a lower-frequency

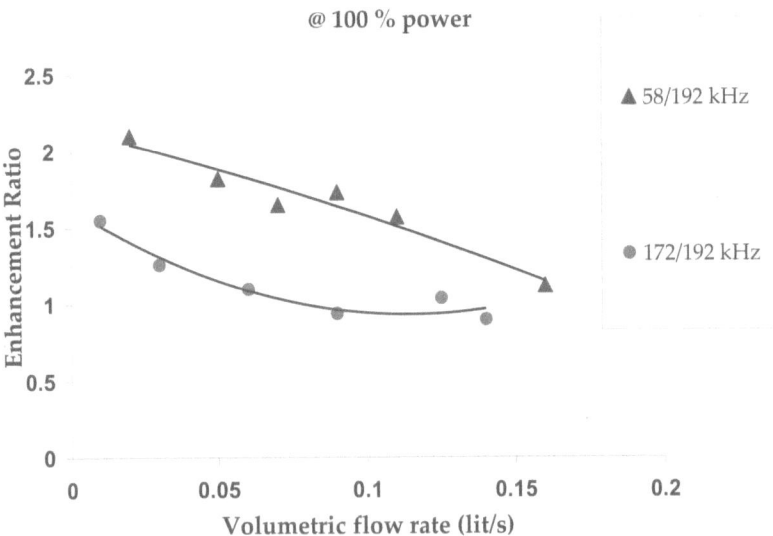

FIGURE 9.13 Heat transfer enhancement in the presence of flow conditions for dual frequency operation.

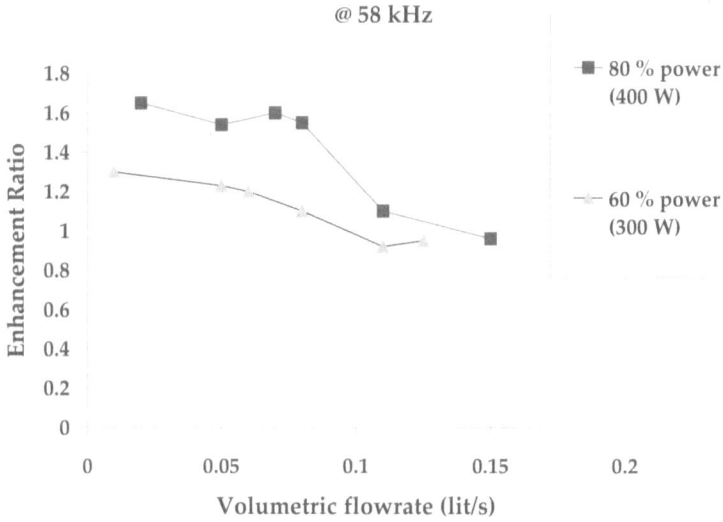

FIGURE 9.14   Effect of power on heat transfer enhancement for flow conditions @ 58 kHz.

operation (58 kHz) and a higher-frequency operation (192 kHz) gives consistently higher enhancement in heat transfer rate compared to 172/192 kHz at all external flow rates. This is in agreement with earlier findings for quiescent conditions, which reinforces the earlier finding that an optimized mix of lower and higher frequencies yields better heat transfer augmentation. This is mainly due to higher cavitation at a lower frequency which adds to the acoustic streaming pattern.

Figures 9.14 and 9.15 show the effect of input ultrasonic power on heat transfer enhancement at various flow conditions.

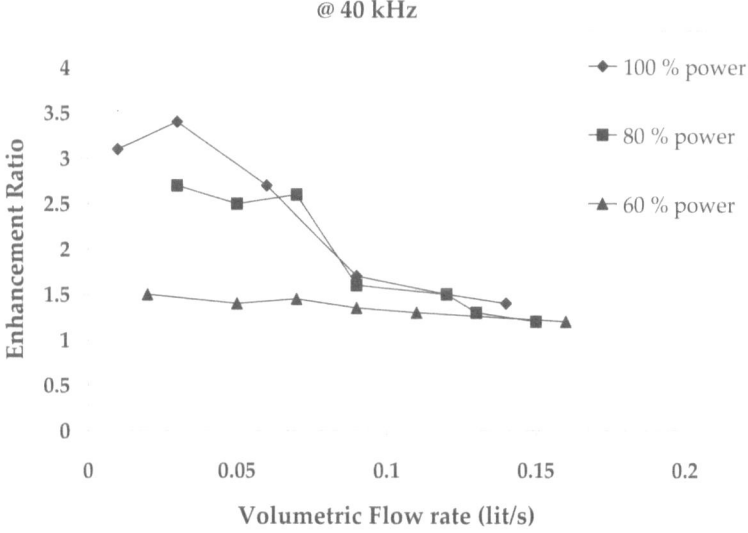

FIGURE 9.15   Effect of power @ 40 kHz.

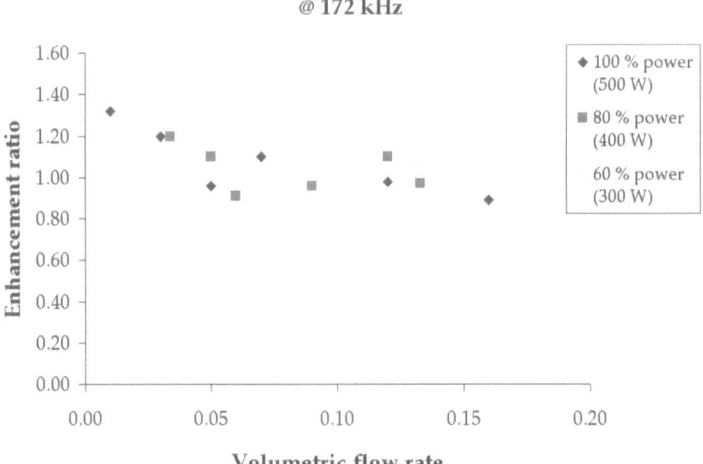

FIGURE 9.16 Effect of ultrasonic power on heat transfer augmentation for 172 kHz operation in the presence of flow conditions.

In Figures 9.14 and 9.15, it is seen that higher input ultrasonic power yields better heat transfer augmentation, which is in agreement with earlier findings for quiescent conditions. It is interesting to note that for 40 kHz at higher flow rates, similar heat transfer enhancements are obtained at all the power levels. Figures 9.16 and 9.17 show the effect of a change of input ultrasonic power on heat transfer augmentation for 172 and 192 kHz frequency of operation. From the figures, it can be concluded that there is hardly any variation with input power—again indicative of the destruction of acoustic streaming pattern due to imposition of external flow.

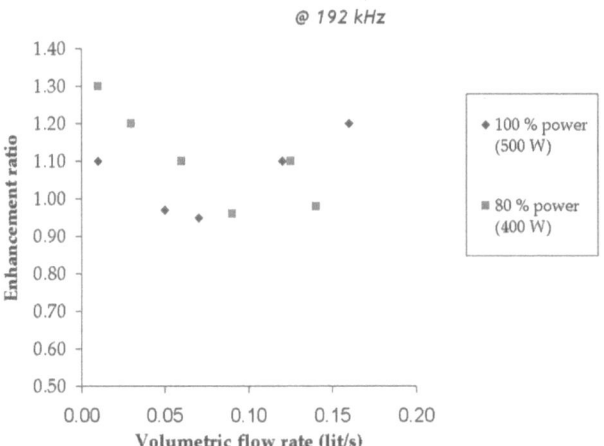

FIGURE 9.17 Effect of ultrasonic power on heat transfer augmentation for 192 kHz operation in the presence of flow conditions.

In summary, there is marked enhancement of heat transfer under quiescent conditions with the application of ultrasound for a typical orthogonal arrangement of the heater and vibrating plates. The enhancement ratio increases with increase in power density and decreases with increase in cavitation intensity. Hence, lower cavitation intensity is required for processes requiring higher heat exchange. The enhancement ratio increases with increase in acoustic streaming velocity. Higher acoustic streaming velocities are obtained for higher frequencies. Hence, for maximum heat transfer enhancement, higher frequency operation is recommended for quiescent conditions.

Flow conditions have a negative effect on heat transfer enhancement due to an imposed acoustic field. Heat transfer enhancement decreases with increase in flow rate. Lower frequency operation, which leverages cavitation as a mechanism to induce localized turbulence, seems to yield better enhancement than higher frequency operation, which is contrary to quiescent conditions. Dual frequency operation yields the best heat transfer enhancement for quiescent as well as flow conditions.

PIF/CIF ANALYSIS

The fact that flow conditions result in the collapse of a cavitational or streaming field is again a severe limitation for intensification of heat transfer using coupled ultrasonic/megasonic systems. Many process tanks used in the industry do involve flow, whether for fresh-fill of chemicals or recirculation. The higher the flow rate, the more effective it is in serving its purpose, typically. This has to be carefully balanced against the benefits of the heat or mass transfer enhancement mechanisms applied.

9.3 NANOFLUIDS

Nanofluids are colloids containing nanosized suspensions of particles in a base fluid. Common base fluids include water, ethylene glycol, and organic liquids. These fluids contain nanoparticles which are typically made of chemically stable metals, and metal oxides such as copper, alumina, etc., in various forms. The size of the nanoparticles imparts some unique characteristics to these fluids, including greatly enhanced energy, momentum, and mass transfer, as well as reduced tendency for sedimentation and erosion of the containing surfaces.

Nanofluids have been investigated for numerous applications, including cooling, manufacturing, chemical and pharmaceutical processes, medical treatments, cosmetics, etc. They are of special interest due to their augmented heat transfer properties (Anusha, 2010). They enhance the thermal conductivity and other heat transfer properties over the base fluid. The usual enhancement techniques for heat transfer can hardly meet the challenge of the ever-increasing demand for heat removal in processes involving electronic chips, laser applications, or similar high-energy devices. The factors which limit the usual techniques are many. One major limitation is the poor thermal characteristics of common heat transfer fluids. They are about two orders of magnitude less efficient in conducting heat compared to metals. This inherent inadequacy of these fluids makes the heat removal mechanism less effective even with the best utilization of their flow properties.

Typically, enhancement of thermal characteristics of nanofluids is in the range of 15–40% over the base fluid, and heat transfer coefficient enhancements have been detected up to 1.5X. An increase in thermal properties of this magnitude cannot be solely attributed to the higher thermal conductivity of the added nanoparticles, leading to speculation regarding other mechanisms that can contribute to the increase in performance. Several have been proposed by the nanofluid research community. But, none has adequately explained the anomalous enhancement in the thermal conductivity of nanofluids. The transition from micron-sized particles to nanoparticles can lead to a number of changes in the physical characteristics of the suspending fluid medium. Two of the major factors are increase in the ratio of particle surface area to volume, and the size of the particles moving into the realm of where quantum effects come into play.

The density of a nanofluid is estimated by the following correlation:

$$\rho = (1-\varphi)\rho_{bf} + \varphi\rho_p \qquad (9.9)$$

where φ is the volume fraction of nanoparticles, and ρ is density of the nanofluid. The subscript "bf" refers to the properties of the base fluid, and the subscript p refers to the properties of the nanoparticles.

Viscosity is an important property of the nanofluid. Prasher et al. (2006) presented experimental results concerning the viscosity of a nanofluid of alumina particles in propylene glycol. Its dependency on particle diameter, nanoparticle volume fraction, and temperature was reported. The viscosity of nanofluids was found to be sensitively dependent on nanoparticle volume fraction, but to be independent of the shear rate, nanoparticle diameter, and temperature—thereby indicating that the nanofluid essentially obeys Newtonian behavior.

Traditional heat transfer fluids such as water, engine oil, and ethylene glycol (EG) are inherently poor heat transfer fluids, and thus, major improvements in cooling capabilities are constrained. The application of nanoparticles provided an effective way of improving heat transfer characteristics of fluids (Eastman et al., 1997). Choi and Eastman (1995) first used the term "nanofluids" to refer to fluids with suspended nanoparticles. Some preliminary experimental results (Eastman et al., 1997) showed that an increase in thermal conductivity of approximately 60% can be obtained for the nanofluid consisting of water and 5-vol% CuO nanoparticles.

According to Xuan and Roetzel (2000), some of the main reasons for substantial heat transfer enhancement are: (i) the suspended nanoparticles increase the surface area and the heat capacity of the fluid, (ii) the suspended nanoparticles increase the effective (or apparent) thermal conductivity of the fluid, (iii) the interaction and collision among particles, fluid, and the flow passage surface are intensified, (iv) the mixing fluctuation and turbulence of the fluid are intensified, and (v) the dispersion of nanoparticles flattens the transverse temperature gradient of the fluid.

Several mechanisms have been proposed for the enhancement of thermal conductivity in nanofluids by researchers. Some of the hypotheses are: Brownian motion, clustering or aggregation of nanoparticles (formation of localized "particle-rich" zones), and formation of a nanolayer (ordered layer of a solid-like structure formed by liquid molecules close to the particles). Tables 9.1–9.3 list some of the proposed mechanisms.

TABLE 9.1
Studies on the Influence of Brownian Motion on Thermal Transport of Nanofluids

Source	Nanofluid	Particle Size (nm)	Influence of Brownian Motion
Evans et al. (2008)	-N/A	10	Not significant, but may have indirect effect by forming clusters
Jang and Choi (2004)	Nanofluids with oxide/ metallic nanoparticles/ carbon nanotubes	6–38.4	Nano convection induced by Brownian motion
Prasher et al. (2006)	Al_2O_3 /H_2O	3–60	Local convection caused by Brownian movement, dependent on particle size
Koo and Kleinstruer (2004)	CuO/ H_2O	10	Strong dependence of the effective thermal conductivity on temperature and material properties of both particle and carrier fluid attributed to the long impact range of the inter-particle potential, which influences the particle motion
Shukla and Dhir (2005)	Al_2O_3/H_2O Au/H_2O Ag/H_2O CuO/H_2O Cu/EG	38.4 17 60 24.2 8	Particle interaction is significant, while Brownian motion contribution from the kinetic part is negligible

TABLE 9.2
Studies on the Influence of Nanolayer on Thermal Transport of Nanofluids

Source	Nanofluid	Nanolayer Thickness (nm)	Nanolayer Conductivity (W m^{-1} K^{-1})	Comments
Evans et al. (2008)		1		May play a role in the enhancement of K, but not solely responsible
Xie et al. (2002)	CuO/Cu/EG Al_2O_3/H_2O	2		Substantial effect on thermal transport
Yu and Choi (2003)	CuO/EG Nanotube/oil	1 & 2 2	k_l, 10 k_l 100 k_l, and k_p k_l, 10 k_l, 100 k_l, and k_p	Enhancement due to increased volume fraction associated with interfacial layer (particularly when particle size is less than 10 nm) Enhancement due to extremely thin needle-shaped ellipsoids
Leong et al. (2006)	Al_2O_3 / H_2O CuO/H_2O	1 nm	2–3 k_l 2–5 k_l	
Shukla and Dhir (2005)	Al_2O_3, CuO, Au, Ag in H_2O	<2 nm	~ 2 k_l	Improves heat transfer only if the distances between the particles are smaller than interfacial layer

k_l = liquid thermal conductivity
k_p = particle thermal conductivity

TABLE 9.3

Studies on the Influence of Clustering on Thermal Transport of Nanofluids

Source	Method: 1—Theoretical 2—Experimental	Nanofluid	Comments
Evans et al. (2008)	1		Clustering of nanoparticles may occur more actively in nanofluids with higher concentration; increased size of the backbone promotes rapid heat conduction across the cluster.
Karthikeyan et al. (2008)	2	CuO/H_2O	Thermal conductivity decreases with elapsed time due to clustering of nanoparticles.
Xuan and Roetzel (2000)	1		Formation of aggregates reduces the efficiency of the energy transport of the suspended nanoparticles.
Xie et al. (2002)	1,2	CuO/H_2O	Derived the fractal dimension of clusters from electron microscopic pictures of nanoparticle clustering; Nanofluids can be treated analogously to a porous medium with the volume of all nanoparticles in the fluid equivalent to the volume of the solid phase in a porous medium.
Prasher et al. (2006)	1		Aggregation can enhance the conduction contribution only if the aggregates are well dispersed, not when one large aggregate is formed.

Preparation of nanofluids is the key step in the use of nanoparticles to improve the thermal conductivity of fluids. Two methods have traditionally been employed in producing nanofluids. One is a single-step method and the other is a two-step method.

The single-step method is a process combining the preparation of nanoparticles with the synthesis of nanofluids, for which the nanoparticles are directly prepared by physical vapor deposition (PVD) technique or liquid chemical method. In this method, the processes of drying, storage, transportation, and dispersion of nanoparticles are avoided, so that the agglomeration of nanoparticles is minimized and the stability of fluids is increased. But, a disadvantage of this method is that only low vapor-pressure fluids are compatible with the process. This limits the application of the method.

The two-step method for preparing nanofluids is a process involving the dispersion of nanoparticles into base liquids. Nanoparticles used in this method are first produced as a dry powder by inert gas condensation, chemical vapor deposition, mechanical alloying, or other suitable techniques, and the nanosized powder is then dispersed into a fluid in a second processing step. This step-by-step method isolates the preparation of the nanofluids from the preparation of nanoparticles. As a result, agglomeration of nanoparticles may take place in both steps, especially in the process of drying, storage, and transportation of nanoparticles. The agglomeration will not only result in the settlement and clogging of microchannels but also decrease the thermal conductivity. Simple techniques such as ultrasonic agitation or the addition of surfactants to the fluids are often used to minimize particle aggregation and improve dispersion behavior. Since nano-powder synthesis techniques have already been scaled up to industrial production levels by several companies, there are potential economic advantages in using two-step synthesis methods that rely on the use of such powders. However, an important problem that needs to be solved is the stabilization of the suspension prepared.

In a study conducted by Anusha (2010), alumina ceramic particles with a median particle size of around 70–80 µm were used as feed material for sono-fragmentation to produce nano-dimensional particles. 5 gm of feed alumina particles were dispersed in 100 ml of ultra-high purity water in a beaker, and sonicated for some time. The contents of the beaker were then decanted and subjected to analyses such as laser particle sizing, TEM, etc. The decanted material was repeatedly sonicated until the desired size was achieved. Sono-fragmentation was enhanced with surfactant (Triton X-100) addition in order to reduce agglomeration tendencies. In a parallel trial, commercially procured nanoparticles, manufactured by a bottom-up technique (vapor phase synthesis) in a comparable size range were also experimented with to provide a comparison with the particles generated via the top-down methodology of sono-fragmentation. The nanofluid was prepared by dispersing the nanoparticles through ultrasonic agitation for 4 hours.

No sedimentation was observed for the fluids for the following 6 hours, and thereafter, minor sedimentation was observed for 4% (volume) suspension but none for 1% and 2% (by volume) suspensions. Thermal conductivity was measured within 2 days of preparation of the nanofluid.

Figure 9.18 shows a comparison of thermal conductivity (in dimensionless units) as a function of temperature for nanofluids prepared with sono-fragmented and commercially procured nanoparticles.

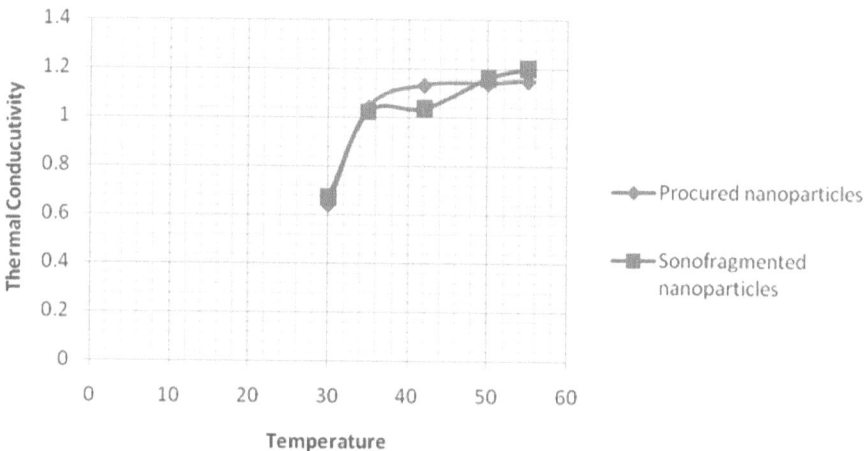

FIGURE 9.18 Comparison of thermal conductivity of nanofluids prepared with procured and sono- fragmented nanoparticles (Anusha, 2010).

As expected, the thermal conductivity of the nanofluid increases with temperature. However, the increase in thermal conductivity of nanofluids shows an irregular trend in the case of sono-fragmented nanoparticles, possibly due to agglomeration. The agglomeration is observed to a relatively lower extent in procured nanoparticles. Particle breakage by a cavitational field does impart additional surface energy to the particles, which can lead to an enhanced tendency to form agglomerates.

Simpson et al. (2019) studied the influence of various parameters, including base liquid, temperature, nanoparticle concentration, nanoparticle size, nanoparticle shape, nanoparticle material, and the addition of surfactant, on nanofluid thermal conductivity. The thermal conductivities of water-based nanofluids containing various nanoparticles are shown in Figure 9.19 as a function of temperature, at a nanoparticle volume concentration of $\phi = 2\%$, and in Figure 9.20 as a function of nanoparticle volume concentration, ϕ, for concentrations below $\phi = 0.01\%$ at room temperature. They observed that for a given temperature, the thermal conductivity of CuO-water nanofluid is higher than the thermal conductivity of Cu-water nanofluid, and that for a given nanoparticle concentration, 5000 nm oxidized graphene nanofluid has higher thermal conductivity than the nanofluid composed of unoxidized graphene nanoparticles having the same size, despite the nanoparticles having the same thermal conductivity.

In their extensive study, Simpson et al. (2019) concluded that the thermal conductivity of nanofluids increases with both temperature and concentration, and the thermal conductivity of nanofluids may increase, decrease, or experience no change with the decrease of nanoparticle size. However, in most cases, the thermal conductivity of nanofluids increased when nanoparticle size was decreased. It was also concluded that the thermal conductivity of nanofluids increases with an increase

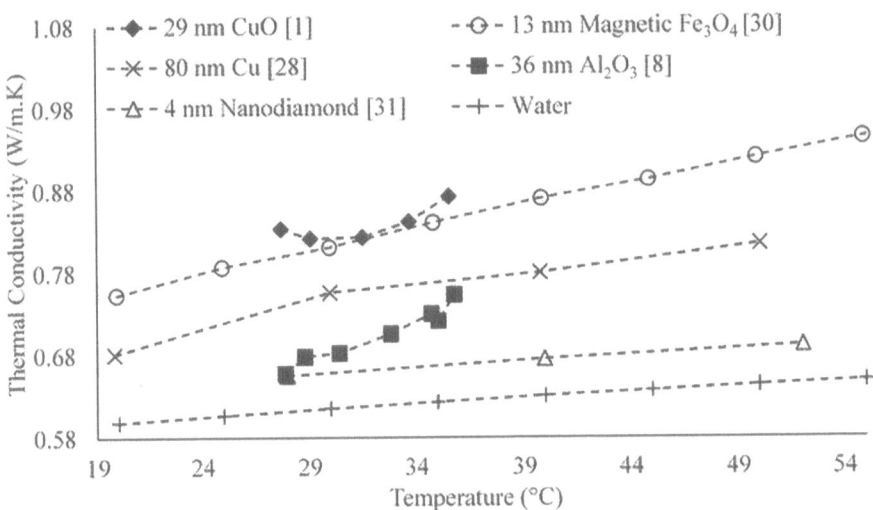

FIGURE 9.19 Variation of water-based nanofluid thermal conductivity as a function of temperature for different nanoparticles at $\phi = 2\%$ (Simpson et al., 2019).

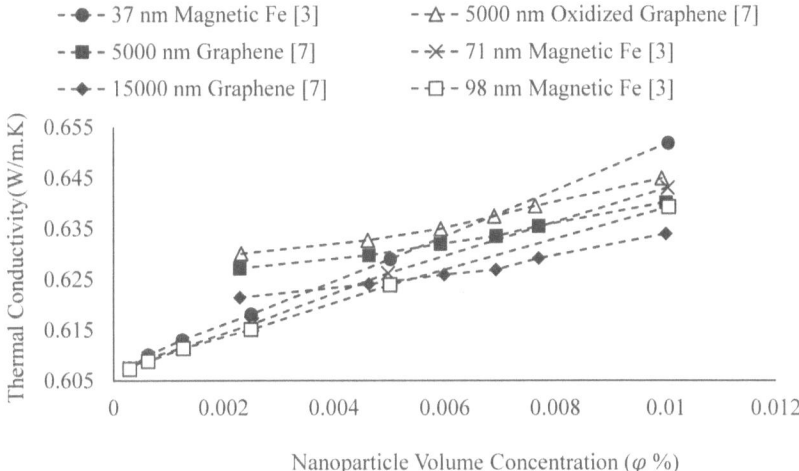

FIGURE 9.20 Variation of water-based nanofluid thermal conductivity as a function of nanoparticle volume concentration, φ, for different nanoparticles at room temperature (Simpson et al., 2019).

in base liquid thermal conductivity, and an increase in nanoparticle surface area. The thermal conductivity of nanofluids containing carbon nanotubes increases with both temperature and concentration, but they recommended further experimentation to determine the best way to prevent nanotube aggregation, in order to maintain nanofluid stability.

Nanofluids intensify heat transfer, and this is not doubted by any researcher. However, the mechanism(s) involved are still a subject of much speculation and fertile ground for further investigation.

PIF/CIF ANALYSIS

The cost associated with nanofluids is primarily contributed by the synthesis or procurement of nanoparticles, and stabilizing them in suspension. The highly nonlinear intensification provided by nanofluids is sufficient to justify the cost involved, in virtually every case where their use is warranted. In flow systems, if the nanoparticles can be captured and reused, it greatly improves their cost-effectiveness.

EXERCISE

Heat transfer enhancement by acoustic fields has been investigated by other researchers: e.g., Tajik et al. (2013), Vainshtein et al. (1995), Franco and Bartoli (2019), and Wang et al. (2020). Review their papers and identify the Process Intensification

Factors obtained in each study. Was there an attempt to estimate the Cost Impact Factor as well? If not, propose a methodology by which this can be done, and an optimization strategy for each application based on both factors.

REFERENCES

Anusha, K.P., "Comparison of Sono-fragmented Nanoparticles and Commercial Nanoparticles in Enhancing Heat Transfer", B.Tech. Project Report, Indian Institute of Technology Madras, 2010.

Awad, S.B., "Ultrasonic Cavitation and Precision Cleaning", *Prec. Clean. J.*, 1996, pp. 12–17.

Busnaina, A.A., G.W. Gale, and I.I. Kashkoush, "Ultrasonic and Megasonic Theory and Experimentation", *Prec. Clean. J.*, 1994, pp. 13–19.

Choi, S.U.S. and J.A. Eastman, "Enhancing Thermal Conductivity of Fluids with Nanoparticles", Report ANL/MSD/CP-84938; CONF-951135-29, ON: DE96004174; TRN: 96:001707, 1995.

Dhanalakshmi, N.P., R. Nagarajan, N. Sivagaminathan, and B.V.S.S.S. Prasad, "Acoustic Enhancement of Heat Transfer in Furnace Tubes", *Chem. Eng. Process.: Process Intensif.*, vol. 59, 2012, pp. 36–42.

Dhanalakshmi, N.P., "Process Intensification via Power Ultrasound - Unifying Principles", Ph.D. Thesis, Indian Institute of Technology Madras, 2013.

Eastman, J.A., S.U.S. Choi, S. Li, L.J. Thompson, and S. Lee, "Enhanced Thermal Conductivity Through the Development of Nanofluids", Proc. Nanophase and Nanocomposite Materials II, Materials Research Society, Boston, 1997, pp. 3–11.

Elsawy, T.M. and A. Busnaina, "Post-CMP Cleaning Using Acoustic Streaming", *J. Electron. Mater.*, vol. 27, 1998, pp. 1095–1998.

Evans, W., R. Prasher, J. Fish, P. Meakin, P.E. Phelan, and P. Keblinski, "Effect of Aggregation and Interfacial Thermal Resistance on Thermal Conductivity of Nanocomposites and Colloidal Nanofluids", *Int. J. Heat Mass Transf.*, vol. 51, 2008, pp. 1431–1438.

Franco, A. and C. Bartoli, "Heat Transfer Enhancement Due to Acoustic Fields: A Methodological Analysis", *Acoustics*, vol. 1, no. 1, 2019, pp. 281–294; https://doi.org/10.3390/acoustics1010016.

Jang, S.P. and S.U.S. Choi, "Role of Brownian Motion in the Enhanced Thermal Conductivity of Nanofluids", *Appl. Phys. Lett.*, vol. 84, no. 21, 2004, pp. 4316–4318.

Karthikeyan, N., J. Philip, and B. Raj, "Effect of Clustering on the Thermal Conductivity of Nanofluids", *Mater. Chem. Phys.*, vol. 109, no. 50, 2008, pp. 50–55; 10.1016/j.matchemphys.2007.10.029.

Koo, J. and C. Kleinstruer, "A New Thermal Conductivity Model for Nanofluids", *J. Nanoparticle Res.*, vol. 6, 2004, pp. 577–588.

Leong, K.C., C. Yang, and S.M.S. Murshed, "A Model for the Thermal Conductivity of Nanofluids - the Effect of Interfacial Layer", *J. Nanoparticle Res.*, vol. 8, no. 2, 2006, pp. 245–254.

Prasher, R., D. Song, and J. Wang, "Measurements of Nanofluid Viscosity and Its Implications for Thermal Applications", *Appl. Phys. Lett.*, vol. 89, 2006, p. 133108.

Shukla, R.K. and V.K. Dhir, "Study of the Effective Thermal Conductivity of Nanofluids", ASME 2005 International Mechanical Engineering Congress and Exposition, November 5–11, 2005, Orlando, Florida, USA, Heat Transfer Division, Heat Transfer, Part B: 2005, pp. 537–541.

Simpson, S., A. Schelfhout, C. Golden, and S. Vafaei, "Nanofluid Thermal Conductivity and Effective Parameters", *Appl. Sci.*, vol. 9, 2019, p. 87; doi:10.3390/app9010087.

Tajik, B., A. Abbassi, M. Saffar-Avval, A. Abdullah, and H. Mohammad-Abadi, "Heat Transfer Enhancement by Acoustic Streaming in a Closed Cylindrical Enclosure Filled With Water", *Int. J. Heat Mass Transf.*, vol. 60, 2013, pp. 230–235.

Thompson, L. and L. Doraiswamy, "Sonochemistry: Science and Engineering", *Ind. Eng. Chem. Res.*, vol. 38, no. 4, 1999, pp. 1215–1249.

Vainshtein, P., M. Fichman, and C. Gutfinger, "Acoustic Enhancement of Heat Transfer between Two Parallel Plates", *Int. J. Heat Mass Transf.*, vol. 38, 1995, pp. 1893–1899.

Vinay, R., "Investigation of High-Frequency, High-Intensity Ultrasonics for Enhancement of Heat Transfer and Gas-Liquid Mass Transfer", B.Tech. Project Report, May 2007.

Wang, X., Z. Wan, B. Chen, and Y. Zhao, "Heat Transfer Enhancement by a Focused Ultrasound Field", *AIP Adv.*, vol. 10, 2020, 085211; https://doi.org/10.1063/1.5133083.

Xie, H., J. Wang, T. Xi, Y. Liu, and F. Ai, "Dependence of the Thermal Conductivity on Nanoparticle-Fluid Mixture on the Base Fluid", *J. Mater. Sci. Lett.*, vol. 21, no. 19, 2002, pp. 1469–1471.

Xuan, Y. and W. Roetzel, "Conceptions for Heat Transfer Correlation of Nanofluids", *Intl. J. Heat Mass Transf.*, vol. 43, 2000, pp. 3701–3707.

Yu, W. and S.U.S. Choi, "The Role of Interfacial Layers in the Enhanced Thermal Conductivity of Nanofluids: A Renovated Maxwell Model", *J. Nanoparticle Res.*, vol. 5, 2003, pp. 167–171.

10 Mass Transfer Rate Enhancement
Case Study

10.1 ACOUSTIC INTENSIFICATION OF DYE UPTAKE BY LEATHER

In many chemical industries, there is a need for some substances to permeate through porous material either naturally or with the aid of some external forces. In leather dyeing, the dye molecules should penetrate through the leather matrix via a diffusion process. Fick's molecular diffusion, in general, is a slow process and needs to be enhanced with an external field. Here, ultrasound is an attractive option for enhancing diffusion. The mechanism of mass transfer is complex and possibly involves the contribution of both cavitation and streaming. Early studies by Javier et al. (1999) show the beneficial effect of ultrasound on mass transport, enhancing mixing and dissolution kinetics at low temperatures. They also found that the diffusion layer thickness decreases with increase in ultrasonic intensity. Ultrasound increases the mass transport with increase in power level, up to a certain level; further increase in power level has a diminishing effect.

The major steps in mass transfer in leather dyeing are:

- from bulk liquid to liquid boundary layer,
- from liquid boundary between leather and bulk liquid to inter-fiber pores, and
- from inter- to intra-pores of fiber matrix.

The process of mass transfer through the matrix will therefore be driven by diffusion, while mass transfer to the matrix will be driven by convection. Since the diffusion process is much slower than convection, the rate-determining step will be diffusion from inter- to intra-pores.

In general, the leather dyeing process depends on initial dye concentration, mechanical agitation, time of dyeing, temperature of the medium, dye particle size, and pH of the medium. To obtain uniformly dyed leather, a higher initial dye concentration should be used with vigorous mechanical agitation; higher temperature and optimum pH should be maintained throughout the process to get the desired result. These techniques will lead to more dye effluent requiring treatment, which is not economical and is hazardous to the environment.

Similarly, using longer mechanical agitation/drumming will lead to surface damage of leather as well as disruption of the fiber bundle. Longer duration and maintaining higher temperatures will lead to more power consumption. To avoid these drawbacks and to enhance the process in a superior fashion, the application of ultrasound on leather dyeing has been studied by many researchers.

The two basic phenomena of ultrasound—acoustic cavitation and acoustic streaming—enhance the diffusion process in different ways. Ultrasound helps in both micro- and macro-mixing of the dyeing process. Early work on leather dyeing of Acid red dye had demonstrated the beneficial effect of ultrasound in achieving more uniform dyeing with less amount of initial dye used, less pollution load as dye uptake is higher, more energy efficiency as it reduces processing time, and enhanced leather properties such as color values and fastness properties without altering the structure of fiber bundle, as verified by SEM analysis. The earlier work motivated a follow-up study of a different dye, i.e., Acid blue 113, on the effect of different frequencies on dyeing. In this study (Dhanalakshmi, 2013), a Taguchi fractional-factorial L9 Design of Experiment (DOE), which is an orthogonal array method, was used to optimize the process, and to identify the significant factor and factor level to achieve a process that consumes less time and is more economical. A dynamic control experiment was performed with the conventional method of dyeing using a mechanical shaker; then, the sonication study was performed to compare with the conventional process. The enhancement was also verified by characterizing the quality of leather dyed with ultrasound and without ultrasound using SEM analyses, which shows the structure of fiber, and color measurement which indicates the depth of dye penetration into the leather matrix. The diffusion process intensification parameter was compared for various frequencies (low to high) and different dyes with their corresponding control experiments, with the kinetics of dye diffusion being studied with respect to the effect of various ultrasonic parameters. The effect of time, stirring speed, initial dye concentration, power level, and frequency were studied using Acid blue 113 on rechromed wet blue fat-liquored crust.

The leather used for this study was a rechromed wet blue fat-liquored cow-calf crust for the Acid blue 113 dyeing process and full chromed cow crust for CI Acid red 119. The entire leather piece was cut into small circles of diameter 9 cm parallel to backbone and numbered sequentially. The left and right side portions of cut leather were used for experiments with and without ultrasound for comparison. Prior to sonication, the weights of the individual leather pieces were measured, and wetting back of the leather pieces was done using 1% SDS and 500% water (based on crust leather weight) for Acid blue 113 dyeing process. The soaked leather was left overnight, ensuring that all the leather pieces are submerged into the solution properly; if required, the solution was stirred. The leather pieces were washed with fresh water after soaking, and were placed in 1000 ml beaker for sonication. The flesh side of the leather-faced the bottom of the transducer for all the experiments. To ensure repeatability, the beaker was placed at the center of the tank and the distance from the bottom of the tank was also noted (Figure 10.1). The experiments were carried out for different concentrations of dye (2, 4, and 6% based on crust weight) with 1000% of distilled water, and for various intervals of time. 2% formic acid was used for dye fixing after a particular time of sonication and was given as three feeds at regular intervals.

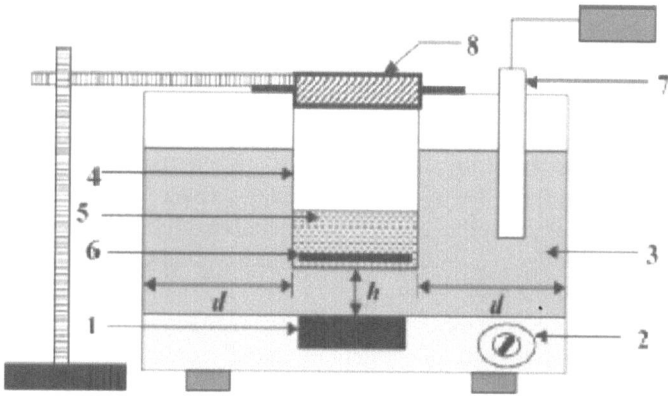

1. Piezo-electric transducer   4. Process beaker   7. Temperature control unit

2. Timer   5. Dye solution   8. Clamp

3. Water   6. Leather Piece

FIGURE 10.1   Leather dyeing: Experimental setup.

The corresponding dye uptake by leather was obtained from the UV-Vis analysis of spent liquor at regular intervals.

The dynamic-control experiments were done using a mechanical shaker at 30°C for different concentrations, stirring speeds, and process times. The sonication experiments were carried out using megasonic and ultrasonic systems with online temperature measurement using a data logger. The leather was placed in a beaker, and the prepared dye solution (1000% water and 2% dye) was transferred to the beaker and irradiated continuously with the sonicator. The initial dye concentration given to the leather was noted as soon as the mass transfer began. The dye uptake by leather was obtained by analyzing the spent liquor concentration using UV-visible spectrophotometric analysis. After the desired time of irradiation, 2% of formic acid was added to the process beaker and the concentration variation was obtained at regular intervals of time. After the desired time of sonication, the leather samples were dried for further qualitative analysis. The same procedure was followed for additional experiments at different frequencies. All the experiments were done in duplicate and averaged values from the trials were used for calculations.

The percentage dye exhaustion was calculated using the formula:

$$\%\text{dye exhaustion} = \frac{\text{dye offered} - \text{dye in the spent liquor}}{\text{dye offered}} \times 100 \qquad (10.1)$$

The Process Intensification Factor, in this case, may then be defined as:

$$\text{PIF} = \left(\% \text{ dye exhaustion with ultrasound}\right) / \left(\% \text{ dye exhaustion without ultrasound}\right) (10.2)$$

The ratio of effective dye diffusivity into the leather matrix with and without sonication may also be taken as a measure of sono-enhancement; a similar definition in terms of the mass transfer coefficient is also possible. However, the definition in Equation (10.2) is the most functionally relevant one, and best suited as a metric to be optimized.

Using a UV-Visible spectrophotometer, the collected samples at each interval were analyzed for unspent dye by measuring the absorbance value at the wavelength λ_{max} (523 nm) of the acid red dye used, after suitably diluting the spent dye liquor. Then, the amount of dye present in the spent liquor was calculated from the calibration graph drawn for the dye.

Quantification of color of the dyed leather was made according to the Commission International de l'Eclairage (CIE) system of color measurement with $10°$ standard observer data. L^*, a^*, and b^* values for both the grain as well as the flesh side of the dyed leathers were obtained using a Reflected spectrophotometer (*Greta Macbeth Spetrolino*, Switzerland).

Leather fiber structure, another important property of final leather, was studied using SEM analysis to detect the influence of ultrasound. Dyed leather samples (58 kHz at 100% power level, 58 + 192 kHz at 60% and 100% power level, and control leather without ultrasound) were cut into uniform sizes and subsequently gold-coated using *Edwards sputtering device*. Analysis was performed using *Leica Cambridge Stereoscan 440* Scanning Electron Microscope. SEM analysis of the leather dyed with ultrasound and without ultrasound was carried out.

Fastness properties of the dyed crust leathers, such as dry-rub fastness, were measured using a *Satra* fastness tester adopting standard *Satra* test procedures. The strength properties of the leather processed with and without ultrasound were also tested. Leather samples for the physical testing were taken parallel to backbone from the circular dyed leather samples following the IUP/1 procedure for sampling and testing. The circular samples taken for ultrasonic experiment and control were within an 18-cm distance from each other at the sampling portion to avoid possible differences due to the variation in location. The following results on dyeing process intensification using ultrasound were obtained and analyzed.

Dye uptake was studied during the course of the dyeing process for a total dyeing time of 2 hours with and without ultrasound for acid red dye. Improvements in the dye exhaustion throughout the dyeing process for acid red at various frequencies and power levels were observed. In general, about 40–85% exhaustion of dye can be achieved in 2 hours of dyeing time using ultrasound, compared to only 20–30% in the absence of ultrasound in stationary conditions (Figures 10.2 and 10.3).

The effects of ultrasonic power input and frequency are captured in Figures 10.4 and 10.5. The effective diffusion coefficient for dye into leather is enhanced by nearly 50–250× under optimized ultrasonic conditions.

As the temperature of the dye bath increases, the percentage exhaustion and dye uptake per gram of leather increases (Figure 10.6). Thus, temperature itself is a process intensifier in this application, albeit to a lower extent compared to sonication.

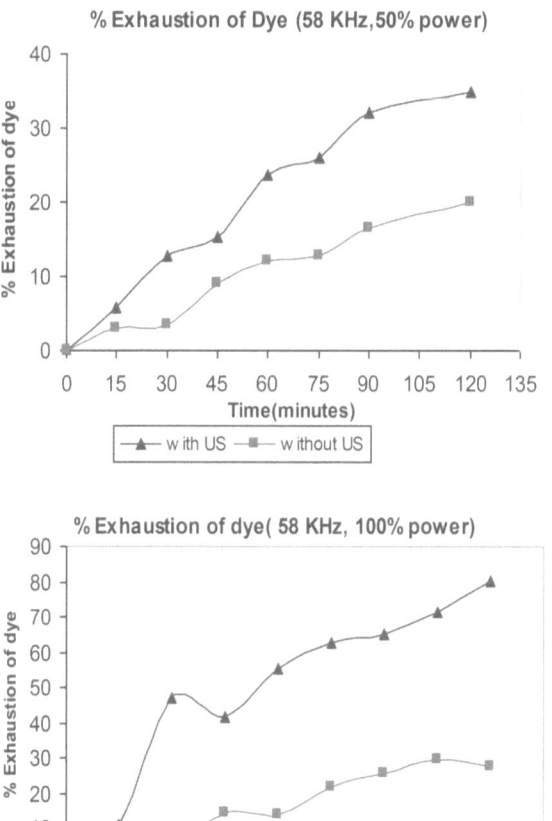

FIGURE 10.2 Effect of time and ultrasonic power input on dye uptake in a 58 kHz ultrasonic field.

The percentage exhaustion of dye by leather when mechanically agitated at 50 rpm during dyeing at room temperature is about 1.2× higher times more than static control at room temperature (Figure 10.7). This shows that agitation also intensifies the process of dye uptake by diffusion, but only by 20%. Also, increasing the stirring speed may result in swirling and vortex formation.

Functional properties of the dyed leather are also improved under ultrasonic exposure (Dhanalakshmi, 2013). Color measurements quantify that leather dyed in presence of ultrasound has greater color values. SEM micrographs indicate a more regular fiber structure for the leathers dyed with ultrasound as compared to that of the control. Fiber bundles are intact and not damaged by exposure to ultrasound. Dry-rub fastness is better for leather dyed in presence of ultrasound when compared with that of control leathers.

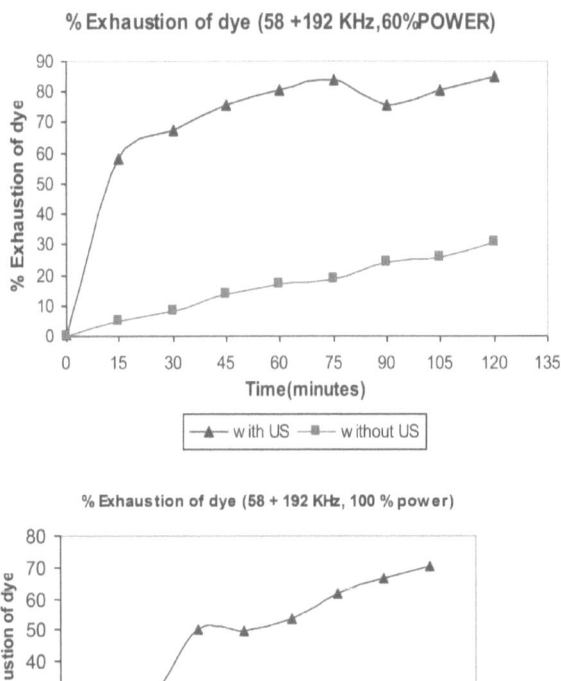

FIGURE 10.3 Effect of time and ultrasonic power input on dye uptake in a 58/192 kHz dual-frequency ultrasonic field.

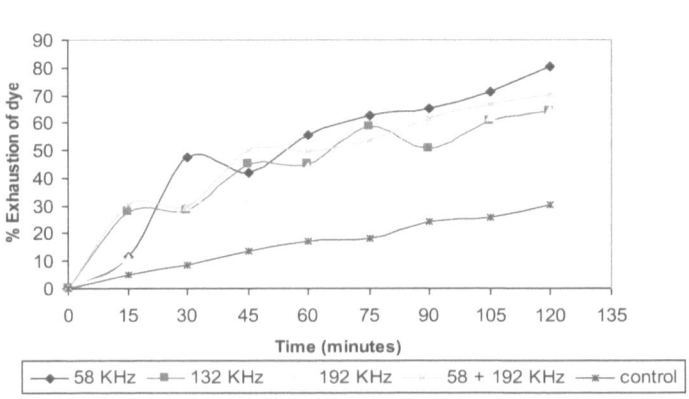

FIGURE 10.4 Effect of time and ultrasonic frequency on dye uptake at 100% input power.

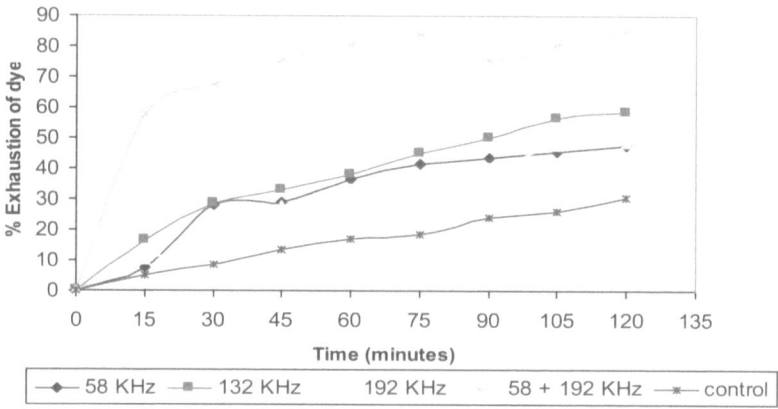

FIGURE 10.5 Effect of time and ultrasonic frequency on dye uptake at 60% input power.

In summary, the dye exhaustion percentage was higher, indicating a higher diffusion rate, in the presence of ultrasound compared to mechanical shaking, even at higher initial dye concentration. Although at high temperatures less energy is converted into kinetic or thermal energy, the drop in viscosity causes mass transfer to increase simultaneously. Therefore, the temperature rise inherent to a sonication system helps in dyeing process.

Use of sonication results in instantaneous expansion and contraction of pores, leading to penetration of dyes through skin\leather. Turbulence induced due to

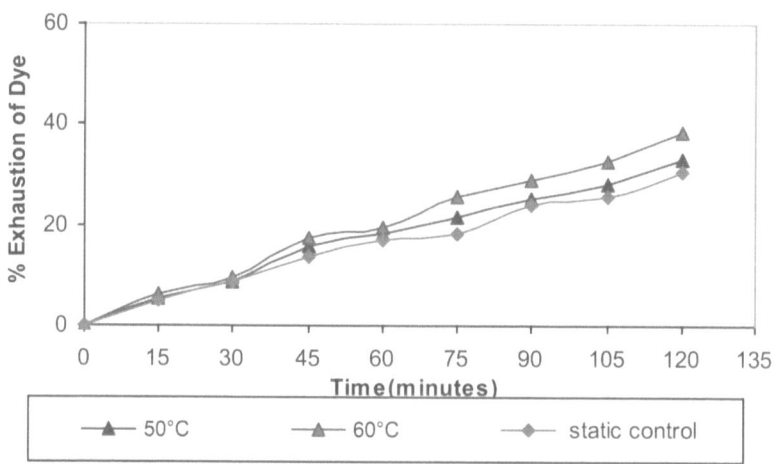

FIGURE 10.6 Effect of time and temperature on dye uptake.

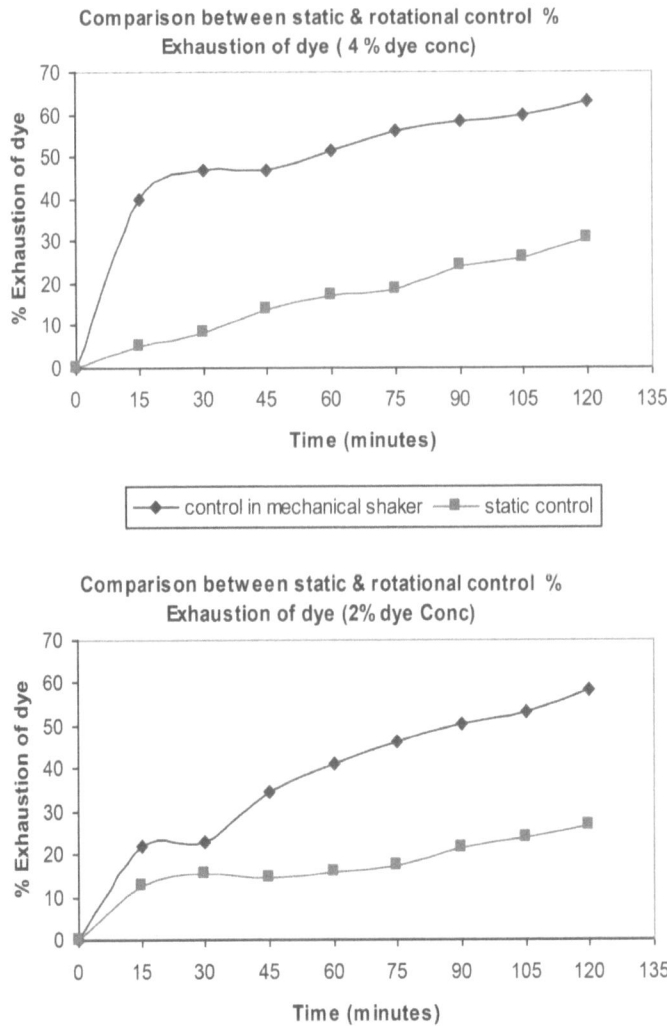

FIGURE 10.7   Effect of mechanical agitation on dye uptake.

microjet formation reduces the concentration gradient in the immediate vicinity of the membrane. Additional internal stirring or turbulence generated due to acoustic streaming may cause higher penetration of dye. Acoustic streaming acts throughout the material leading to a higher average internal flow rate, i.e., higher diffusion rate. It also increases the velocity of particles in the medium. Based on these results, for a given dye concentration, the rate at which dyeing proceeds depends on frequency, and follows the order: lower frequency>dual frequency>higher frequency>control.

PIF/CIF ANALYSIS

Effluents from leather tanneries are a severe health hazard. Their treatment before release, and requirements for zero discharge, impose high costs on the industry. Intensification of dye transfer helps reduce processing time and dye volume, thereby alleviating the environmental impact and reducing operational costs for the tannery. Given the inherent diffusional limitations to dye transfer, the use of acoustic fields appears to be a smart innovation.

10.2 ACOUSTIC INTENSIFICATION OF GAS-LIQUID MASS TRANSFER

Gas-liquid mass transfer is of great importance in bioprocessing because of the requirement for oxygen in aerobic cell cultures. In non-Newtonian liquids, examples abound of gas absorption in pseudoplastic flow relevant to industrial processes such as a fermentation broth, slurry, fluidized bed, etc. One common method for improving the mass transfer is by carrying out the reaction in a stirred vessel. The energy input from the stirrer device serves to break up the bubbles, leading to an increased contact area between gas and liquid.

Many chemical (gas-liquid) and biochemical reactions are limited by mass transfer of the reactants to the reaction system. In this regard, the transfer of mass from a swarm of bubbles into a turbulent liquid has been a growing area of interest. It is well known that the application of ultrasound results in the intensification of chemical processes, and that the presence of dissolved gases is crucial as they either act as nuclei for cavitation or as reactants for chemical or biochemical reactions. Hence, transfer of gas into the solution in the presence of ultrasound warrants further study. Conventionally, gas-liquid mass transfer processes have always involved the use of mechanically stirred contractors or continuously stirred tank reactors. These have good mixing and mass transfer characteristics. The ultrasonic operation is identified as an alternate means of agitating the medium and its mass transfer characteristics have been studied by Vinay (2007).

Only a few such studies have been reported in the literature (Gondrexon et al., 1997; Kumar et al., 2004, 2005). Kumar et al. (2004) have quantified the volumetric oxygen uptake coefficients for both ultrasonic bath- and horn-type sonicators in terms of operating parameters of the system. Very little focus, however, has been accorded to the effect of frequency on the observed mass transfer enhancement. There are virtually no reports on effect of dual frequency operation on gas-liquid mass transfer in open literature. Frequency plays an important role in deciding the distribution of input ultrasonic power between the cavitation and acoustic streaming phenomena; hence, frequency effect warrants a detailed study. The study by Vinay (2007) investigates the relationship between the local cavitation intensity, acoustic streaming, and gas-liquid mass transfer in a typical ultrasonic bath setup. The observed increase in volumetric mass transfer coefficients is interpreted through an idealized eddy structure of turbulence in the vicinity of inlet sparged bubbles

(Linek et al., 2004) incorporating two models, viz., "the slip velocity" model (Calderbank and Moo-Young, 1961), and "the eddy cell" model (Lamont and Scott, 1970).

Mass transfer experiments were conducted by Vinay (2007) using a beaker (containing completely degassed solution) which is clamped at the central position within an ultrasonic tank, with air being bubbled at a constant rate of 27.77 cm^3/s by means of air pump. Sodium carbonate was used to prepare completely degassed solutions. The dissolved oxygen concentration was monitored using a dissolved oxygen probe (Horiba™, WX – 22, 23D series. time constant ~ 10 s) which uses an oxygen electrode. The cavitation intensity was measured using a cavitation meter (ppb500™). The ultrasonic operation was done in pulsed mode. Continuous mode results in an increase in temperature, and simultaneous determination of dissolved oxygen (DO) concentration also becomes difficult due to damage to the oxygen electrode tip by cavitation erosion. DO concentration is measured at intervals of ½ minute (for some experiments at 1 minute), and the aeration was continued till a constant value of dissolved oxygen concentration is achieved.

The volumetric mass transfer coefficient is obtained for oxygen uptake by degassed solution by dynamic oxygen electrode method (Gogate and Pandit, 1999). Transient mass balance for oxygen within the beaker yields:

$$\frac{\partial C}{\partial t} = k_L a \left(C^* - C \right) \tag{10.3}$$

where C^* is the equilibrium concentration of oxygen at the bubble surface, and C is the oxygen concentration in the bulk of the liquid; $k_L a$ is the volumetric mass transfer coefficient (s^{-1}).

This equation on integration yields:

$$\ln \left(\frac{C^*}{C^* - C} \right) = k_L a t \tag{10.4}$$

From this relation, the volumetric oxygen transfer coefficient $k_L a$ can be computed by plotting ln $(C^*/(C^* - C))$ vs. time (t), with the slope of the resultant straight line yielding $\mathbf{k_L a}$.

This method has its advantages and disadvantages. It is not suitable for the determination of mass transfer coefficients for large-scale cavitational reactors, as it uses the assumption of constant gas hold-up and steady $k_L a$ value during the whole aeration process, hence resulting in lower uptake coefficients. Knowledge of flow pattern becomes necessary for exact determination. Gogate and Pandit (1999) have conducted an excellent survey of the various methods used in the determination of the volumetric mass transfer coefficients in bioreactors and enumerated the advantages, disadvantages, and scope of applicability of each method in various situations. The percentage increase in volumetric mass transfer coefficients, the PIF in this case, is obtained as:

$$\text{Increase}(\%) = \left(\frac{k_L a_{\text{withUS}} - k_L a_{\text{withoutUS}}}{k_L a_{\text{withoutUS}}} \right) * 100 \tag{10.5}$$

Though there is a facility in the equipment which directly indicates the input ultrasonic power, for the dual-frequency system, the input power was obtained calorimetrically (Kumar et al., 2004) by operating the tank in continuous mode and noting the temperature rise in the tank by means of a digital thermometer system. When the accuracy of the power indicator was verified calorimetrically for single-frequency operations, there was only a deviation of about 5–10% when compared with the calorimetrically obtained power. This gives an energy efficiency of 90–95% for the ultrasonic bath. Calorimetric determination of ultrasonic power dissipated in the medium was also performed for a locally sourced horn-type sonicator operating at 33, yielding an energy efficiency of only 10–15%. For efficient transfer of ultrasonic power to the beaker, the coupling fluid between the vibrating plate and the beaker must be maintained at isothermal conditions (Kumar et al., 2004), which is difficult to maintain while operating in continuous mode as this results in bulk fluid heating. This is avoided by the use of a pulse mode of operation which results in negligible heating effects.

The volumetric mass transfer coefficients were obtained both in presence of ultrasound (for varying frequencies and power densities) as well as in its absence. The volumetric mass transfer coefficient was obtained as $3.25 * 10^{-3}$ (s^{-1}) at a constant inlet air flow rate (27.77 cm^3/s), in the absence of ultrasound. The percentage increase in volumetric mass transfer coefficients for oxygen uptake in the presence of ultrasound was then calculated.

Volumetric mass transfer coefficients showed an increase with increase in input power densities for a constant frequency of operation (Figure 10.8). This was true for all the frequencies of operation. The effect of power was more pronounced for

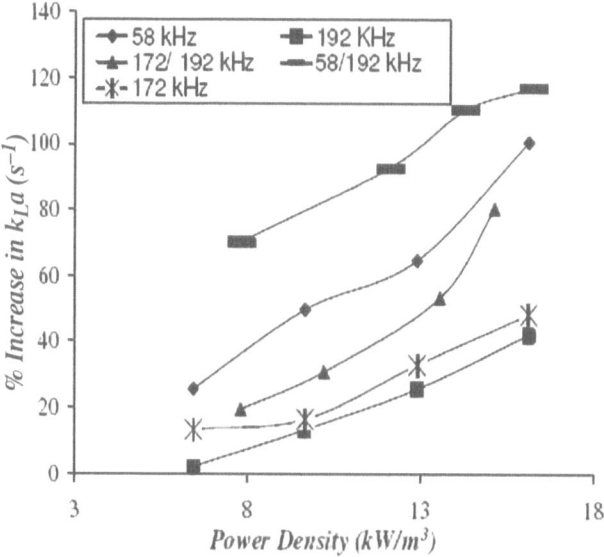

FIGURE 10.8 Variation of % increase in volumetric mass transfer coefficients with input power density.

frequencies of 58, 172, 192, and 172/192 kHz. Even though 58/192 kHz yielded the highest volumetric mass transfer coefficients for all power levels, the corresponding rate of increase of volumetric mass transfer coefficients was the least among all. About a 27% increase in volumetric mass transfer coefficients was observed for an increase in power density from 7.8 to 16.2 kW/m^3.

Highest rate of increase was observed for 132 kHz, which is about a 66% increase in volumetric mass transfer coefficient for increase in power density from 10 to 25 kW/m^3. The observed trend can be explained on the basis of increased cavitational activity at higher ultrasonic power dissipation rates. Cavitation enhances the gas-liquid mass transfer by physically breaking the inlet-sparged bubbles, thereby increasing the interfacial area for gas-liquid mass transfer.

In addition, the eddies that result in surface renewal of liquid at the bubble surface are more in number when the cavitational activity is higher. In other words, the intensity of turbulence is higher when the cavitational activity is higher, and this plays a dominant role in enhancing the gas-liquid mass transfer (Kumar et al., 2004; 2005). It is interesting to observe that in a dual-frequency operation involving 58/192 kHz, the additional frequency (192 kHz) does not merely superimpose its own effect on the other frequency (58 kHz), but instead the enhancement achieved is greater than the sum of enhancements achieved by the individual frequencies of 58 and 192 kHz.

This reinforcing effect occurs for 172/192 as well. The dual frequency 172/192 kHz follows the same trend as the individual frequencies 172 and 192 kHz, and the observed enhancements are almost equal to the sum of enhancements due to the individual frequencies at lower power levels; at higher power levels, the enhancement due to dual-frequency exceeds the sum of individual enhancements. This may indicate the existence of a "threshold" power level beyond which cavitation becomes significant even at higher frequencies.

Volumetric mass transfer coefficients show a decrease with increase in frequency at constant specific power dissipation (Figure 10.9). This is observed for all the power levels of operation. The highest volumetric mass transfer coefficients are observed for 58 kHz (for single frequency operation) and 58/192 kHz (for dual frequency operation). Overall, dual-frequency operation gives the highest volumetric mass transfer coefficient at comparable power densities.

The volumetric mass transfer coefficients are significantly affected by cavitation intensity, which appears to play a dominant role in determining effective gas-liquid mass transfer. Higher volumetric mass transfer coefficients are obtained for higher cavitation intensities at constant input power densities (Figure 10.10). Cavitation intensity, though, is not an independent variable, but one dependent on frequency and power of operation. Higher cavitation intensity is obtained at a lower frequency and higher ultrasonic power dissipation. The cavitation intensity scales as $1/f^3$ and is directly proportional to the power (Nagarajan, 2006).

Higher percentage increase is obtained at higher cavitation intensity. The percentage increase seems to saturate toward the higher end of cavitation intensity. This indicates that for a given inlet air flow rate, there is a limiting value of cavitation intensity above which there is no further increase in volumetric mass transfer coefficient. Cavitation intensity is a direct measure of cavitational activity or local turbulence intensity in the medium. As discussed earlier, higher cavitational activity leads

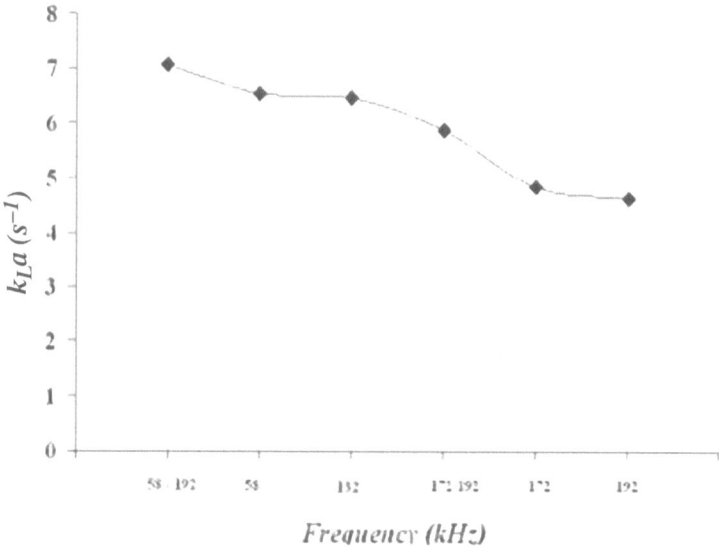

FIGURE 10.9   Variation of volumetric mass transfer coefficient with frequency for 100% power level.

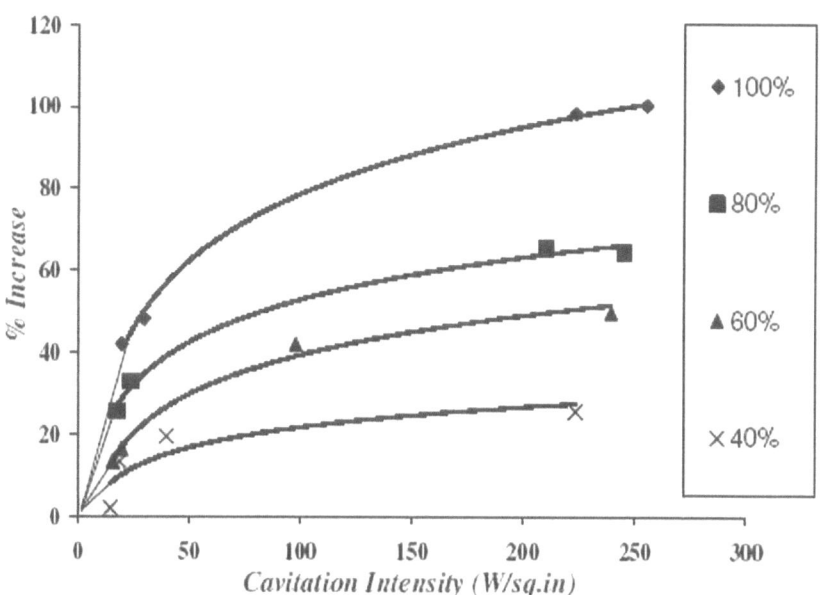

FIGURE 10.10   Variation of % increase in volumetric mass transfer coefficient with local cavitation intensity.

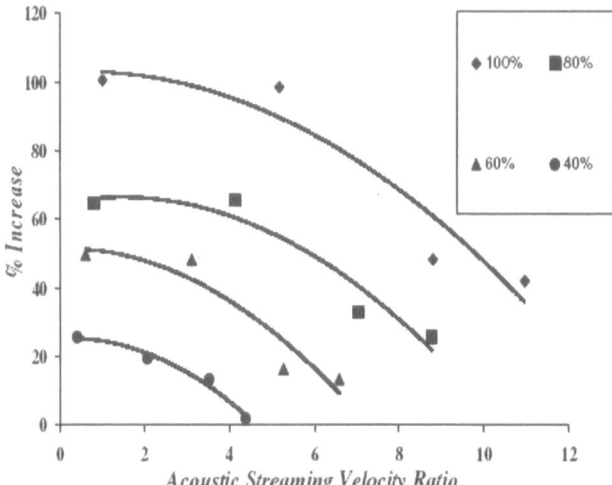

FIGURE 10.11 Variation of percentage increase in volumetric mass transfer coefficient with acoustic streaming velocity ratio for different power levels.

to higher gas-liquid mass transfer. This can be explained on the basis of the eddy cell model (Lamont and Scott, 1970). Cavitational activity is directly related to the intensity of turbulence in the system. When it is higher, the intensity of turbulence in the system is higher, which results in renewal of bubble surface by eddies that bring in fresh parcels of liquid to the bubble surface. This increases the concentration driving force at the bubble-liquid interface, hence resulting in higher gas-liquid mass transfer.

As far as the gas-liquid mass transfer is concerned, cavitation is the dominant factor. Acoustic streaming seems to have a negative effect on gas-liquid mass transfer at a given power level. Higher volumetric mass transfer coefficients are obtained for lower acoustic streaming velocity ratios (Figure 10.11). At higher acoustic streaming velocities, a higher gross mean flow of bulk fluid relative to the bubble interface is obtained; this is expected to result in the higher surface renewal of liquid at the bubble surface (slip velocity model), and hence, in higher gas-liquid mass transfer. However, experimental results are contradictory to this model.

Two dual-frequency operations, viz., 58/192 and 172/192 kHz, were conducted at different ultrasonic energy dissipation rates. A higher volumetric mass transfer coefficient was obtained for 58/192 (yielding the highest volumetric mass transfer coefficient among all frequencies tested) than for 172/192 kHz operation. This can be explained on the basis of cavitation and acoustic streaming phenomena which are instrumental in deciding the rate of gas-liquid mass transfer. However, the former is the more dominant factor than the latter as far as the gas-liquid mass transfer is concerned. Higher cavitation intensity is obtained for 58/192 kHz due to the use of lower frequency 58 kHz as one frequency component. Hence, 58/192 scores over 172/192 kHz.

Even though 58 and 58/192 kHz have comparable cavitation intensities at higher ultrasonic energy dissipation rates, a higher volumetric mass transfer coefficient is obtained for the latter because of the use of an additional higher frequency. This facilitates a higher bulk fluid motion (due to acoustic streaming), thereby aiding in surface renewal of liquid at the air bubble surface. The use of higher frequency results in higher gross mean flow of fluid relative to the bubble surface, and as predicted by slip velocity model, higher volumetric mass transfer coefficient would be obtained at higher gross mean flow of fluid. This is evident from experimental results for dual-frequency operations, even though for single-frequency completely contradictory results were obtained. This would appear to indicate that acoustic streaming does have a secondary effect on gas-liquid mass transfer. Moreover, an optimized mix of lower frequency (58 kHz) and higher frequency (192 kHz) yields the best results.

The rate of gas-liquid mass transfer increases when there is sufficient driving force for mass transfer. The large number of eddies that are created due to cavitation implosions results in the surface renewal of the liquid at the bubble interface. Since resistance to mass transfer is only in the liquid phase, this surface renewal registers a significant increase in the rate of mass transfer.

Higher turbulence is achieved at higher cavitation intensity which is, in turn, obtained at lower frequencies and higher power levels. Though the bulk fluid motion is important in disseminating oxygen throughout the tank, it is only of secondary importance and does not seem to result in efficient surface renewal by gross mean flow relative to the bubble surface. Cavitation is thus the dominant factor for gas-liquid mass transfer enhancement.

Ultrasound has two distinct influences on gas-liquid mass transfer. First, it increases the liquid phase mass transfer coefficient by increasing the turbulence in the medium, behaving much the same way as a normal stirrer agitator. Second, high cavitation intensity ensures the breakage of inlet air bubbles which results in an increase in interfacial area for mass transfer. Thus, it prevails over conventional stirrer agitators by virtue of the cavitation phenomenon. Also, strong convective currents developed within the tank (due to acoustic streaming) further aid the mass transfer process. Nevertheless, acoustic streaming is only a reinforcing effect as opposed to cavitation which is the dominant factor as far as gas-liquid mass transfer is concerned.

The volumetric mass transfer coefficient is correlated to specific ultrasonic power dissipation or power density and frequency in the following manner:

$$k_l a = k_l a^0 \left(1 + C_1 (P/V)^{C2} f^{C3}\right) \tag{10.6}$$

This is a generalized correlation in which the volumetric mass transfer coefficient obtained without ultrasound is, in turn, correlated in the following manner (in case of other means of agitation):

$$k_l a^0 = g_1 (V_s) g_2 (\epsilon) \tag{10.7}$$

where $g_1 (V_s)$ is a function of gas inlet superficial velocity, and $g_2 (\epsilon)$ is function of specific power input to the liquid medium by any other means of agitation such as stirrer etc. Here, ϵ refers to the gassed power input to tank and is correlated to the stirrer speed. Since we do not use any other means to agitate the liquid apart from ultrasound, g_2 ($\in$) is constant in this case. Also, g_1 (V_S) is constant as the superficial gas inlet velocity is kept constant in all the experiments.

Thus, equation (31.4) reduces to:

$$k_1 a = A\left(1 + C_1 \left(P / V\right)^{C2} f^{C3}\right) \tag{10.8}$$

where A is a constant. The parameters C_1, C_2, C_3 are obtained using non-linear regression analysis using POLYMATH, as $C_1 = 2.32$, $C_2 = 0.61$, $C_3 = -0.73$, with $A = 3.25*10^{-3}(s^{-1})$.

There is marked enhancement in gas-liquid mass transfer using power ultrasound under quiescent conditions. A higher percentage increase in volumetric mass transfer coefficients is obtained at a lower frequency, and hence at higher cavitation intensity. Volumetric mass transfer coefficient increases with increase in power density. The basic mechanisms responsible for the increase in gas-liquid mass transfer are: (1) Increase in interfacial area, (2) Creation of intense local turbulence due to cavitation implosions, and (3) Enhanced convective flow pattern due to acoustic streaming effects.

Dual-frequency modes that combine both effects register a higher gas-liquid mass transfer enhancement than single-frequency operation at comparable power densities and cavitation intensities. Cavitation is the more dominant factor than acoustic streaming effects as far as gas-liquid mass transfer enhancement (under quiescent conditions) is concerned. Eddy cell model gives satisfactory explanations for the observed gas-liquid mass transfer enhancement. Dual-frequency operation is explained well on the basis of both eddy cell and slip velocity models. For simultaneous heat and mass transfer enhancement, a trade-off between cavitation intensity and acoustic streaming is necessitated, which is achievable by dual-frequency operation involving an optimized mix of low-frequency and high-frequency transducers.

PIF/CIF ANALYSIS

Gas-liquid mass transfer is a relatively slow process that requires intensification in many applications. In some ways, the use of a cavitational field for this purpose may be an overkill and one that should be adopted only under specific circumstances. Simple mechanical agitation will, in most cases, serve the purpose. In bioprocessing, ultrasonic intensification should be given careful consideration prior to implementation since many bio-materials can suffer cavitational damage.

EXERCISE

Mass transfer enhancement by acoustic fields has been investigated by several researchers: e.g., Xie et al. (2015), Hasan and Farouk (2013), Yao (2016), Komarov et al. (2007), and Sujith and Zinn (2000). Review their papers and identify the

Process Intensification Factors obtained in each study. Was there an attempt to estimate the Cost Impact Factor as well? If not, propose a methodology by which this can be done, and an optimization strategy for each application based on both factors.

REFERENCES

Calderbank, P.H. and M.B. Moo-Young, "The Continuous Phase Heat and Mass Transfer Properties of Dispersions", *Chem. Eng. Sci.*, vol. 16, 1961, pp. 39–61.

Dhanalakshmi, N.P., "Process Intensification via Power Ultrasound - Unifying Principles", Ph.D. Thesis, Indian Institute of Technology Madras, 2013.

Gogate, P.R. and B. Pandit, "Survey of Measurement Techniques for Gas–Liquid Mass Transfer Coefficient in Bioreactors", *Biochem. Eng. J.*, vol. 4, 1999, pp. 7–15.

Gondrexon, N., V. Renaudin, P. Boldo, Y. Gonthier, A. Bernis, and C. Petrier, "Degassing Effect and Gas Liquid Transfer in a High-Frequency Sonochemical Reactor", *Chem. Eng. J.*, vol. 66, no. 21, 1997, pp. 21–26.

Hasan, N. and B. Farouk, "Mass Transfer Enhancement in Supercritical Fluid Extraction by Acoustic Waves", *J. Supercrit. Fluids*, vol. 80, 2013, pp. 60–70.

Javier, D.C., A. Neudeck, C. Richard, and F. Marken, "Low-Temperature Sonoelectrochemical Processes Part 1. Mass Transport and Cavitation Effects of 20 KHz Ultrasound in Liquid Ammonia", *J. Electroanal. Chem.*, vol. 477, 1999, pp. 71–78.

Komarov, S.V., N. Noriki, K. Osada, M. Kuwabara, and M. Sano, "Cold Model Study on Mass-Transfer Enhancement at Gas-Liquid Interfaces Exposed to Sound Waves", *Metallurg. Mater. Transact. B.*, vol. 38, no. 5, 2007, pp. 809–818.

Kumar, A., P.R. Gogate, A.B. Pandit, H. Delmas, and A.M. Wilhelm, "Gas-Liquid Mass Transfer Studies in Sonochemical Reactors", *Ind. Eng. Chem. Res.*, vol. 43, 2004, pp. 1812–1819.

Kumar, A., P.R. Gogate, A.B. Pandit, H. Delmas, and A.M. Wilhelm, "Investigation of Induction of Air Due to Ultrasound Source in the Sonochemical Reactors", *Ultrason. Sonochem.*, vol. 12, 2005, pp. 453–460.

Lamont, J.C. and D.S. Scott, "An Eddy Cell Model of Mass Transfer into Surface of a Turbulent Liquid", *AIChE J.*, vol. 16, 1970, pp. 513–519.

Linek, V., M. Kordac, M. Fujasová, and T. Moucha, "Gas–Liquid Mass Transfer Coefficient in Stirred Tanks Interpreted Through Models of Idealized Eddy Structure of Turbulence in the Bubble Vicinity", *Chem. Eng. Process.*, vol. 43, 2004, pp. 1511–1517.

Nagarajan, R., "Use of Ultrasonic Cavitation in Surface Cleaning: Mathematical Model to Relate Cleaning Efficiency and Surface Erosion Rate", *J. IEST*, vol. 49, 2006, pp. 40–50.

Sujith, R.I. and B.T. Zinn, "A Theoretical Investigation of Enhancement of Mass Transfer from a Packed Bed Using Acoustic Oscillation", *Can. J. Chem. Eng.*, vol. 78, 2000, pp. 1145–1150.

Vinay, R., "Investigation of High-Frequency, High-Intensity Ultrasonics for Enhancement of Heat Transfer and Gas-Liquid Mass Transfer", B.Tech. Project Report, May 2007.

Xie, Y., C. Chindam, N. Nama, S. Yang, M. Lu, Y. Zhao, J.D. Mai, F. Costanzo, and T.J. Huang, "Exploring Bubble Oscillation and Mass Transfer Enhancement in Acoustic-Assisted Liquid-Liquid Extraction With a Microfluidic Device", *Sci. Rep.*, vol. 5, 2015, p. 12572. https://doi.org/10.1038/srep12572.

Yao, Y., "Enhancement of Mass Transfer by Ultrasound: Application to Adsorbent Regeneration and Food Drying/Dehydration", *Ultrason. Sonochem.*, vol. 31, 2016, pp. 512–531.

11 Pollution Abatement and Microcontamination Control

Case Study

11.1 METHYL VIOLET DEGRADATION

The chemical and mechanical effects of ultrasound caused by cavitation bubble implosion result in "hot spots" where the temperature is extremely high. Even though the temperature is extraordinarily high, the local hot spot region itself is so small that the heat dissipates quickly. Thus, at any given time, the bulk liquid remains near ambient temperature. This local heat from implosion decomposes water into extremely reactive hydrogen atoms and hydroxyl radicals. However, during the quick-cooling phase, H^+ and OH^- radicals recombine to form hydrogen peroxide and molecular hydrogen. Some other compounds may be added to the system irradiated with ultrasound, and cause a wide range of secondary reactions to occur. Most of the organic compounds are degraded in this environment.

One chemical of interest is methyl violet (MV) dye, a mixture of tetramethyl, pentamethyl, and hexamethyl pararosanilines. MV is mainly used in textile dyeing to obtain purple and also give deep violet colors, especially in printing inks and paints. These colored dyes from industrial effluent are characterized by high values of suspended solids (SS), chemical oxygen demand (COD), and biochemical oxygen demand (BOD), as well as heat, acidity, basicity, and presence of other soluble substances. These dyes, even in small amounts, are highly visible and toxic to the aquatic environment, as dyes absorb and reflect sunlight, thereby affecting bacterial growth and photosynthesis in aquatic environments. Traditional chemical and physical processes, such as elimination by adsorption on activated carbon, coagulation by chemical agents, oxidation by ozone or hypochlorite, and electrochemical methods are not cost-effective.

Recently, solid wastes, such as banana stalk, beer brewery discharge, agricultural waste, passion fruit, fly ash, etc., are being used as low-cost absorbents (Chang-qi et al., 2008). These methods do not eliminate the colors completely, are sometimes expensive, and usually cause other waste pollutants as secondary products. Therefore, the purification of colored water discharges remains a problem for many industries, and it is necessary to develop new technologies to treat these waters. Among several physicochemical processes, Fenton's oxidation is one of the oldest of advanced oxidation processes but continues to be used successfully, as it is comparatively cheap, of low toxicity, and easy to handle as a reagent (Fe^{2+} and H_2O_2) with relatively simple

DOI: 10.1201/9781003283423-11

technology (Walling, 1975). Fenton's reagent, a mixture of hydrogen peroxide and ferrous iron, is effective for color, COD, and TOC (total organic carbon content) removal from dye effluent. In the Fenton system, the free radicals generated are considered as dominant species with the potential ability to oxidize almost all organic contaminants in an aqueous solution into carbon dioxide and water. For the decolorization of azo dye, destruction of dye to mineral state is not necessary since the removal of color is associated with the breaking of the chromophores, i.e., conjugated unsaturated bond (–N=N–) in molecules. However, the end products formed are of concern due to their toxicity.

The main aim of the study undertaken by Dhanalakshmi and Nagarajan (2011) was to evaluate the enhancement of Fenton's oxidation process by using ultrasound to achieve complete decolorization of azo dye. The process intensification due to ultrasound is defined by the parameter:

$$PIF_{DA} = (DA)_{with US} / (DA)_{w/o US} \qquad (11.1)$$

where DA represents "Decolorization Activity", and is given by $(A_i-A_f)/A_i$. and A_i and A_f are the initial and final activities, resp.

The sono-intensified process results in mineralization of solute without producing sludge or some other material that must then be discarded, and this constitutes a major advantage of this process. In this study, both US and Fenton's reagents were used in combination to achieve complete decolorization of methyl violet. Both horn-type and tank-type sonicators at various power levels (250, 400, and 500 W) and frequencies ranging from 20 to 1000 kHz were investigated. The control experiments were performed at room temperature without sonication. The temperature effect on degradation of methyl violet was studied for constant bath temperatures (30, 40, 50, 60, and 80°C). A parametric study on process intensification (PIF_{DA} %) was conducted.

Initially, the water in the ultrasonic tank was degassed to remove the small suspended gas bubbles and dissolved gases. Methyl violet of 12 mg quantity was added to 400 ml of distilled water in a 1000 ml beaker and was homogenized in an ultrasonic bath for half an hour for good dissolution. The dye oxidant, Fenton's reagent, consists of $FeSO_4.7H_2O$ and H_2O_2 (30% v/v) prepared by dissolving equivalent amount (10 mg) of weighed $FeSO_4$ in 10 ml of distilled water, 100 ml of H_2O_2 was measured and kept aside. After dissolution, 3 ml of 0.1 N H_2SO_4 was added to maintain optimum pH of 3. At the start of the reaction, all the reactants were mixed simultaneously and irradiated continuously with US for 150 minutes. The kinetics of oxidation enhanced by sonication were tracked by taking samples at regular intervals from 0 to 150 minutes and analyzing them using UV-Vis spectrophotometer (580 nm). The residual concentration of dye was deduced from a calibration curve at a wavelength of maximum absorbance (580–581 nm). The cells used were 1 cm thick quartz. The same procedure was followed to study the kinetics of dye degradation for various frequencies (25, 40, 58, 68, 132, 58 + 192, 430, 470 kHz and 1 MHz) and at different power levels (250,400, and 500 W).

The pH was measured using pH meter at regular intervals to ensure the optimum pH of 3. The UV-Vis spectra of MV solution were recorded from 400 to 650 nm using

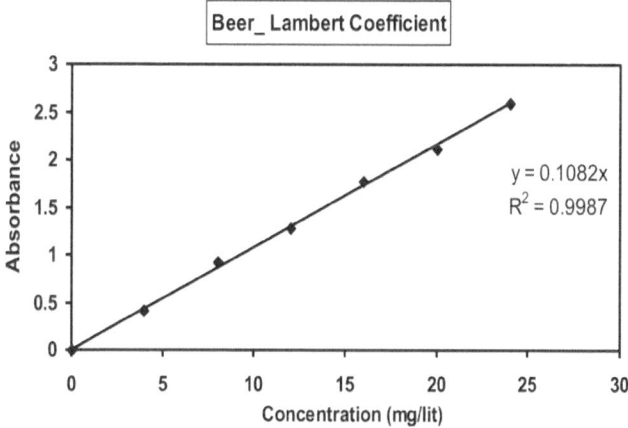

FIGURE 11.1 Calibration curve for methyl violet.

a spectrophotometer, showing maximum absorbance wavelength at 581 nm. The calibration curve obtained for different known concentrations of dye and their corresponding peak absorbance at wavelength of maximum absorbance (580–581 nm) was plotted. With this, the concentration of MV with time can be determined by measuring the absorption intensity, and hence, concentrations from the calibration curve. The calibration curve for the methyl violet dye is shown in Figure 11.1.

Figure 11.2 shows the effect of three different processes, viz., sonication, Fenton/agitation, and sonication/Fenton, carried out for the degradation of methyl violet. Here, 40 kHz at 500 W sonication was used, and a laboratory stirrer with propeller blade at 2000 rpm was used for agitation. Figure 11.3 shows the effect of Fenton's reagent in degradation process. It was observed that there is no decrease in concentration without Fenton's reagent, and a clear decrease in presence of Fenton's reagent. With excess Fenton, there appeared no peak in the absorption spectrum as it reached

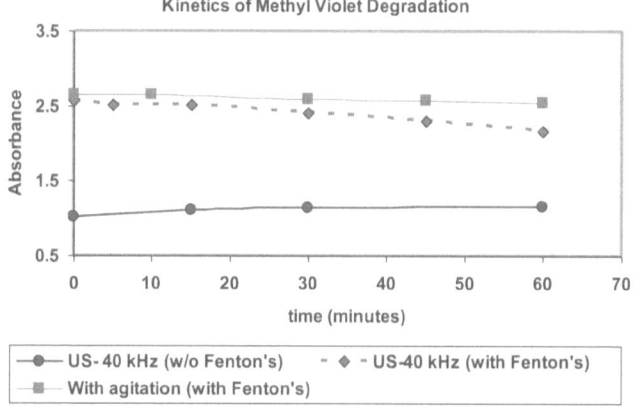

FIGURE 11.2 Degradation study of MV using various processes.

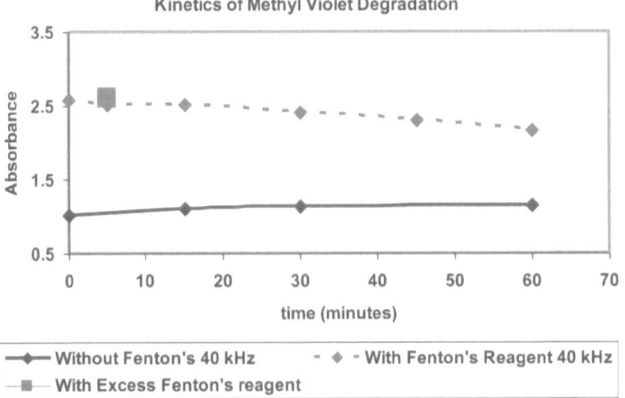

FIGURE 11.3   Effect of Fenton's reagent on reaction kinetics.

the very high absorbance value immediately. Hence, excess Fenton's is undesirable for this degradation process.

To study the effect of agitation on degradation of methyl violet, experiments conducted at a fixed rpm were performed at room temperature and the kinetics of discoloration were obtained. From the results (Figure 11.4), the reaction rate constant was estimated to be 0.03 min^{-1} assuming first-order kinetics. The reason for lowered degradation might be CO_2 dissolved in the solution that produces HCO_3^- and CO_3^{2-} ions. These would trap hydroxyl radicals and reduce the reaction kinetics.

Reaction:

$$HO^\bullet + CO_3^{2-} \rightarrow OH + CO_3^{\bullet-}$$

Dye degradation process is carried out in the presence of hydroxyl radicals, which are formed from the reaction between H_2O_2 and Fe^{2+}. At higher concentrations of

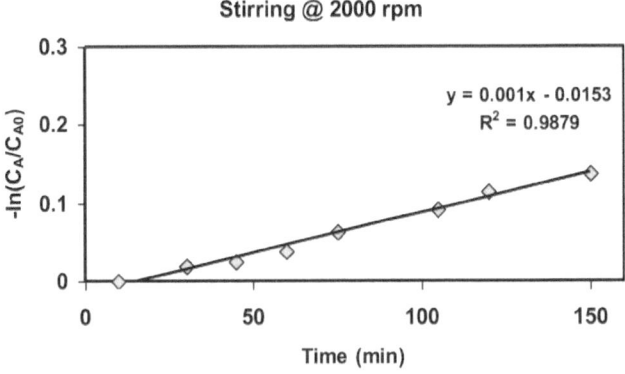

FIGURE 11.4   Effect of stirring on dye degradation.

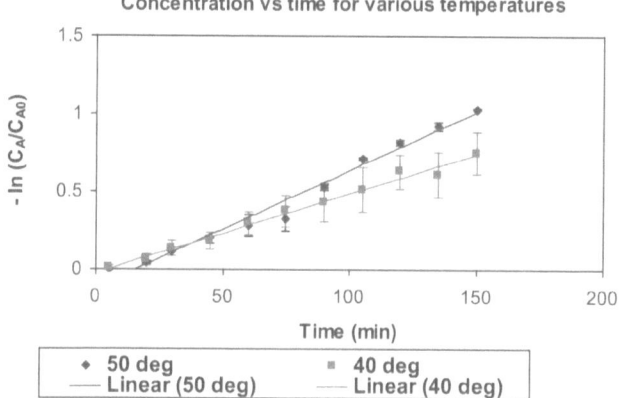

FIGURE 11.5   Temperature effect on degradation of MV.

H_2O_2, its molecules themselves will trap hydroxyl radicals by the reaction given below, thereby countering the degradation effect.

$$H_2O_2 + HO^{\bullet} \rightarrow H_2O + HO_2^{\bullet}$$
$$HO_2^{\bullet} + HO^{\bullet} \rightarrow H_2O + O_2$$

Reaction:

$$HO^{\bullet} + HO^{\bullet} \rightarrow H_2O_2$$

From Figure 11.5, it is seen that the reaction between H_2O_2 and Fe^{2+} is affected by the temperature, and hence, so are the kinetics of dye degradation. Experiments were conducted for temperatures of 40°C and 50°C, and the results indicate that the temperature has a significant effect on reaction rate. The initial 50 minutes of reaction is affected by low temperatures, but after 50 minutes, the reaction yield is high at higher temperatures. Also around 60°C, there is a slight reduction in the yield, due to the decomposition of H_2O_2 at the relatively higher temperature. Hence, the optimum temperature for maximum degradation by Fenton/US is around 50°C.
Reaction:

$$2H_2O_2 \rightarrow 2H_2O + O_2$$

From Figures 11.6–11.8, the effect of ultrasonic frequency and input power on degradation may be understood. In all cases, the dual frequency setting appears to be more effective, and higher power causes an earlier degradation of MV.

The MV degradation data show that at higher cavitation intensity, the reaction is accelerated as the OH radicals at the bubble interface are more in number, resulting in faster degradation of azo dyes. The dye degradation process is intensified in the mutually reinforcing presence of oxidizing agent and ultrasound, resulting in complete decomposition of dye in less time compared to control and benchmark of mechanical agitation.

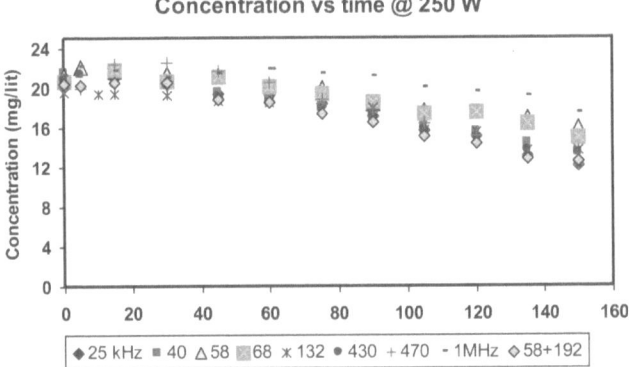

FIGURE 11.6   Effect of frequency at 250 W.

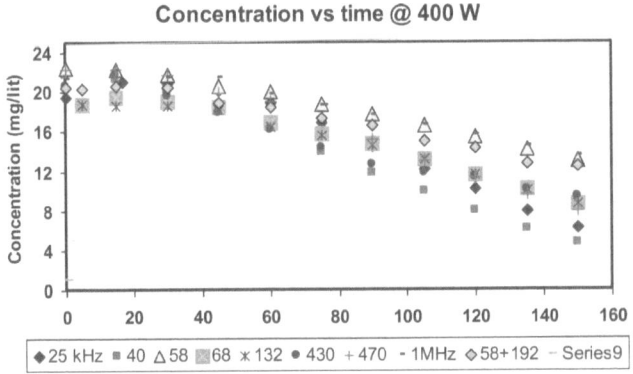

FIGURE 11.7   Effect of frequency at 400 W.

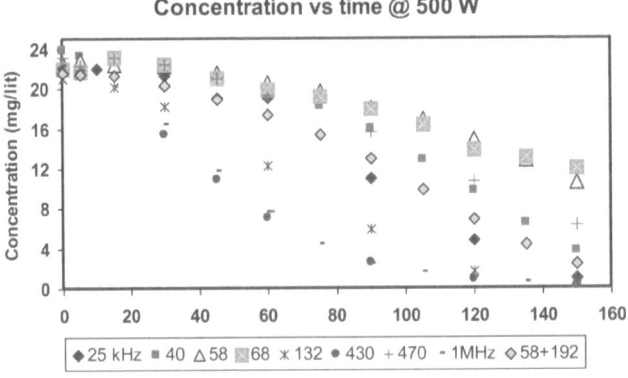

FIGURE 11.8   Effect of frequency at 500 W.

MV degradation is temperature-dependent, and as temperature increases, the reaction proceeds faster. The temperature rise associated with sonication is therefore beneficial. In both tank-type and probe-type sonicators, for a given frequency, increase in power level increases the decomposition rate. This is due to an increase in the number density of cavitation bubbles and the associated implosion intensity. The consequent rise in reaction rates of hydroxyl ion and dye at the bubble interface results in an enhancement of degradation.

In probe-type ultrasonic operation, pulsed mode proves to be more cost-effective than continuous. Though probe-type sonicator is more intense and effective in degradation, tank-type is preferred from a scalability point of view as the volume treated per batch can be ten times higher compared to probe-type.

Higher frequencies also enhance sonochemical degradation. As the number density of bubbles formed increases, so does the gas-bubble interfacial area. Hence, the high local concentration of OH radical in the enhanced interfacial region helps in oxidation reaction causing higher degradation of dyes. Higher frequencies showed higher reaction rates and less half-life time than lower frequencies. Based on these results, higher initial dye concentration can also be treated completely using ultrasonics. The rate at which the reaction proceeds depends on frequency, and follows the order: higher frequency > dual frequency > lower frequency.

11.2 COMPONENT SURFACE CLEANING IN MICROELECTRONICS MANUFACTURING

It has often been estimated that the single most important source of contamination in a manufacturing or precision-assembly cleanroom is the piece parts themselves, along with the packaging material in which they arrive. Any contamination arriving in the cleanroom on the product or its packaging material is most proximate to the material most at risk. By comparison, the workstations and garments of workers in the cleanroom are not likely to come in contact with the product. In addition, the probability that contaminants from the cleanroom will find their way to the product is similarly remote.

Parts cleanliness is an issue for designers, manufacturing engineers, cleaning process engineers, procurement engineers, and quality assurance personnel and their management. Indeed, because of the dominant importance of contamination on piece parts and the subsequent attempts to remove this contamination and preserve the cleanliness of the parts after cleaning, virtually every discipline will become involved. Chemists, chemical engineers, mechanical engineers, facilities engineers, etc. will contribute to the multidisciplinary solutions required.

The types of products adversely affected by contamination vary widely by industry and by volume of production. High-volume products such as semiconductors, disk drives, flat panel displays, and CDROM or DVD are affected. Low-volume manufacturers, such as equipment fabricators or manufacturers in the aerospace industry, are equally concerned, though they may choose different (e.g., batch-oriented) approaches to deal with contamination issues.

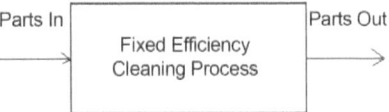

FIGURE 11.9   A simplified fixed-efficiency cleaning process (Welker et al., 2006).

In the manufacture of hard disk drives, particulate contamination can cause disk damage, disk scratching, and head crashes: magnetic contamination can result in unrecoverable data erasure: organic contamination can cause stiction failure, where the motor current is not sufficient to lift a head stuck to the disk upon landing; and ionic contaminants can cause corrosion of the head and/or disk. These defects may be detected at the manufacturing line or may escape to the field, thereby impacting customers. While there are many contributors to the total contamination inventory in disk drives, the component parts, including heads and disks, are certainly one of the major sources. With the increased leveraging of low-cost manufacturing into high-capacity, high-performance products, contamination issues have come to the forefront industry-wide. Increasingly, drive manufacturers are expecting component suppliers, external and internal, to characterize the cleanliness of their parts, and to achieve optimal cleanliness levels on these components. With the simultaneous environmental regulatory pressure to eliminate the use of certain solvents, the industry and its vendor base have been forced to be creative in developing methods for cleaning (and for cleanliness measurement) that are both cost-effective and functionally satisfactory (Nagarajan, 1997).

Early in the manufacture of high-technology products, little attention was paid to the surface cleanliness of parts. Even to this day, some high technology manufacturers use relatively primitive methods to characterize and control contamination on piece parts. In the 70s and 80s, many manufacturers assumed their in-house cleaners provided adequate protection to prevent contamination incoming on piece parts from affecting their processes. The view was the result of a lack of quantitative measurement of parts' cleanliness. Cleaning processes can be viewed as being relatively fixed in efficiency. The cleaning process removes a fraction of the contamination on the parts going into it. Figure 11.9 illustrates a simplified cleaning process (Welker et al., 2006).

This should not be taken to suggest that the cleaning efficiency of all cleaning processes does not vary. Quite to the contrary, the fraction of contamination removed is a function of the contamination level on the parts going into the cleaning process. If parts put into the process are very dirty, then a very high percentage of the incoming contamination will be removed. Conversely, if relatively clean parts are put into the process, then the percent contamination removal will be less. This is illustrated in Figure 11.10.

Parts with high incoming contamination levels are relatively dirty. Parts with large amounts of contamination can be cleaned to apparently very high cleaning efficiency. This may be misleading. Much of the contamination on heavily contaminated parts may be relatively easy to remove. Parts with low incoming contamination levels can be relatively clean. The contamination on these surfaces may be

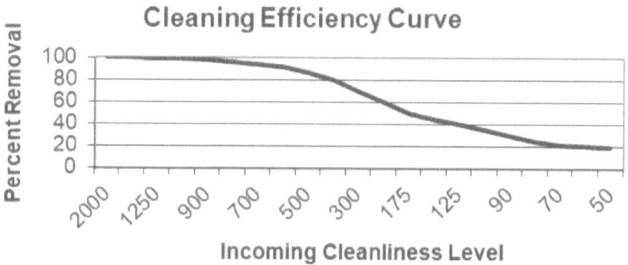

FIGURE 11.10   A cleaning efficiency curve is showing the relationship between incoming cleanliness level and percent removal (Welker et al., 2006).

extremely difficult to remove, resulting in apparently low cleaning efficiencies. An interesting extrapolation here is that very clean parts can actually get dirtier going through a cleaning process—i.e., negative cleaning efficiency. This can be observed frequently in practice when clean plastic parts of an HDD (hard disk drive) assembly are washed right after a basket of dirty screws in the same cleaner.

The key to the development and implementation of effective cleaning processes is the ability to quantitatively assess the cleanliness of the product produced by the cleaning process. The acceptance criteria for a cleaning process can be based on a number of criteria, choosing among approaches to specifying the cleanliness level and among the various classes of contaminants to control. The process of cleaning deployed clearly establishes surface cleanliness levels that may be achieved. There are thus many requirements for "intensification" of surface cleaning, and many avenues to pursue in this context.

11.2.1   Spray Cleaning

Spray cleaning and rinsing are widely employed methods. In spray cleaning, particle removal is entirely by shear stress, and removal forces are therefore proportional to the square of particle diameter. As a result, finer particles are increasingly difficult to remove by spray. It is to be borne in mind that particle adhesion forces in the van der Waals regime typically scale as particle diameter; thus, for most conventional cleaning processes, the cleaning efficiency, i.e., the ratio of particle removal to particle adhesion forces, decreases with particle size. Depending on the nature of the liquid being sprayed, varying degrees of organic and inorganic contamination removal also are afforded. However, the use of surfactants in sprays is found to interfere with cleaning efficiency. This is believed to be due to the formation of foam on the surface of the parts being cleaned. The foam on the surface acts as a cushion and disperses some of the fluid velocity, decreasing cleaning effectiveness. Since spray cleaning is almost entirely due to shear stress, maximizing the fluid velocity on the surface increases cleaning efficiency. For this reason, with a given volume of fluid, solid jets are more efficient than fan jets. In addition, the stream of liquid should not break up into discrete droplets before the surface is impacted. Spray break up before hitting the surface decreases cleaning efficiency.

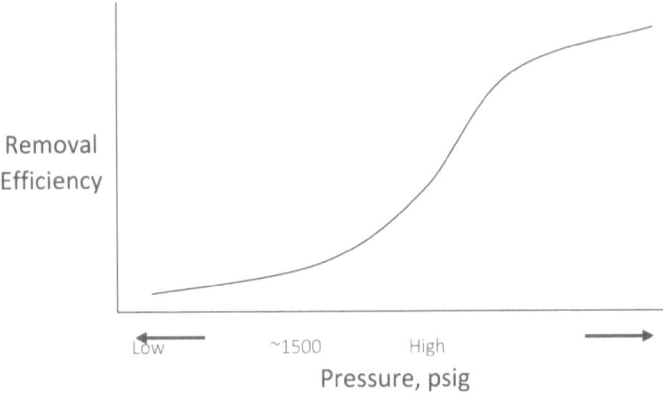

FIGURE 11.11 Characteristic S-shaped cleaning efficiency curve for spray cleaning (Nagarajan and Welker, 1992).

In the late 1970s and through the 1980s spray cleaning using solvents was popular (Musselman and Yarbrough, 1987). Nonflammable, high-volatility solvents, especially CFCs (chlorofluorocarbons), were the most popular solvent spray applications. Today, with the ban on the use of CFCs due to their high ozone depletion potential, water is largely being used in place of organic solvents. In general, fluid velocity is directly proportional to spray pressure. Many low-pressure spray processes are used, where the pressure is 90 psig or less. Higher spray pressures are often used in spin rinse dryers. A spin rinse dryer for cleaning and drying a cassette of parts usually operates at less than 400 psig. A spin rinse dryer for single wafers or disks can operate at typically up to 3,000 psig. High-pressure spray cleaning for precision cleaning of metal parts typically ranges up to about 10,000 psig (Nagarajan and Welker, 1992). Figure 11.11 shows the characteristic S-shaped removal efficiency curve seen with high-pressure spraying. The transition at about 1500 psig presents the demarcation point between low- and high-pressure cleaning efficiency regimes.

11.2.2 Spin-Rinse-Dryer Cleaning

Spin rinse dryers are an attractive cleaning alternative for rotationally symmetrical parts. Where surfaces are relatively smooth, fan nozzles are more efficient than needle jets. Where surfaces are not flat, needle jets can be more effective than fan jets. In the spin rinse dryer, the shear stress cleaning action of the spray is enhanced by the increased velocity of the fluid over the surface of the part due to high-velocity rotation of the part. High pressure has been shown to improve part cleaning dramatically. Parts that are not perfectly rotationally symmetrical can be cleaned on rotors designed with counterweights built-in to provide balance. Both one-part-at-a-time and batch spin rinse dryers are available (Welker et al., 2006).

Spin rinse dryers are among the most energy efficient at drying. Drying is accomplished by high-speed rotation of the substrate being cleaned. Heated, filtered dry nitrogen is often used as a drying gas. High-speed rotation during part drying has been observed to lead to electrostatic charging of surfaces. The source of this electrostatic charging has not been analytically proven. The most likely explanation is that some part charging occurs when the liquid is mechanically spun off the surface. It is doubtful that friction between the air and the part contributes significantly to this charging directly. It is more likely that collisions between aerosolized droplets recirculating in the high-velocity airflow induced by the high-speed rotation of the part and the spinning part are contributing to charging. Air ionization of the heated, filtered gas pumped through the chamber during drying can aid in reduction of the level of charge on the dried parts.

11.2.3 Vapor Degreasing

Vapor degreasing continues to be an important cleaning process, despite the current unavailability of CFCs, which previously were the dominant chemicals used in vapor degreasing. Today, several other solvent systems have emerged as suitable replacements. In vapor degreasing, the cold part is immersed in the hot vapors over a boiling bath of liquid. The solvent condenses on the cold parts and drips off. Vapor degreasing is very efficient at removing organic contaminants and can be effective at flushing away large particles, but not small particles. Depending on the solvent chosen, it can also be effective at removing the ionic contaminants associated with solder residues. Alcohol vapor degreasers which boil a water/alcohol azeotrope can be effective for removal of flux residues.

Hydrofluorocarbons (HFCs), alcohols, acetone, and even water are used as vapor degreasing agents today with great success. Unfortunately, the clock is ticking and the time period for the use of HFCs is running out. Many regional authorities are also increasing restrictions on the use of volatile organic compounds, which continue to contribute locally to photochemical smog and global warming.

11.2.4 Chemical Cleaning

Exceptionally effective chemical cleaning methods that are more often thought of as surface treatment methods are electropolishing and bright dipping. Electropolishing is normally thought of as a surface treatment for stainless steels, but can also be used on aluminum, brass, and other alloys. In electropolishing, the roughness of the surface is reduced. This eliminates hiding places for particles and chemicals, making subsequent cleaning processes more efficient. However, there is a note of caution about electropolishing and bright dipping. Careful and thorough rinsing is mandatory to eliminate electropolishing chemicals and substrate residues. Dimensional tolerances are also a critical consideration (Welker et al., 2006). Alkaline etch has also been used successfully as a chemical cleaning process. An alkaline etch is generally effective at removal of residual soaps and organic contaminants prior to subsequent plating processes. Again, a note of caution is appropriate. An alkaline etch can produce a heavy oxide film on reactive metals such as aluminum. This oxide film may not be tightly adhered to and can result in subsequent particle-shedding problems if not properly sealed or removed.

11.2.5 SOLVENT CLEANING

The most widely used solvent in precision cleaning is water. Pure water has high surface tension, about 70 dyne-cm, and thus is a poor wetting agent for hydrophobic surfaces. Since many surface contaminants are hydrophobic, modifying chemicals are needed to make water suitable for wetting hydrophobic surfaces. Chemicals available include simple surfactants, formulated detergents, solvents such as alcohols and emulsion-based hydrocarbons or other organic chemicals. These are added in varying quantities, depending on the chemistry of the system, in order to reduce the surface tension of the mixture to typically 35 dyne-cm or less. Water is a low-cost solvent compared with other alternatives. However, the chemistries added to improve wetting can increase direct costs and can complicate waste disposal.

One of the biggest advantages of the use of water as a solvent is its widespread acceptance as an alternative to CFCs for precision cleaning. It has been estimated that 90 to 95% of all precision cleaning can be done using water. As a consequence, the variety of equipment available as water-based cleaners is wider than for any other solvent. One of the biggest disadvantages of water comes in drying. Because of the relatively high boiling point and heat of vaporization of water, costs associated with its drying process are relatively high.

One of the work-horse chemicals in the cleaning process industry has been normal methyl pyrollidone (NMP). When used at elevated temperatures (about 170°F), NMP shows a remarkable ability to dissolve polymers, such as polycarbonates, polyacrylates, and polyurethanes. However, it has no tendency to dissolve polyolefins. This chemical selectivity can be used to advantage in product and process design. The primary drawbacks include the cost of heating the chemical, some safety concerns about burns from splashes, and the rather strong odor. Though low in toxicity, many find the odor offensive and cannot tolerate working around it for extended periods of time.

A few critical (strategic) applications have been granted exceptions to the international agreement to eliminate ozone-depleting chemicals. These chemicals include chlorinated hydrocarbons, such as 1,1,1, trichloroethane and hydrochlorofluorocarbons (HCFCs). For those applications that have not been granted such an exception, alternative solvents are required. Fortunately, there are several alternatives currently available. Among these alternatives are HFCs. HFCs are attractive because they have relatively low toxicity, are chemically stable, and are non-flammable. They exhibit low viscosity, low surface tension, and low boiling point, all of which are ideal properties for cleaning complex parts. HFCs are among the few solvents that can effectively dissolve the perfluorinated lubricants used extensively in the disk drive and aerospace industries. One drawback is the high cost of HFCs. A second is their high global warming potential. HFCs released into the atmosphere will remain there for hundreds to thousands of years.

Several solvents based on natural plant products are also receiving widespread acceptance for niche processes. Among these chemicals are terpenes, derived from pine trees, and limonenes, derived from citrus fruits. All have zero GWP (global warming potential), zero ODP (ozone depletion potential), low toxicity, and complete water solubility. The primary disadvantage of the use of these chemicals is the odor, which at low concentrations or for short exposure times is pleasant, but with prolonged exposure can become annoying.

11.2.6 Mechanical Agitation Cleaning (Undulation and Sparging)

Mechanical agitation-based cleaning processes are immersive in nature, and do not employ ultrasonic agitation. Generally, parts are immersed in a bath and either undulated, rotated, or sparged. In sparging, the parts are generally held stationary in the liquid and are showered from below with jets of liquid containing air bubbles. All mechanical agitation-based cleaning processes have relatively low particle removal efficiencies, but can be moderately effective for removal of organic or inorganic contaminants, depending on the chemistry of the bath. Advantages include relatively low equipment cost, low maintenance, and a reduced tendency to cause damage to parts being cleaned (Welker et al., 2006).

11.2.7 Manual Cleaning (Swabbing and Wiping)

Swabbing and wiping are remarkably effective cleaning processes. It has been estimated that the particle removal efficiency of wiping is equivalent to high-pressure spray at several thousand pounds per square inch. Thus, if the surfaces to be cleaned are not too geometrically complex, wiping may be an excellent choice. Indeed, nearly every high-technology manufacturer depends on wiping and swabbing to prepare their tools and workstations for use in processing.

In the aerospace industry, many of the structures that must be cleaned are so large that manual wiping is often the only cost-effective alternative. Several cleaning techniques have been evaluated for their ability to achieve the required cleanliness levels. Vacuum brushing was not found to be capable of achieving MIL-STD-1246B Level 100 cleanliness. Solvent flushing was rejected for large surfaces as being likely to involve the use of large quantities of solvent having cost and environmental impacts. Thus, the cleaning methods of choice for large, complex surfaces are limited to wiping and swabbing with solvents. Multiple solvent wipes typically are capable of achieving Level 100 cleanliness.

A primary consideration is the selection of tools provided for personnel who will perform the wiping operation. In addition to having the correct wipers, a variety of swabs may be needed to get into the hard-to-reach spaces. The selection of solvents for large complicated structures must consider the variety of materials used in the assembly. Manual wiping can also result in prohibitive costs in high labor-rate countries, such as Japan or Singapore. As a result, manual wiping operations are frequently subcontracted to lower labor-rate countries, such as China.

11.2.8 Specialty Cleaning (Plasma, UV, and UV/Ozone, CO_2 Snow, Supercritical Fluid, etc.)

Plasmas have often been described as the fourth state of matter. Gas, such as oxygen, is exposed to high-energy radiofrequency radiation in a low-pressure chamber. This causes the oxygen molecules to dissociate into ions, which, because of their charge, can be accelerated toward a surface. When these impinge on the contaminants, the contaminants undergo chemical reactions. Many of the reaction products produced are harmless gasses, which are pumped away by the chamber vacuum system. Plasma cleaning is most effective for removal of thin layers of low molecular weight organic contaminants. It is ineffective at particle removal or removal of inorganic contaminants. The equipment is relatively expensive and requires a good deal of maintenance. In addition, it is a batch process.

Ultraviolet light, either alone or in combination with ozone, can be an effective cleaning process for removal of organic contaminants. In ultraviolet cleaning, high energy (short wavelength) light illuminates organic molecules on a surface. Ultraviolet energy is absorbed by organic contaminants: the absorbed energy breaks chemical bonds and produces volatile reaction products. Because the reaction products are volatile, no residues are left on the surface. This can be enhanced by illuminating with high-power ultraviolet laser. Both UV and laser UV cleaning are ineffective in removing particles or ionic contaminants. Both are line-of-sight cleaning processes: only areas illuminated are cleaned. Areas in shadows will not be cleaned at all.

In UV ozone cleaning, ultraviolet light is used to convert oxygen into ozone. The ozone diffuses around the part and promotes reactions in organic contaminants that produce volatile reaction products, just as in UV-only cleaning. Just as with UV cleaning, UV ozone cleaning is not effective at removal of particles or ionic contaminants. However, the ozone is free to diffuse in the bath, and hence, UV ozone cleaning is not limited to line-of-sight.

11.2.9 SUPERCRITICAL FLUID CLEANING

Supercritical (SC) fluid cleaning is an interesting niche cleaning process. A supercritical fluid exists when the pressure is greater than the critical pressure and the temperature is greater than the critical temperature. Table 11.1 shows the typical physical properties of supercritical fluids (density, diffusivity, and dynamic viscosity) compared to the gas or liquid phases of the same material.

SC fluids have unusual properties: they have the density of a liquid and the viscosity of a gas, and are best suited for removal of organic contaminants. The selectivity of the SC fluid in removing organic compounds can be adjusted by adjusting the temperature and pressure of the SC fluid. The most common of the SC fluids used

TABLE 11.1
Physical Properties of SCFs

Property	Symbol	Units	Gas Phase	Liquid Phase	Supercritical Fluid Phase
Density	ρ	g/cm^3	0.001	0.8–1	0.2–0.9
Diffusivity	D	cm^2/s	0.00005–0.00035	5×10^{-6}	10^{-3}
Dynamic viscosity	μ	g/cm s	10^{-4}	10^{-2}	10^{-4}

for precision cleaning is carbon dioxide (CO_2). $SCCO_2$ has zero ODP and zero GWP since the carbon dioxide used comes from the atmosphere, to begin with. Using $SCCO_2$ has an advantage over the use of other solvents in that no solvent residues remain on the parts after cleaning, and therefore, an additional drying step is not required. The drawback to $SCCO_2$ cleaning is that the operating conditions require high pressures, and consequently, the equipment can be heavy and expensive. In addition, $SCCO_2$ cleaning is a batch process. Removal efficiencies for particle and inorganic contaminants are low as well.

11.2.10 CO_2 Snow Cleaning

Carbon dioxide snow is another niche cleaning process that has found several applications. In the CO_2 snow cleaning process, liquid CO_2 is sprayed through specially designed nozzles where the heat of expansion freezes out tiny solid particles of dry ice. The expanding gas accelerates these particles so that they can be dispensed like a spray. The snow is relatively effective at removing particles and low-molecular-weight oils. It is less effective at removal of inorganic contamination and higher-molecular weight oils and greases.

Several problems were noted with early designs of CO_2 snow cleaners. One was the tendency to freeze the parts being cleaned. The chief disadvantage created was the tendency of the parts to condense moisture from the atmosphere. In a few cases, differential expansion induced mechanical problems, particularly with rigid epoxy bonds. The friction between the high-velocity snow particles and the surface being cleaned sometimes led to electrostatic charging of the parts.

Several of these problems have been overcome by more advanced designs. In advanced CO_2 snow cleaners, an auxiliary heated air supply is added to reduce the cooling effect of the snow. The addition of in-line air ionizers to this auxiliary heated air supply has greatly reduced the tendency for an electrostatic charge to be generated. One lasting drawback of these snow cleaners is the relatively small area over which they operate. Basically, they form a pencil-thin spray of snow. They are thus a line-of-sight cleaning process. If the area to be cleaned is reproducible from part to part, fixtures can be designed to automate the cleaning process to some degree.

Another niche cleaning process similar to CO_2 snow is argon/nitrogen snow cleaning (Banerjee, 2015). The argon/nitrogen snow cleaning approach is particularly attractive because the cleaning medium is volatile, as in CO_2 snow cleaning, and the cleaning chemicals are particularly inert.

11.2.11   Ultrasonic Cleaning

One of the most popular cleaning processes is ultrasonic cleaning. Ultrasonic cleaning can be very effective, but a number of performance features of ultrasonic cleaning need consideration. The tendency of the ultrasonic energy to damage parts is a consideration in the selection of frequency and power density in the cleaning tank and can influence the design of the equipment and design of the process.

Low-frequency ultrasonic cleaning (less than about 200 kHz) is relatively omnidirectional. As frequency increases above 200 kHz, ultrasonic cleaners tend to become more directional. Very-low frequency ultrasonic cleaning (less than 30 kHz) can produce sub-harmonics that can be heard by workers and can be a source of irritation. These can also result in severe damage to surfaces. Ultrasonic cleaners with frequencies in the 30–70 kHz frequency range tend to produce less mechanical damage to parts and usually operate quieter than low-frequency ultrasonic cleaners.

The susceptibility of materials to ultrasonic damage is complex. For example, non-porous metals with high surface hardness (for example, T6 hardness in aluminum) are not often damaged by ultrasonic cleaning below 200 kHz. Conversely, porous cast aluminum can easily be damaged by 68 kHz and lower frequencies. Mechanical damage to delicate structures also is well documented. Examples include breakage of wire bonds, delamination of adhesive bonds, changes to shapes of fine metal parts, etc. Some strategies have been implemented to try and minimize the damage from ultrasonic cleaners. The theory behind these is that the damage occurs because of standing waves in the ultrasonic tank that stay in fixed locations with respect to the parts being cleaned.

One technique is to vary the frequency of the ultrasonic energy. The frequency cannot be varied far from the center of resonance or the power level in the ultrasonic tank drops markedly. For example, in sweep frequency systems where ultrasonic energy is operated at a resonant frequency of 47 kHz, the sweep is less than +/– 2 kHz. A second technique is to move the parts slowly in the tank during the cleaning process, a procedure termed undulation. Undulation is often used together with sweep frequency to reduce damage.

One final consideration is the damage that occurs when parts are drawn through the fluid/air interface. The energy density at the interface is higher than within the bulk of the cleaning fluid. Drawing a damage-sensitive part through the fluid/air interface while the ultrasonics are running can result in severe parts damage. Thus, many processes are designed to turn off the ultrasonic energy as parts are passing through the interface: this is often referred to as employing the quiet interface.

Ultrasonic cleaners work by the two principal mechanisms detailed in earlier chapters: cavitation and acoustic streaming. These work together in all forms of ultrasonic cleaning, but the relative contribution of each is a function of frequency. At low ultrasonic frequencies, cavitation is very strong and dominates the cleaning process. At high ultrasonic frequencies, cavitation bubbles are very small, but acoustic streaming velocities can be very high. Thus, at high frequencies, acoustic streaming dominates the cleaning process and less of the cleaning occurs due to cavitation.

The way cavitation and acoustic streaming combine to clean is important to understand. When acoustic waves constructively combine, the resulting decrease in pressure creates a localized bubble. In properly degassed solutions, this bubble almost entirely consists of solvent vapors. When the ultrasonic pressure waves separate, the localized pressure drops and the bubble collapses. When this occurs, a microscopic jet of liquid is formed, jetting from the bubble wall into the volume of the bubble. This high-velocity jet scours the surface of parts it comes in contact with, knocking loose material from the surface. This cavitation action preferentially takes place at discontinuities on surfaces. Discontinuities can be scratches, pin-holes in paint, and preexistent pits, among other features. For this reason, ultrasonic erosion that is so common below 70 kHz is almost always associated with these kinds of surface features.

The second mechanism of contamination removal in ultrasonic cleaning is acoustic streaming. Acoustic streaming carries contamination away from surfaces into the bulk of the liquid so that the contaminants cannot redeposit on the surfaces. The acoustic stream cannot penetrate through the boundary layer of motionless fluid that surrounds all of the surfaces in the ultrasonic tank. Particles dislodged from the surface by cavitation action are not swept away from the surface and become reattached. At high frequencies (>200 kHz), the acoustic streaming is highly directional, and hence, the orientation of the part to be cleaned becomes critical. At low ultrasonic frequencies, the acoustic streaming is randomized and not highly directional.

The size and number of cavitation bubbles produced in ultrasonic cleaning are a function of a large number of parameters. Frequency is the most important. As frequency increases, the number of cavitation bubbles increases, but the size of the cavitation bubbles decreases. Other factors affecting the degree of cavitation action include the vapor pressure of the liquid, the temperature of the liquid, the amount of dissolved gas in the liquid, the surface tension of the liquid, and the presence of contaminants in the liquid. At any fixed frequency and power level, the greater the number of bubbles produced, the smaller the bubbles will be and vice versa.

As the surface tension of the liquid decreases, the ease of bubble formation increases. This results in an increase in the number of bubbles, but a decrease in bubble size. Similarly, as the temperature of the liquid gets closer to the boiling point of the liquid, the ease of bubble formation increases. Thus, at higher liquid temperatures, the number of cavitation bubbles increases and the size of cavitation bubbles decreases.

Dissolved gases have an adverse effect on cleaning since ultrasonic energy is expended in generating gas bubbles and moving them around. Similarly, particles in the liquid decrease cleaning efficiency because energy is lost in moving the particles around.

Figure 11.12 illustrates a comparison of 40 kHz ultrasonic versus 400 kHz megasonic cleaning. The material is a bare, A300 cast and machined part: a soft aluminum alloy. The parts were measured using 40 kHz ultrasonic extraction followed by liquid-borne particle count. Megasonic cleaning is not as efficient at the removal of small particles (less than 25 μm). Ultrasonic cleaning tends to cause more erosion damage, as evidenced by relatively larger numbers of large particles (larger than 25 μm).

Several other factors can affect cleaning efficiency. Ultrasonic energy can be absorbed by many polymers. For this reason, plastic containers are not used as cleaning containers in ultrasonic cleaning processes. Plastic coatings, liners, or spacers may be used when metal-to-metal contact between cleaning inserts and the parts being cleaned causes, or can potentially cause, physical damage.

Excessive mechanical agitation can cause the ultrasonic cavitation to completely collapse. This introduces a problem. In order to remove the contamination from the fluid, it must be recirculated through filters and reintroduced into the ultrasonic cleaning tanks. If the recirculation rate is too small, say a tiny fraction of the tank volume, clean-up times will be prolonged and the tank will continue to accumulate

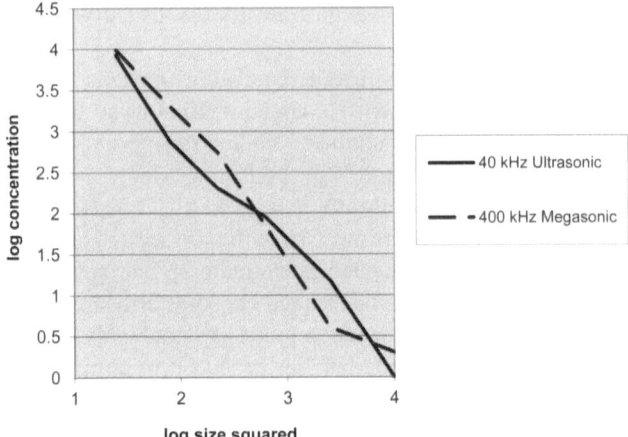

FIGURE 11.12   40 kHz ultrasonic versus 400 kHz megasonic cleaning (Welker et al., 2006).

contamination with each successive load to be cleaned. The effect of the recirculation rate on clean-up time is illustrated in Figure 11.13. Conversely, too high a recirculation rate can cause the cavitation to completely collapse. Most manufacturers of ultrasonic cleaning tanks recommend a maximum recirculation rate of no more than from 25% to 40% of the tank volume per minute.

It is equally important to control the way the fluid is re-introduced into the tank. For example, if the return fluid is allowed to jet into the tank it will cause extensive stirring, causing the ultrasonic cavitation to collapse, even at recirculation rates as little as 5% of tank volume per minute. It is best to use diffuser screens to reduce the velocity of the returning fluid to prevent this collapse from occurring.

Tank Clean-Up Curves as a Function of Recirculation Rate

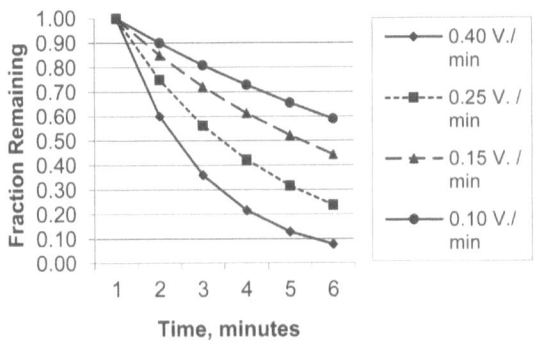

FIGURE 11.13   Tank clean-up rate as a function of volume fraction recirculation rate through theoretical, 100% efficient filters (Welker et al., 2006).

11.2.12   COMPONENT CLEANING IN THE HDD INDUSTRY

Cleaning of the mechanical components of a hard disk drive is typically accomplished in two stages (Nagarajan, 1997): the component supplier performs a rough-clean, and the drive manufacturer then does a final-clean prior to introducing the parts into a Class 100 (or better) cleanroom assembly environment. Recently, the trend has been toward outsourcing many assembly steps that used to be done at the drive manufacturing facility. As a result, more sub-assemblies, that are inherently uncleanable, are now being procured for direct entry into the drive assembly area. This has shifted the emphasis on component cleaning out to the supplier. The generic, all-purpose cleaning equipment favored in the past by drive manufacturers for in-house cleaning of multiple parts is being supplanted by more customized cleaners at supplier locations offering a wider variety of chemistries and mechanisms. In the late 80s and early 90s, many disk drive manufacturers made a major change away from using chlorofluorocarbons (CFCs) in cleaning to using water-based systems.

Phillips (1991) describes the challenges of making this transition. He lists steps in the evaluation process that remain relevant any time a cleaning process change is being considered:

- Review process steps where cleaning occurs.
- Eliminate any unnecessary cleaning steps.
- Identify constraints.
- Define cleanliness requirements.
- Select and implement alternative process.

A high-volume precision aqueous cleaning system, installed at Digital Equipment Corporation to clean hard disk electromechanical components, was developed In three phases—process and chemistry definition, prototyping, and production implementation. Its eight-station system contained a holding stage, an immersion ultrasonic wash, a rough spray rinse, an immersion ultrasonic rinse, a final spray rinse, clean air blow-off, and infrared radiation heating and exit staging. A high-purity water production and reclamation system was integrated within the cleaner. This configuration is fairly representative of most aqueous cleaners used in the disk drive industry, though final drying is more frequently accomplished by convective air drying or vacuum drying.

Gibbons (1987) discusses the then state-of-the-design art in parts cleaning technology throughout IBM. Among the forms of cleaners detailed are vapor degreasers, ultrasonic cleaners and non-immersion cleaners such as high-pressure spray washers and atmospheric-pressure atomizer sprayers. Solvent and aqueous cleaning media are contrasted. Drying methods, filtration needs, cleanliness measurements, and automation trends are reviewed as well. Although predominantly driven by environmental considerations, water cleaning has also, in general, resulted in cleaner parts and higher product yields (Nusbaum, 1991)—clearly a situation where environmental and manufacturing objectives were not mutually exclusive. In IBM San Jose, an aqueous cleaner with ultrasonics for disk drive parts was first implemented in 1989. Larger, more aggressive aqueous ultrasonic cleaners for the same purpose went online in July 1992 (Nagarajan and Welker, 1992)—the date when IBM San Jose ceased to use CFCs in cleaning—2.5 years ahead of the EPA mandate and 1.5 years ahead of the Corporate directive. In addition, IBM San Jose worked extensively with their supplier base to ensure that manufacturing processes were compatible with water-based cleaning, and with their own design and development teams to ensure that parts were designed for solvent-free cleaning (Nagarajan, 1991).

Thus, the implementation of aqueous cleaning in the microelectronics and precision-assembly industries intensified not only the elimination of ozone-depleting substances but also resulted in lower operating costs and higher process yields—a win-win scenario that is not always achievable. But when it does happen, oh, baby!

EXERCISE QUESTIONS

1. When an acoustic field is coupled to a liquid medium for the purpose of process intensification, the associated physical and chemical effects can sometimes be in conflict. For example, lower frequencies can intensify the physical effects due to higher cavitation intensity, but the reduction in bubble density can hinder the interfacial reactions promoted by the bubbles. Thus, an optimum frequency would need to be selected that would present the perfect balance. Explore this in the context of the MV-degradation degradation and other applications where the conflict may be present.

2. Component cleaning is not regarded as a value-add activity by microelectronics manufacturers, and rightly so. If contaminants were not present on surfaces to begin with, there would be no need to expend time and energy to remove them. "Clean manufacturing", however, also adds cost due to more stringent environmental controls, superior material selection, etc. This again is a classical optimization problem that lends itself to quantitative analysis. Assuming that a Class 10 cleanroom deposits 30% fewer particles on a critical surface than a Class 100 cleanroom, a linear relationship between surface contamination levels and process yield, and a cleaning process cost increase of 5% for every 10% increase in surface contamination levels, derive an optimization strategy by defining appropriate PIF and CIF parameters. Make any other necessary assumptions and justify them in your final resolution.

REFERENCES

Banerjee, S., "Cryoaerosol Cleaning of Particles from Surfaces", in *Particle Adhesion & Removal*, K.L. Mittal and R. Jaiswal, Editors, Scrivener Publishing LLC, Beverly, MA, 2015, Chapter 12, pp. 453–476.

Chang-qi, C., C. Jian-ping, and T. Zhiguo, "Degradation of Azo Dyes by Hybrid Ultrasound-Fenton Reagent", *IEEE Trans.*, 2008, pp. 3600–3603.

Dhanalakshmi, N.P. and R. Nagarajan, "Ultrasonic Intensification of Chemical Degradation of Methyl Violet: An Experimental Study", *World Acad. Sci. Eng. Technol.*, vol. 59, 2011, pp. 537–542.

Gibbons, J., "Cleaning Plant Design", *J. Soc. Environ. Eng.*, March 1987, pp. 7–12.

Musselman, R.P. and T.W. Yarbrough, "Shear Stress Cleaning for Surface Departiculation", *J. Environ. Sci.*, 1987, pp. 51–56.

Nagarajan, R. and R.W. Welker, "Precision Cleaning in a Production Environment With High-Pressure Water", *J. IES*, 1992, pp. 34–44.

Nagarajan, R., "Guidelines for Design of Machining Processes to Eliminate Solvent Cleaning", Proc. 11th IEEE/CHMT International Electronics Manufacturing Technology Symposium, IEEE (Institute of Electrical and Electronics Engineers), San Francisco, CA, 1991, pp. 300–306.

Nagarajan, R., "Survey of Cleaning and Cleanliness Measurement in Disk Drive Manufacture", *Precis. Clean. J.*, February 1997, pp. 13–22.

Nusbaum, V.A., "Are Manufacturing and Environmental Objectives Compatible?", Proc. 11th IEEE/CHMT International Electronics Manufacturing Technology Symposium, 1991, pp. 275–277.

Phillips, Q.T., "Meeting the Challenges of CFC Elimination from the Cleaning Process", *Microcontamination*, February 1991, pp. 51–56.

Walling, C., "Fenton's Reagent Revisited", *Acc. Chem. Res.*, vol. 8, 1975, pp. 125–131.

Welker, R.W., R. Nagarajan, and C.E. Newberg, *Contamination and ESD Control in High-Technology Manufacturing*, Wiley-Interscience, 2006.

12 Formulation of Nanoemulsion
Case Study

Emulsions are generally defined as a heterogeneous mixture of (at least) two immiscible liquids, in which one constituent liquid is oil and the other is water. They can be either oil-in-water (O/W) emulsions or water-in-oil (W/O) emulsions, where water is a continuous phase in the former, and oil is the continuous phase in the latter (Figures 12.1 and 12.2). Emulsions can be broadly classified as microemulsions and nanoemulsions. Emulsions with droplets large enough to settle in the gravitational field, and size as small as 1 μm, are generally called microemulsions. Those with droplet sizes below 0.5 μm (Capek, 2004; Solans et al., 2003) are called ultrafine emulsions/miniemulsions (El-Aasser and Sudol, 2004)/nanoemulsions. Thus, nanoemulsions consist of very fine emulsions with a droplet diameter of preferably less than 100 nm (Rao and McClements, 2012).

The essential difference between a microemulsion and a nanoemulsion is that microemulsions are thermodynamically stable, i.e., they are equilibrium systems, while nanoemulsions are not thermodynamically stable. Nanoemulsions are non-equilibrium systems, thermodynamically unstable in nature (Anton et al., 2008; Mason et al., 2006), which have a propensity to separate into constituent phases (Gutierrez, 2008). Also, microemulsion formulation requires a higher amount of surfactant, while nanoemulsions can be produced with a lower amount of surfactant in the range of 5–10% (w/w) (Tadros et al., 2004).

Nanoemulsions appear to be almost transparent or translucent to the naked eye due to their nano-sized droplets. They are metastable systems whose structure depends on the methods by which they have been prepared. They are very fragile in nature with low viscosity, high kinetic stability, and optical transparency (Wang et al., 2007). They are usually fluid, and at the slightest hint of destabilization, their transparency decreases, and creaming becomes visible (Sonneville-Auburn at al., 2004). The advantageous properties of nanoemulsions are mainly due to their small droplet size—for example, the long-term physical stability of the nanoemulsions can be attributed to the fact that the Brownian motion of the particles in a nanoemulsion is sufficient to overcome the gravitational force which causes sedimentation. The nano-range size of the droplets also prevents their flocculation as well as their coalescence (Tadros et al., 2004).

Hence, it can be concluded that the stability of nanoemulsion increases with decrease in droplet size, sometimes referred to as "approaching thermodynamic stability" (Tal-Figiel, 2007). Reducing droplet sizes in nanoemulsions to nano-scale offers some very interesting physical properties such as optical transparency and

DOI: 10.1201/9781003283423-12

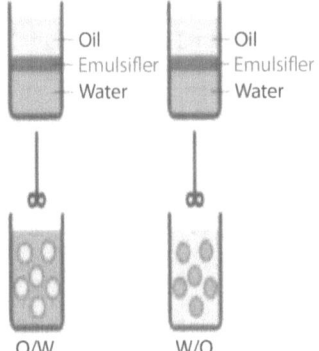

FIGURE 12.1 Schematic diagram for o/w and w/o emulsions (Bhattacharya, 2013).

unusual elastic behavior. Smaller droplet size in nanoemulsions makes them highly stable, increases their fluidity, and enhances their textural and aesthetic properties, which are very desirable in commercial applications. Small droplet sizes lead to transparent emulsions so that the product appearance is not altered by the addition of any oil phase (Leong et al., 2009).

Nanoemulsions form attractive systems in products such as polymerization reaction media to obtain nanoparticles in chemical industries, as nano-encapsulators to

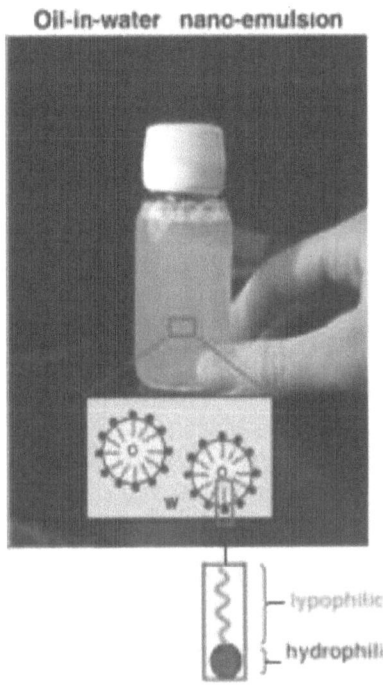

FIGURE 12.2 An O/W nanoemulsion system (Sharma et al., 2010).

harness controlled delivery of additives such as flavors, colors, micronutrients, etc., in processed food in food industries, as potent nano-colloidal drug carriers, ophthalmic delivery systems, and coating material for tablets in pharmaceutical industries, in cosmetics as personal care formulations, in fragrance and perfumes as the matrix to encapsulate the volatile compounds, as eco-friendly pesticide delivery systems in agro-industries, as larvicidal and anti-microbial agents, as disinfectants, in adjuvant vaccine development, etc. (Bhattacharya, 2013).

Smaller droplet-sized nanoemulsions are preferred in food and beverage industries since they lead to a creamier mouthfeel and more storage efficiency. Specifically, nanoemulsions can deliver high concentrations of oil-soluble food supplements with different functional attributes into a range of water-based food products and beverages without affecting the visual quality of the product. These supplements could be triglycerols (clouding agents, carrier oils, nutrients, and bioactive lipids), citrus oils (flavoring agents), phytosterols (nutraceuticals), carotenoids (colorants and anti-oxidants) and Vitamin E (oil-soluble vitamins). Nanoemulsions provide a lipophilic environment that facilitates high drug (e.g., Camptothecin–an anti-cancer agent) loading, appreciable stability of the drug within the droplets before its release, reductions in dosage, and a decrease in systemic toxicity. The capacity of nanoemulsions to dissolve large quantities of hydrophobics, along with their mutual compatibility and ability to protect drugs from hydrolysis and enzymatic degradation, makes them ideal vehicles for parenteral transport. Nanoemulsions can serve as ocular eye drops by virtue of the sustained release of the drug applied to the cornea, high penetration in the deeper layer of the ocular structure as well as ease in sterilization (Bhattacharya, 2013).

Nanoemulsions are desirable in skin care products due to their high penetration power through body membranes, merging textures, bioavailability, and hydrating power. Nanoemulsions have anti-bacterial and anti-larvicidal properties (e.g., nanopermethrin). In addition to that, pesticides, when used in nanoparticulate form, are required in much lower concentrations as compared to bulk pesticides and can be formulated without the use of organic solvents, thus saving many non-target organisms and causing no harm to the environment. The antimicrobial activity of the nanoemulsions was also investigated with water-in-oil nanoemulsions that are diluted in water just before application to oil-in-water nanoemulsions. The efficiency of nanoemulsions in inactivation of the ebola virus was investigated with promising results (Bhattacharya, 2013). Due to these characteristics of nanoemulsions and due to the innumerable possibilities they offer, they have gained unparalleled importance in research.

12.1   FORMULATION PROCESS

Stability and characteristics of the nanoemulsions largely depend on the methods by which they have been prepared; nanoemulsions cannot form spontaneously (Solans et al., 2003). Energy input either from the chemical potential of the components or from some mechanical device, is required to form stable nanoemulsions. These are generally prepared by high-energy emulsification (Wang et al., 2007) as well as low-energy emulsification methods (Uson et al., 2004). Low-energy techniques which take

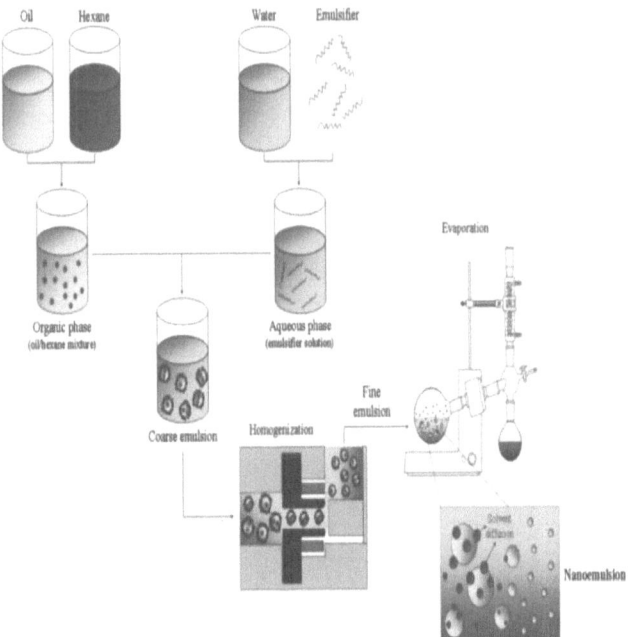

FIGURE 12.3 Schematic diagram for a nanoemulsion formulation by homogenizer (Troncoso et al., 2012).

advantage of the physicochemical properties of the system that affect the hydrophilic-lipophilic balance (HLB) include condensation methods such as phase inversion temperature (PIT) emulsification (Izquierdo et al., 2002, 2004) and phase inversion composition (PIC) (Sajjadi, 2006). However, several limitations were found in using low-energy emulsification methods, such as the requirement of a large number of surfactants, difficulty in temperature control, and in selecting the combination of surfactant and co-surfactant (Wulff-Perez et al., 2009). High-energy methods include high-pressure homogenization, use of acoustic fields, etc. (Kentish et al., 2008).

Advantages of using high-energy emulsification techniques include flexible control of droplet size distribution, and ability to produce fine emulsions from a variety of materials (Seekkuarachchi et al., 2006). The traditional method for the preparation of nanoemulsions was homogenization (Figure 12.3), first used in food industries (McClements, 2004). Homogenization, being an energy-intensive process, paved the way for a high-energy emulsification device, i.e., microfluidizer since the 1990s (Strawbridge et al., 1995). But the recent surge in the cost of energy has led to the use of ultrasonics in the preparation of nanoemulsions (Tal-Figiel, 2007).

12.2 ULTRASOUND APPLICATION

Ultrasonic processors are used as homogenizers, to reduce and distribute small particles in a liquid to improve uniformity and stability. These particles (dispersed phase) can be either solids or liquids. This process is termed as "ultrasonic emulsification".

Emulsification was one of the first applications of ultrasound. Ultrasound emulsification was reported for the first time by Woods and Loomis (1927). The first patent for ultrasonic emulsification was granted in the year 1944 in Switzerland. Since then, extensive research is ongoing in the field of ultrasonic emulsification, carried out by many eminent scientists. A wide range of intermediate and consumer products, such as cosmetics and skin lotions, pharmaceutical ointments, varnishes, paints and lubricants, and fuels are based wholly or in part on emulsions.

Emulsions are dispersions of two or more immiscible liquids. High-intensity ultrasound supplies the power needed to disperse a liquid phase (dispersed phase) in small droplets in a second phase (continuous phase). In the dispersing zone, imploding cavitation bubbles cause intensive shock waves in the surrounding liquid and result in the formation of liquid jets of high liquid velocity. In order to stabilize the newly formed droplets of the dispersed phase against coalescence, emulsifiers (surface active substances, surfactants), and stabilizers are added to the emulsion. As coalescence of the droplets after disruption influences the final droplet size distribution, efficiently stabilizing emulsifiers are used to maintain the final droplet size distribution at a level that is the same as the distribution immediately after the droplet disruption in the ultrasonic dispersing zone. At appropriate energy density levels, ultrasound can well achieve a mean droplet size in the micrometer range resulting in the formation of micro-emulsions. Advanced technologies have used ultrasonic emulsification to produce emulsions with even finer droplets, i.e., below 1 μm, which are termed as "nanoemulsions".

Extremely high shearing capability of ultrasound or acoustic fields has been exploited in recent years to formulate different varieties of nanoemulsions. Ultrasonic emulsification is found to have superior efficiency in producing nanoemulsions as compared to mechanical agitation and traditional homogenization in terms of better characteristic properties. It is again found more suitable as compared to microfluidization in terms of commercial feasibility. The use of an acoustic field in nanoemulsion formulation has helped in the optimal utilization of resources. The ultrasonic emulsification technology is rapidly attracting interest due to its inherent advantages, which include low cost, minimal byproduct formation and ability to be scaled up for bulk production for industrial applications.

An emulsion is formed when the interfaces of two immiscible liquids are irradiated with ultrasound waves. During the formation of an emulsion, tiny droplets of one liquid, i.e., the dispersed phase, get scattered in the other liquid—which is the constituent phase or the dispersion medium. The size of the droplets of the dispersed phase dictates the formation of either a microemulsion or a nanoemulsion. Notable observations about the earlier work done on the emulsification process can be summarized as follows (Gaikwad and Pandit, 2008):

1. Emulsification process starts only after a certain threshold of cavitation intensity is reached.
2. The stability of an emulsion in terms of droplet size and distribution increases with an increase in the irradiation power.
3. The dispersed phase droplets tend to be in a lower size range in proportion to the emulsification time, while the fractional dispersed phase hold-up decreases with an increase in emulsification time.

Nanoemulsions produced by ultrasonics have smaller droplet sizes, which has the effect of increased interfacial contact area, and hence, higher emulsification efficiency. Two mechanisms are believed to be involved in ultrasonic emulsification. First is acoustic cavitation, and the second is development of interfacial capillary waves (Li and Fogler, 1978a; Tal-Figiel, 1990; Tal-Figiel, 2005), i.e., acoustic streaming. Cavitation dominates at low frequency, while streaming dominates at high frequency. An increase in ultrasonic frequency causes an increase in the amplitude of the interfacial capillary waves making them unstable. These unstable capillary waves then eventually result in droplets of oil breaking away from the crests of the wave and their dispersion in the water medium (Tal-Figiel, 2007). Acoustic cavitation results in turbulent implosions which act as a very effective method for breaking up the primary droplets of oil into nanosize (Li and Fogler, 1978a,b).

Studies comparing ultrasonic emulsification with rotor-stator dispersing have found ultrasound to have superior efficiency in terms of mean droplet size and energy efficiency (Canselier et al., 2002). Ultrasonic emulsification was also found to be more advantageous from the industrial point of view as compared to microfluidization technique in terms of cleaning, equipment maintenance, operation, and cost (Jafari et al., 2006). The nanoemulsions produced by ultrasonic emulsification require less amount of surfactant for a desired mean particle diameter, are less polydispersed, and are more stable in nature as compared to those prepared by microfluidization. Also, the energy consumption through heat loss was found to be much lower in ultrasonic emulsification than in microfluidization (Tadros et al., 2004). Two mechanisms, acoustic streaming, and acoustic cavitation, are mainly responsible for acoustically induced effects. Acoustic streaming, resulting in development of unstable interfacial capillary waves with increase in acoustic frequency, leads to the eruption of dispersed phase droplets into the continuous phase. Acoustic cavitation resulting in turbulent implosions acts as a very effective method of breaking up the primary droplets of oil into nano-size by pressure fluctuations. Although excellent shear force is generated by ultrasound for droplet breakage, the rate of droplet coalescence is determined by the surface activity at interface of the dispersed phase and dispersion medium. Therefore, along with the ultrasonication, the addition of surfactant to the oil-water mixture plays a very important role in the formation of a stable nanoemulsion by lowering the interfacial tension (σ).

Tal-Figiel (2007) considered Weber number (We) as an important parameter that describes droplet fragmentation. Droplet deformation increases with an increase in Weber number. Weber number is defined as the ratio of the external stress, $G\eta$, to the Laplace pressure, P.

$$We = (G\eta / P) = (G\eta r / 2\sigma) \tag{12.1}$$

As clearly seen, a decrease in surface tension (σ) will cause an increase in Weber number and thus droplet deformation. It has already been proved in earlier research work that for the disruption of liquid into droplets using ultrasonic waves, the amplitude must be above a threshold value.

This threshold amplitude (A_{crit}) is given by-

$$A_{crit} = (2\eta_c / \rho_c)(\rho_c / \pi \sigma f)^{0.333} \tag{12.2}$$

Thus, the droplet size of the nanoemulsion also depends on the frequency of the ultrasonic waves. A relation between specific volume power density (ϵ) and residence time in the ultrasonic field with droplet size has also been reported in the literature. Droplet size decreases with increase in ultrasonication time until a system-specific minimum droplet size is obtained. For the investigated systems, the relation is:

$$d_{32} \propto d_{max} \propto \left[\epsilon(t)^{-b} \right] \text{where } b = 0.18 - 0.29 \tag{12.3}$$

From the equations above, it can be concluded that the main parameters which govern the droplet size in a nanoemulsion are oil:surfactant ratio, ultrasonication time, power, and frequency (Tal-Figiel, 2007).

Surfactants aid in droplet break-up by decreasing the oil-water interfacial surface tension. Surfactants quickly adsorb on the surface of the newly formed droplets in nanoemulsions, which prevents the coalescence of droplets. Surfactants are known to reduce interfacial tension, causing a reduction in the droplet size of the nano-emulsions, and, in turn, make the nanoemulsion systems kinetically stable. This phenomenon depends on the surfactant concentration used while formulating the nanoemulsion (McClements, 2004; Tadros et al., 2004). At low surfactant concentration, enough surfactant is not present in the system to adsorb over the newly formed droplets which causes droplet coalescence. The concentration of the surfactant determines the total droplet surface area, the rate of diffusion, and the adsorption phenomena (Li and Chiang, 2012).

There is a decrease in mean droplet size with increase in surfactant concentration up to a certain point, after which the rate of decrease in droplet size is minimal or almost negligible with further increase in surfactant concentration (Kentish et al., 2008). This can be attributed to the fact that in presence of a very high concentration of surfactants, micelles play a very important role in determining the stability of a nanoemulsion system (Dickinson, 2003; Dickinson et al., 1998).

Once the newly formed dispersed phase droplets are completely coated with surfactants, the excess surfactant in the system forms micelles that do not adsorb on the already surfactant-coated droplets. As a result of this, an attractive force develops between the dispersed phase droplets by a depletion mechanism. Thus, when two droplets approach in a solution of non-adsorbing micelles, the latter are expelled from the gap, generating a local region with almost a pure solvent.

This osmotic pressure in the liquid surrounding the droplet pair exceeds that between the drops and consequently compels the droplets to aggregate (Napper, 1983). The rate of adsorption of surfactants on the droplet surface and droplet size distribution of the newly formed droplets is influenced by the time an emulsion has been exposed in the ultrasonic field (Jena and Das, 2006). Hence, emulsification time or irradiation time was also considered to be an important factor that governs the thermodynamic equilibrium in an oil-in-water nanoemulsion system. The requirements of both droplet break-up and coalescence dictate that small molecule surfactants are more suited to producing stable nanoemulsions compared to macromolecular surfactants because of their greater ability to rapidly adsorb to droplet interfaces and due to their much lower dynamic interfacial tensions (Kralova and Sjoblom, 2009; Leong et al., 2009).

The primary objective of the work by Bhattacharya (2013) was to determine a commercially viable, intensified process for nanoemulsion formulation and understand its mechanism of operation at the nano-scale. The effects of acoustic fields on this process were extensively studied with a focus on the use of very high-frequency, dual-frequency as well as low-frequency acoustic waves for nanoemulsion formulation. In this regard, megasonic and ultrasonic systems were used. The second objective of this study was to show the superior efficiency of tank-type sonicators over probe-type ultrasonicators in producing stable neem oil nanoemulsions, which enables scale-up at an industrial level while consuming less time. This was mainly done by size analysis of the various nanoemulsion samples produced by probe-type ultrasonicators as well as by tank-type sonicators. The third objective of the study was to optimize by quantitative characterization the concentration of surfactant Tween-20 to be used in producing a stable neem oil nanoemulsion, thus making the process more cost-effective.

An important step is to identify the factors, both process parameters and compositional parameters, which could have a significant influence on the formulation of "best" nanoemulsions. Various quantitative as well qualitative characterizations were done to assess the "best" nanoemulsions. The final objective of the work was to optimize the experimental conditions to formulate a better quality, i.e., stable neem oil nanoemulsion by a designed experiment based on the Taguchi method. Finally, optimization results were validated by confirmation experiments.

12.3   EXPERIMENTAL SETUP AND PROCEDURE

The oil which has been used for the preparation of nanoemulsion is neem oil. Neem has been used in India since 2000–4000 BC, and was referenced in ancient Indian texts. Neem oil has been recognized for more than 2,000 years to possess powerful medicinal properties. Its use originated in ancient India and neighboring countries, where the neem tree *(Azadirachta indica)* was deeply revered as one of the most versatile medicinal plants known. Neem oil is made from the extract of the plant *Azadirachta indica* and generally is light to dark brown in color, bitter in taste, and has a strong odor. It comprises mainly of triglycerides and large amounts of triterpenoid compounds. It is hydrophobic in nature, and in order to emulsify it in water, formulation with the appropriate amount of surfactant is a must. Neem oil in this study was obtained from a local oil mill shop in Chennai, Tamil Nadu, India.

Surfactant used in this study was Tween-20, i.e., polyoxyethylen-20-sorbitan-monolaurate which has a monolaurate tail ($C_{12:0}$) and a molecular weight of 1228 g/mol. Tween-20 surfactant was used as it is readily miscible with plant oils. It is non-ionic in nature, and less sensitive to pH and ionic changes. The molecular weight of the surfactant plays an important role in the emulsification process since the kinetics of the surfactant movement from the dispersed phase to the continuous phase decrease with increase in molecular weight of the surfactant, thereby altering the formation of oil droplets at phase boundary (Ostertag et al., 2012). Small-molecule surfactants, when dissolved in water, were seen to decrease the surface tension of pure water (from 72.46 to ~40 mN/m) and showed lower interfacial surface tension when mixed with oil, which is favorable for the formation of small oil droplets in the

emulsion system (Mao et al., 2009). Also, small molecules such as nonionic surfactants are known to reduce the interfacial surface tension more, thus causing increased
stability (Kralova and Sjoblom, 2009), as compared to polymeric surfactants having
large molecules (Tadros et al., 2004). Therefore, a small molecule non-ionic surfactant, Tween-20, was chosen as the emulsifier in Bhattacharya's (2013) study for stabilizing the neem oil nanoemulsions. The aqueous medium or the dispersion phase used
was purified water from a Milli-Q™ integral water purification system.

Three tank-type sonicators and a single probe-type ultrasonicator were used for
equipment optimization in this study. Tank-type sonicators were provided by Crest
Ultrasonics (Trenton, New Jersey), while probe-type ultrasonicator operating at
20 kHz, 750 W was supplied by Sonics and Materials. Inc., USA. Out of the three
tank-type sonicators, one was operated in the megasonic range, i.e., at a frequency of
1 MHz with 1200 W power input, and the other two were dual-frequency and single-
frequency ultrasonicators. The dual-frequency ultrasonicator works at 58 + 192 kHz,
500 W, and the single-frequency ultrasonicator works at 25 kHz, 500 W.

Neem oil, Tween-20 surfactant, and purified water were used as components for
preparing the nanoemulsion samples by sonication. To compare the efficiency of
tank-type sonicators with that of probe-type sonicators, neem oil nanoemulsions
were prepared by using different oil:surfactant ratios, i.e., 1:0.3, 1:1.5, and 1:3, making the total volume up to 100 ml with a sonication time of three hours, at 100%
operational power. A blue cap bottle was used as a sample holder if a tank-type ultrasonicator is used, while a simple glass beaker was used in case of probe- type ultrasonicator. Size analysis was done by an Acoustic Particle Sizer instrument supplied
by Matec Applied Sciences, USA. Each measurement was performed in triplicate.

For optimizing the Tween-20 surfactant concentration, various oil:surfactant
ratios were tried. The ratios chosen for the optimization study of surfactant concentration are 1:0.05, 1:0.1, 1:0.2, 1:0.3, 1:0.4, and 1:0.5. Total volume of the neem oil
nanoemulsion samples, in this case, was again 100 ml. Quantitative measurements
(size analysis, stability test, and surface tension analysis) were then performed to
optimize the amount of Tween-20 surfactant to be used in nanoemulsion formulation.
A 90 plus-Particle Size Analyzer supplied by Brookhaven Instrument Corporation,
USA, was used for size analysis and stability tests. Each of the measurements was
done in duplicate.

For the Taguchi optimization, neem oil nanoemulsions were formulated as per the
conditions given in each trial of the L_{27} orthogonal array (Table 12.1). For example,
the neem oil nanoemulsion of Trial 1 was prepared by mixing neem oil and Tween-
20 surfactant in the ratio of 1:0.2, making the total volume up to 100 ml by using
purified water, keeping the initial sample temperature at room temperature, and
emulsifying it for two hours in tank-type sonicator operating at 25 kHz and 70%
operational power.

The sizes of the droplets and their distribution in the neem oil nanoemulsions were
determined by a 90 plus-Particle Size Analyzer supplied by Brookhaven Instrument
Corporation, USA. The 90-plus-Particle Size Analyzer performs fast, routine submicron particle size measurements on a wide variety of samples and concentrations.
It is an ideal instrument for measuring colloids, latexes, micelles, proteins, and other
nanoparticles. This instrument is based on the principles of dynamic light scattering.

TABLE 12.1
Design Parameters, Their Values at Three Levels, and the Taguchi L_{27} Matrix

Controlled Parameters	Level 1	Level 2	Level 3
Frequency of acoustic field	25 kHz	58+192 kHz (Dual)	1 MHz
Operational power	100% Power	85% Power	70% Power
Oil:surfactant ratio	1:0.2	1:0.3	1:0.5
Temperature of sample	35°C	45°C	55°C
Time of emulsification	2 hours	3 hours	4 hours

Trials	(A) Frequency	(B) Power	(C) Oil:Surfactant	(D) Temperature	(E) Time
1	1(25kHz)	1(70%)	1(1:0.2)	1(35°C)	1(2hrs)
2	1(25kHz)	1(70%)	2(1:0.3)	2(45°C)	2(3hrs)
3	1(25kHz)	1(70%)	3(1:0.5)	3(55°C)	3(4hrs)
4	1(25kHz)	2(85%)	1(1:0.2)	1(35°C)	2(3hrs)
5	1(25kHz)	2(85%)	2(1:0.3)	2(45°C)	3(4hrs)
6	1(25kHz)	2(85%)	3(1:0.5)	3(55°C)	1(2hrs)
7	1(25kHz)	3(100%)	1(1:0.2)	1(35°C)	3(4hrs)
8	1(25kHz)	3(100%)	2(1:0.3)	2(45°C)	1(2hrs)
9	1(25kHz)	3(100%)	3(1:0.5)	3(55°C)	2(3hrs)
10	2(Dual)	1(70%)	1(1:0.2)	2(45°C)	1(2hrs)
11	2(Dual)	1(70%)	2(1:0.3)	3(55°C)	2(3hrs)
12	2(Dual)	1(70%)	3(1:0.5)	1(35°C)	3(4hrs)
13	2(Dual)	2(85%)	1(1:0.2)	2(45°C)	2(3hrs)
14	2(Dual)	2(85%)	2(1:0.3)	3(55°C)	3(4hrs)
15	2(Dual)	2(85%)	3(1:0.5)	1(35°C)	1(2hrs)
16	2(Dual)	3(100%)	1(1:0.2)	2(45°C)	3(4hrs)
17	2(Dual)	3(100%)	2(1:0.3)	3(55°C)	1(2hrs)
18	2(Dual)	3(100%)	3(1:0.5)	1(35°C)	2(3hrs)
19	3(1 MHz)	1(70%)	1(1:0.2)	3(55°C)	1(2hrs)
20	3(1 MHz)	1(70%)	2(1:0.3)	1(35°C)	2(3hrs)
21	3(1 MHz)	1(70%)	3(1:0.5)	2(45°C)	3(4hrs)
22	3(1 MHz)	2(85%)	1(1:0.2)	3(55°C)	2(3hrs)
23	3(1 MHz)	2(85%)	2(1:0.3)	1(35°C)	3(4hrs)
24	3(1 MHz)	2(85%)	3(1:0.5)	2(45°C)	1(2hrs)
25	3(1 MHz)	3(100%)	1(1:0.2)	3(55°C)	3(4hrs)
26	3(1 MHz)	3(100%)	2(1:0.3)	1(35°C)	1(2hrs)
27	3(1 MHz)	3(100%)	3(1:0.5)	2(45°C)	2(3hrs)

The sample materials (emulsions) scatter incoming laser light. Due to the random motion of these particles, the scattered light intensity fluctuates in time. Processing the fluctuating signal with a state-of-the-art digital auto-correlator yields the particle's diffusion coefficient, from which the equivalent spherical particle size is calculated using the Stokes-Einstein equation. The neem oil nanoemulsion samples were diluted with purified water in a 3 ml cuvette in order to reduce the effects of multiple scattering of light.

The stability of the neem oil nanoemulsions was assessed by determining the change in mean droplet size as a function of time. It was done by measuring the mean size and the size distribution of the nanoemulsion samples with the 90 plus-Particle Size Analyzer, at different time intervals from the time of formation of the nanoemulsions.

10 ml of neem oil with a defined concentration of surfactant and water was used to prepare all 27 nanoemulsion samples (as per the trials in Taguchi L_{27} design matrix). The mixing was made inside a laminar flow enclosure to avoid any contamination of the bulk mixture. The bulk mixtures were then subjected to different frequencies of the acoustic field at different operational powers for a varied time of emulsification (as per the Taguchi L_{27} design matrix).

12.4   RESULTS AND DISCUSSION

12.4.1   EMULSION DROPLET SIZE DISTRIBUTION

Neem oil, Tween-20 surfactant, and purified water were used as components for preparing the nanoemulsion samples by sonication. To compare the efficiency of tank-type sonicators with that of probe-type sonicators, neem oil nanoemulsions were prepared by using different oil:surfactant ratios, i.e., 1:0.3, 1:1.5, and 1:3, making the total volume up to 100 ml with a sonication time of 3 hours. Size analysis was done by an Acoustic Particle Sizer instrument supplied by Matec Applied Sciences, USA. Each measurement was performed in triplicate.

Figures 12.4–12.7 show the normalized particle size distribution of the neem oil nanoemulsion samples for all the oil:surfactant ratios of 1:0.3, 1:1.5, and 1:3, for a 25 kHz, 500 W tank-type sonicator, for dual frequency (58 + 192 kHz), 500 W tank-type sonicator, for 1 MHz, 1200 W tank-type sonicator and for a 20 kHz, 750 W probe-type sonicator, respectively. From the normalized particle size distribution graphs plotted with particle size (μm) on the x-axis and normalized particle count

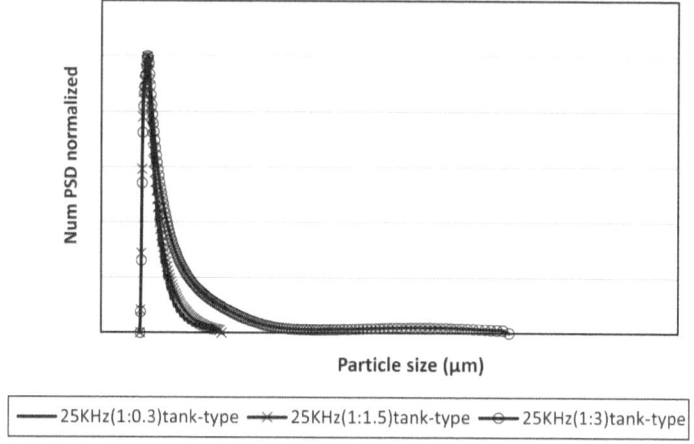

FIGURE 12.4   Normalized particle size distribution graph for 25 kHz tank-type sonicator.

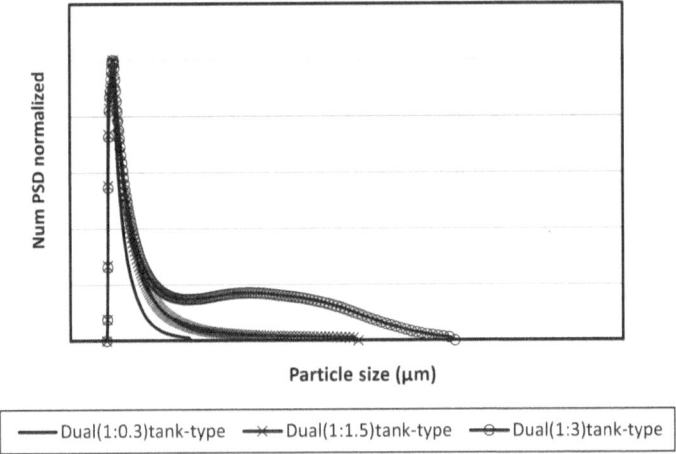

FIGURE 12.5 Normalized particle size distribution graph for dual frequency (58 + 192 kHz) tank-type sonicator.

on the y-axis, the minimum and maximum particle size in a nanoemulsion sample is calculated. The mean particle size of a nanoemulsion sample is given by the APS instrument directly.

From the normalized PSD graphs of all the neem oil nanoemulsion samples, it was observed that the minimum particle size in all the samples ranges from 10 to 12 nm, while a large variation is seen in case of maximum particle size. In nanoemulsions prepared from tank-type ultrasonicators, the maximum size of the particle varied from 20 to 170 nm. On the other hand, the maximum size of the particles is much larger in nanoemulsions prepared from probe-type ultrasonicators, i.e., 96 nm–1.4 μm. Similarly, the mean size of the particles in nanoemulsions prepared

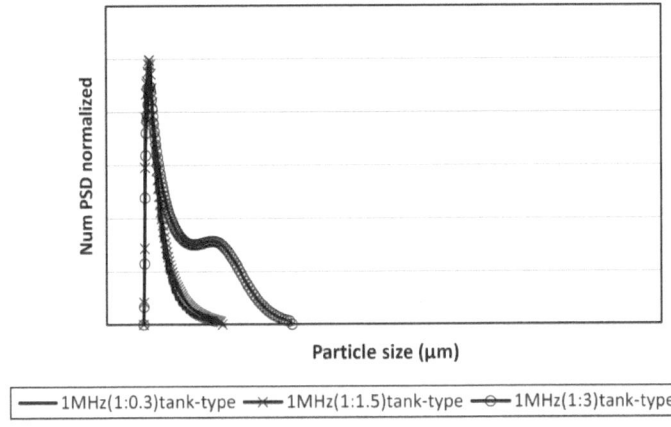

FIGURE 12.6 Normalized particle size distribution graph for 1 MHz tank-type sonicator.

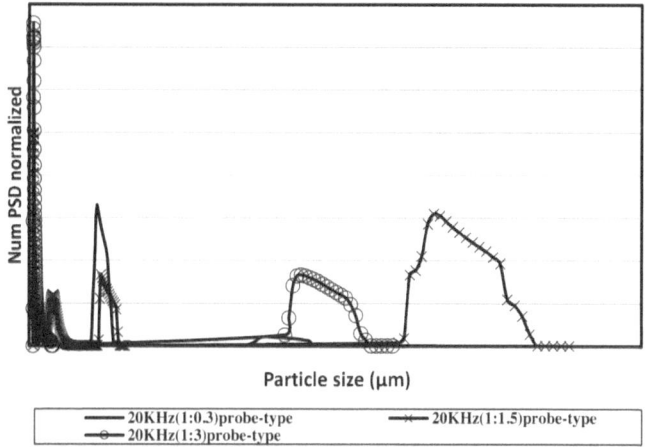

FIGURE 12.7   Normalized particle size distribution graph for 20 kHz probe-type sonicator.

from tank-type ultrasonicators (12–80 nm) was much lower than the mean size of the particles in nanoemulsions prepared from probe-type ultrasonicator (150–670 nm).

Figures 12.8–12.11 show the cumulative particle size distribution of the neem oil nanoemulsion samples for all the oil:surfactant ratios, i.e., 1:0.3, 1:1.5, and 1:3, for a 25 kHz, 500 W tank-type sonicator, for dual frequency (58 + 192 kHz), 500 W tank-type sonicator, for 1 MHz, 1200 W tank-type sonicator and for a 20 kHz, 750 W probe-type sonicator, respectively. From the cumulative particle size distribution graphs plotted with particle size (um) on x-axis and cumulative particle count on the y-axis, the D_{90} and D_{50} values are calculated for each nanoemulsion sample.

From the cumulative PSD graphs, D_{90} and D_{50} values were then calculated for all the samples in four replicates, and average was taken. D_{90} value signifies that particle diameter where 90% of the particle distribution in a particular nanoemulsion sample

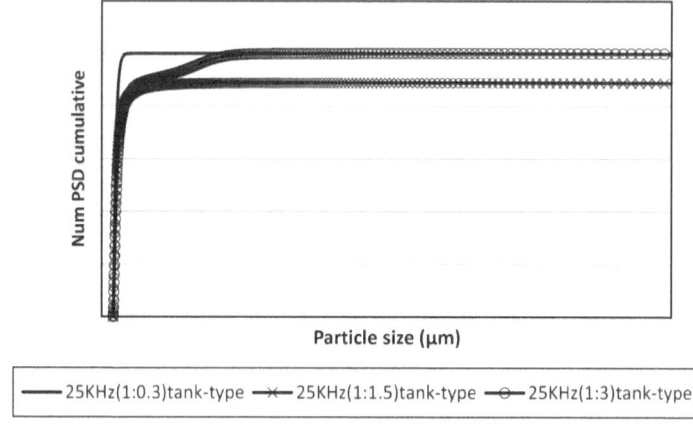

FIGURE 12.8   Cumulative particle size distribution for 25 kHz tank-type sonicator.

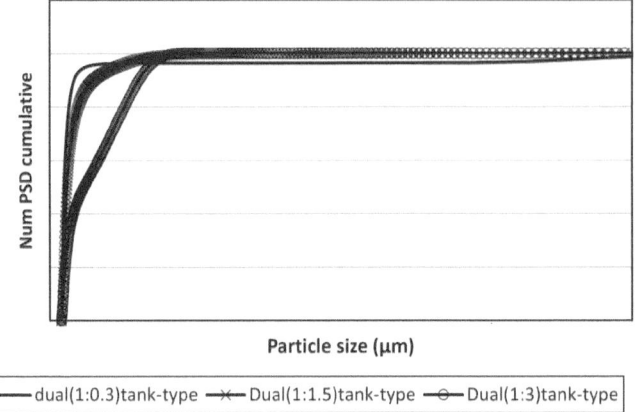

FIGURE 12.9   Cumulative particle size distribution for dual frequency (58 + 192 kHz) tank-type sonicator.

is smaller than that value. D_{50} is also known as the median diameter or medium value of particle diameter and is the particle diameter value for which the cumulative distribution percentage reaches 50%. Therefore, the smaller the value of D_{90} and D_{50} the better it is since it shows that the sample has more particles in the lower size range. Again, it can be observed that nanoemulsion samples prepared from tank-type ultrasonicators have lower D_{90} and D_{50} values as compared to the nanoemulsions prepared from probe-type ultrasonicator.

To compare the efficiency of a tank-type sonicator and a probe-type sonicator, sonicators operating at a comparable frequency are taken into consideration. The nanoemulsions prepared from a tank-type sonicator operating at 25 kHz were compared to those prepared by 20 kHz probe-type sonicator in terms of droplet size and distribution. From the normalized particle size distribution graphs, it was observed

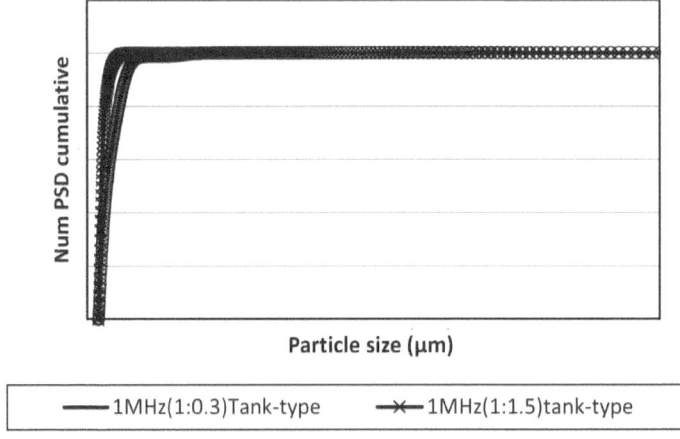

FIGURE 12.10   Cumulative particle size distribution for 1 MHz tank-type sonicator.

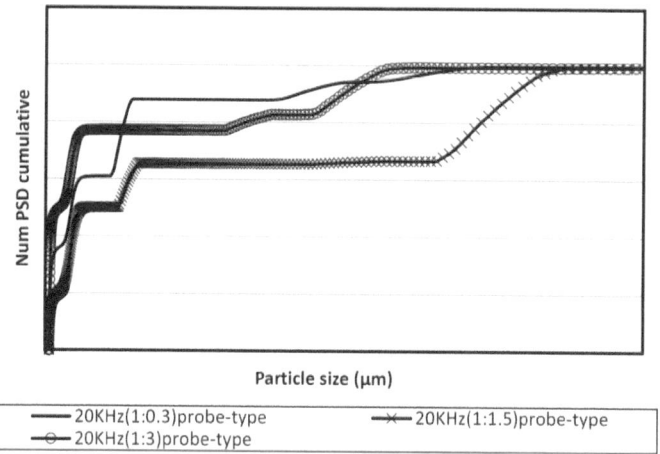

FIGURE 12.11   Cumulative particle size distribution for 20 kHz probe-type sonicator.

that in nanoemulsions prepared from tank-type sonicators operating at 25 kHz, 500 W, the maximum size of the particle varied from 30–130 nm.

On the other hand, the maximum size of the particles is much larger in nanoemulsions prepared in probe-type ultrasonicators, i.e., 96 nm–1.4 µm. Similarly, the mean size of the particles in nanoemulsions prepared from 25 kHz tank-type sonicator (25–40 nm) was much lower than the mean size of the particles in nanoemulsions prepared in probe-type ultrasonicator (200–670 nm). From cumulative particle size distribution graphs, it can be observed that nanoemulsion samples prepared from tank-type sonicators have lower D_{90} and D_{50} values compared to the nanoemulsions prepared from probe-type ultrasonicator.

In a 25 kHz tank-type sonicator, the D_{90} value varies from 18 to 110 nm for all the different oil:surfactant ratio samples. This means that 90% of the particles in these nanoemulsion samples are smaller than 110 nm in size, thus indicating a more stable nanoemulsion. Similarly, 50% of the particles in samples prepared in tank-type sonicators lie below 50 nm in size. On the other hand, a large D_{90} and D_{50} value obtained in case of nanoemulsion samples prepared from probe-type ultrasonicators indicate that the size of most of the particles in these samples lies in the higher size range of 100 nm to 1.3 µm. Thus, based on the minimum, mean, and maximum particle size measurement as well as on the D_{90} and D_{50} values, it may be concluded that for a comparable frequency, tank-type sonicators are more efficient in preparing nanoemulsions of smaller particle size in nano range compared to probe-type ultrasonicators, facilitating easy industrial scale-up consuming less time.

Size analysis was done for neem oil nanoemulsion samples prepared by three tank-type sonicators operating at 25 kHz, dual (58 + 192 kHz) and 1 MHz, respectively, at 100% operational power for an emulsification time of three hours. Nanoemulsions with oil:surfactant ratios of 1:0.05, 1:0.1, 1:0.2, 1:0.3, 1:0.4, and 1:0.5 were formulated and size analysis was done by dynamic light scattering instrument, i.e., 90-plus particle sizer was used for this purpose. From the size analysis results, it was found that for an oil:surfactant ratio of 1:0.05, the mean diameter of the droplet of neem

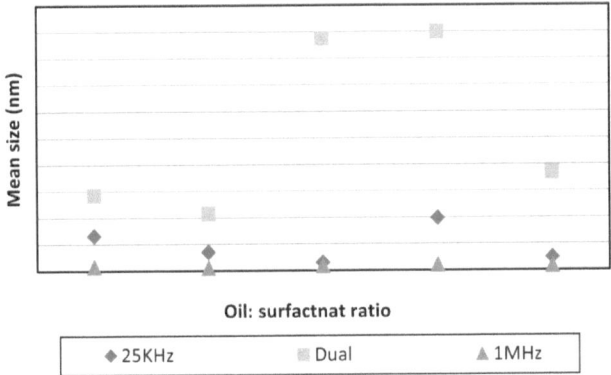

FIGURE 12.12 Mean droplet size as a function of oil:surfactant ratio.

oil nanoemulsion samples prepared by all the three tank-type sonicators was in the micro-range, indicating that at this particular oil:surfactant ratio, the surfactant concentration is too low to form a nanoemulsion. Size analysis results for neem oil nanoemulsions with oil:surfactant ratio of 1:0.1, 1:0.2, 1:0.3, 1:0.4, and 1:0.5 is shown here in the form of graphs. Figure 12.12 shows the variation of the number-mean size of the droplets of neem oil nanoemulsions for all five oil:surfactant ratios, i.e., 1:0.1, 1:0.2, 1:0.3, 1:0.4, and 1:0.5, for three different frequencies.

This graphical representation (Figure 12.12) indicates that nano-scale droplets of neem oil nanoemulsion can be prepared by using tank-type sonicators at all the ratios. The mean droplet size of the neem oil nanoemulsions with an oil:surfactant ratio of 1:0.1, 1:0.2, 1:0.3, and 1:0.5 is seen to be much smaller than the neem oil nanoemulsions with oil:surfactant ratio of 1:0.4 at all the three frequencies. Figure 12.13 shows the variation of mean droplet size with different acoustic field frequencies for all oil:surfactant ratios (1:0.1, 1:0.2, 1:0.3, and 1:0.5).

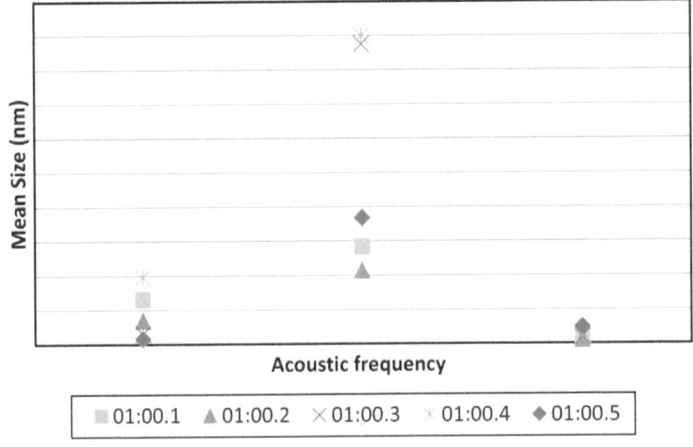

FIGURE 12.13 Mean droplet size as a function of acoustic frequency.

From Figure 12.13, it can be inferred that the mean droplet size of the neem oil nanoemulsion prepared at 25 kHz and 1 MHz is much smaller than that prepared at dual-frequency (58 + 192 kHz). The mean droplet size of the neem oil nanoemulsions prepared from 25 kHz tank-type sonicator and 1 MHz tank-type sonicator is below 100 nm, while the mean droplet size prepared by dual-frequency tank-type sonicator is below 500 nm. This can be attributed to the fact that at 1 MHz, generation of unstable interfacial capillary waves, i.e., streaming, dominates, and at the lower frequency of 25 kHz, cavitation dominates; neither dominates at the intermediate dual-frequency, resulting in larger droplet size of the neem oil nanoemulsions prepared from dual (58+ 192 kHz) tank-type sonicator.

From Figures 12.12 and 12.13, it can be observed that the droplet size of the neem oil nanoemulsion prepared from 1 MHz tank-type sonicator is even smaller than that of the neem oil nanoemulsions prepared in a 25 kHz tank-type sonicator. The mean droplet size of the neem oil nanoemulsions prepared from 1 MHz tank-type sonicator is below 50 nm, while the mean droplet size prepared by 25 kHz tank-type sonicator is below 100 nm. The high cavitation intensity prevailing in a tank-type sonicator operating at 25 kHz, 500 W indicates that the reduction in droplet size of neem oil nanoemulsion in this type of sonicator is mainly due to the mechanism of cavitation. The larger-sized droplets in neem oil nanoemulsion prepared by dual-frequency tank-type sonicator can be attributed to the low cavitation intensity in this type of ultrasonicator. Streaming dominates over cavitation as a size-reduction mechanism at a very high acoustic frequency, i.e., in 1 MHz, 1200 W tank-type sonicator.

Earlier, cavitation was assumed to be the prevalent mechanism for size reduction and disruption caused by the application of acoustic field (Richards, 1929). But, the results of the study by Bhattacharya (2013) suggest that the streaming phenomenon is more effective in causing disruption and size reduction, i.e., high-frequency acoustic waves cause size reduction more effectively than low-frequency acoustic waves. This is probably associated with the generation of shear forces when the bubbles move in the direction of the wave, at very high acoustic field frequencies. When these shear forces become large, they can influence particle aggregation by sweeping single particles away or limiting the size of dispersed phase particles by drawing particles off their perimeter. Ultrasonic frequencies ≥ 1 MHz are preferred for particle manipulation, so that high-pressure amplitudes can be employed without inducing ultrasonic cavitation, with its associated vigorous order-disrupting bubble cavity (Coakley et al., 2000). Streaming, in general, causes drag and stress on particles and particle clusters. Large-scale streaming, in particular, may drive particles out of desired positions, prevent particle aggregation or even cause disruption (Spengler and Jekel, 2000).

12.4.2 STABILITY ANALYSIS

The main difference between a nanoemulsion system and microemulsion system is that microemulsions are thermodynamically stable, while nanoemulsions are not. Nanoemulsions are non-equilibrium systems in which the droplets tend to agglomerate over time, resulting in coalescence due to Ostwald ripening. This then gives rise

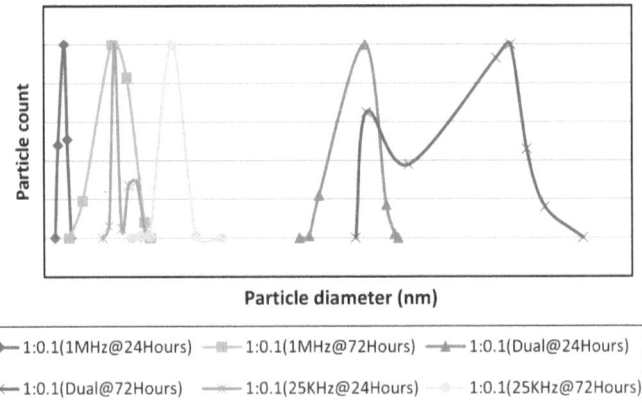

FIGURE 12.14a   Particle count in neem oil nanoemulsion with oil:surfactant ratio of 1:0.1 as a function of time (24 hours, 72 hours).

to unstable nanoemulsions with higher droplet sizes (mean diameter) whose physical and chemical properties degrade from those of stable nanoemulsions. Therefore, in order to check the stability of the neem oil nanoemulsion samples, particle size was measured with a time gap of 24 hours and 72 hours from the time of formulation. If the mean diameter of each neem oil nanoemulsion sample is found to be almost the same in both measurements, the nanoemulsion sample is considered to be stable. Figures 12.14a-e show the particle size distribution graphs for the neem oil nano-emulsions with all the oil:surfactant ratios, i.e., 1:0.1, 1:0.2, 1:0.3, 1:0.4, and 1:0.5, respectively, at two-time intervals, at all frequencies.

From Figures 12.14a and 12.14d, it can be observed that mean droplet diameter and size distribution for neem oil nanoemulsion samples with oil:surfactant ratio of 1:0.1 and 1:0.4 vary over a period of 72 hours, i.e., mean droplet diameter and size

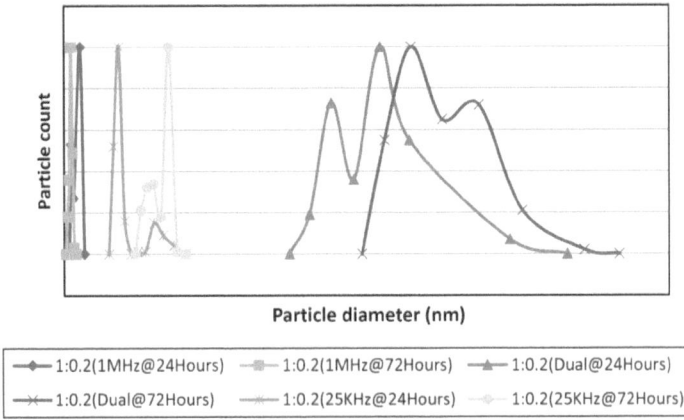

FIGURE 12.14b   Particle count in neem oil nanoemulsion with oil:surfactant ratio of 1:0.2 as a function of time (24 hours, 72 hours).

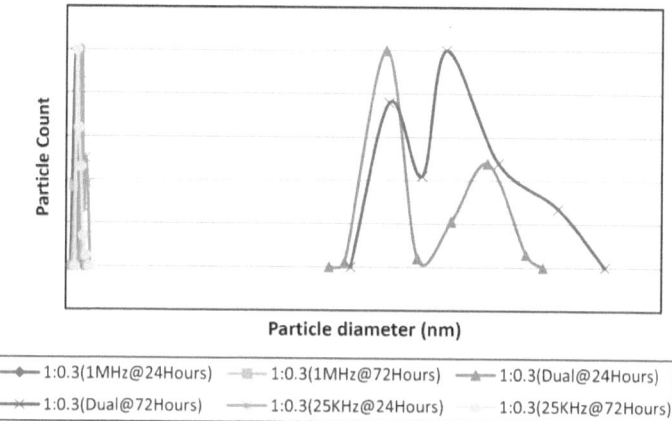

FIGURE 12.14c Particle count in neem oil nanoemulsion with oil:surfactant ratio of 1:0.3 as a function of time (24 hours, 72 hours).

distribution for neem oil nanoemulsion samples with oil:surfactant ratio of 1:0.1 and 1:0.4 do not remain constant or nearly constant with time, resulting in the formation of unstable neem oil nanoemulsions. The main reason behind such instability in the neem oil nanoemulsion sample with oil:surfactant ratio of 1:0.1 is due to the low concentration of the surfactant, which is not sufficient to form a thin film covering the oil droplets completely. As a result, the interfacial surface tension tends to increase over time and finally, the droplets coalesce to form larger size droplets, thus causing a higher mean size. It could also be observed in Figures 12.14(b), 12.14(c), and 12.14(e) that the mean droplet diameter and size distribution for neem oil nanoemulsion samples with oil:surfactant ratio of 1:0.2, 1:0.3, and 1:0.5 do not vary over a period of 72 hours, i.e., mean droplet diameter and size distribution for neem oil

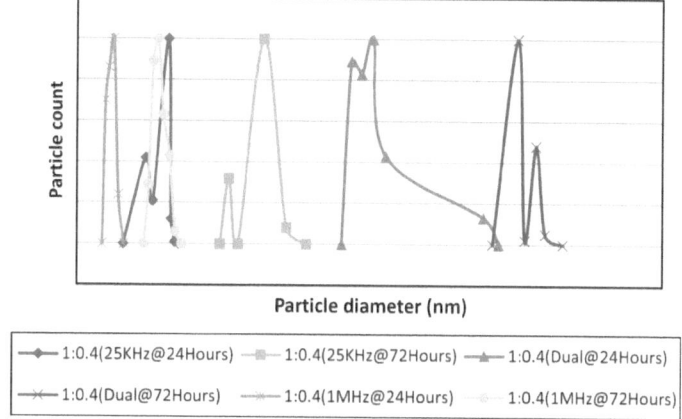

FIGURE 12.14d Particle count in neem oil nanoemulsion with oil:surfactant ratio of 1:0.4 as a function of time (24 hours, 72 hours).

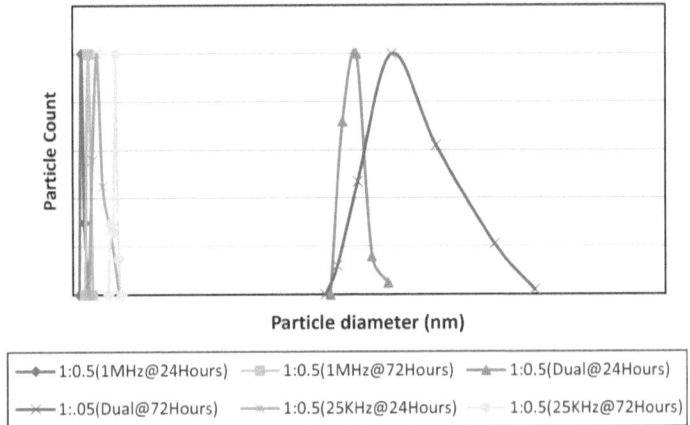

FIGURE 12.14e Particle count in neem oil nanoemulsion with oil:surfactant ratio of 1:0.5 as a function of time (24 hours, 72 hours).

nanoemulsion samples with oil:surfactant ratio of 1:0.2, 1:0.3, and 1:0.5 remain constant or nearly constant with time, resulting in the formation of stable neem oil nanoemulsions. Thus, the stability test results (Figure 12.14(a–e)) show that the neem oil nanoemulsions with oil:surfactant ratio of 1:0.1 and 1:0.4 do not form stable nanoemulsions, while neem oil nanoemulsions with oil:surfactant ratios of 1:0.2, 1:0.3, and 1:0.5 form stable nanoemulsions.

Morphology and structure of neem oil nanoemulsion at a micro-scale was studied by simple optical microscope. Figure 12.15(a) shows the microstructure of neem oil nanoemulsion with a mean droplet size of around 200 nm, captured with an objective lens of resolution 100X. Figure 12.15(b) shows the image of the same neem

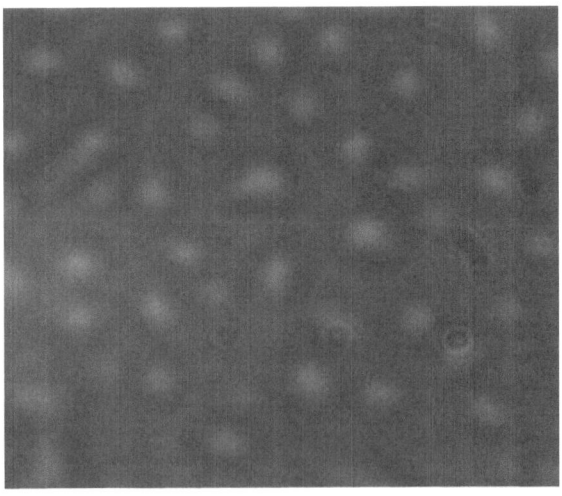

FIGURE 12.15a 100X image of neem oil nanoemulsion.

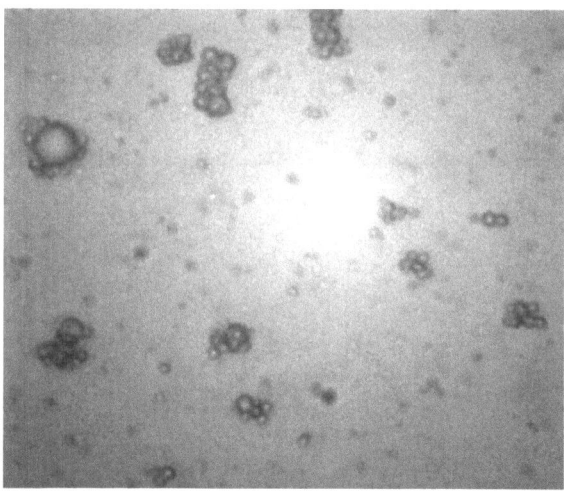

FIGURE 12.15b   50X image of neem oil nanoemulsion.

oil nanoemulsion sample seen with an objective lens of resolution 50X. From these images, it can be seen that the droplets are spherical in shape and are quite uniformly distributed in the sample.

Morphology and structure of neem oil nanoemulsion at a nano-scale was studied by Atomic Force Microscope (AFM). Only one neem oil nanoemulsion sample was used for AFM analysis to confirm the size range given by DLS. Figures 12.16a and 12.16b show the 1D AFM image and 3D AFM image, respectively, of a neem oil nanoemulsion sample whose, mean droplet size was estimated to be 154.4 nm by dynamic light scattering instrument. The AFM analysis revealed that nanoemulsion droplets were spherical in shape and discrete with size in nanometer range (~141 nm).

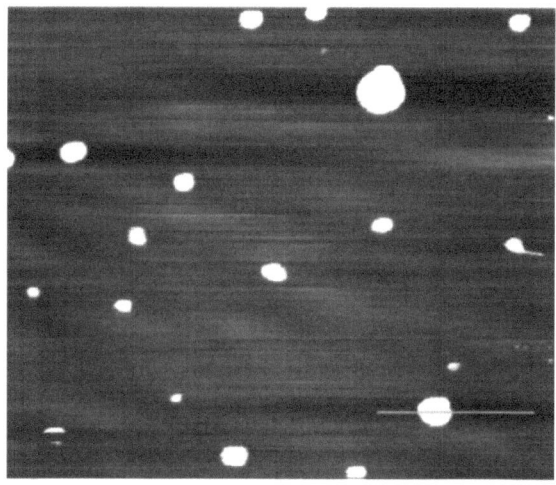

FIGURE 12.16a   AFM image of neem oil nanoemulsion.

FIGURE 12.16b   AFM 3D image of neem oil nanoemulsion.

Optimization results suggested that the frequency at which the acoustic field is applied plays a dominating role, and is the most significant parameter in causing size degradation of neem oil nanoemulsion droplets, followed by operational power, oil:surfactant ratio, time of emulsification and temperature of neem oil nanoemulsions. The optimum experimental condition for obtaining a neem oil nanoemulsion with minimum mean droplet size was found to be acoustic field frequency of 1 MHz, 100% operational power, i.e., 1200 W, oil:surfactant ratio of 1:0.5, emulsification time of 4 hours and initial temperature of neem oil nanoemulsion of 45°C.

The results also indicated that oil:surfactant ratio, frequency of the acoustic field, and the operational power are all significant parameters in reducing the oil-water interfacial surface tension in neem oil nanoemulsions, followed by a much smaller influence of time of emulsification and initial temperature of neem oil nanoemulsion. Among all the three parameters, oil:surfactant ratio is the most influencing parameter in causing reduced surface tension. The optimum experimental condition for obtaining a neem oil nanoemulsion with minimum surface tension is an acoustic field frequency of 1 MHz, 100% operational power, i.e., 1200 W, oil:surfactant ratio of 1:0.5, emulsification time of 4 hours, and initial temperature of neem oil nanoemulsion as 55°C.

The optimum condition for the zone of inhibition against *E. coli* was determined as: oil:surfactant ratio of 1:0.5, 85% operational power, dual (58 + 192 kHz) acoustic field frequency, emulsification time of 3 hours and initial sample temperature of 45°C. Similarly, the optimum condition for the zone of inhibition against *S. aureus* was determined as: acoustic field frequency at 1 MHz, oil:surfactant ratio of 1:0.5, 70% operational power, initial sample temperature of 55°C and emulsification time of 2 hours.

Finally, the results of the confirmatory experiments by Bhattacharya (2013) for quantitative and qualitative characterizations (mean droplet size and surface tension) and functional characterization (antimicrobial activity) were in good agreement with the expected values at the optimum conditions.

REFERENCES

Anton, N., J. Benoit, and P. Saulinder, "Design and Production of Nanoparticles Formulated from Nano-Emulsion Templates", *J. Controll. Release.*, vol. 128, no. 3, 2008, pp. 185–199.

Bhattacharya, S., "Acoustically-Enhanced Formulation of Neem Oil Nanoemulsions: Optimization of Process Parameters by Quantitative and Qualitative Characterization", M.S. Thesis, Indian Institute of Technology Madras, 2013.

Canselier, P., H. Delmas, A.M. Wilhelm, and B. Abismail, "Ultrasonic Emulsification: An Overview", *J. Disper. Sci. Technol.*, vol. 23, no. 1–3, 2002, pp. 333–349.

Capek, I., "Degradation of Kinetically-Stable O/W Emulsions", *Adv. Colloid. Interfac.*, vol. 107, 2004, pp. 102–110.

Coakley, W.T., J.J. Hawkes, M.A. Sobankski, C.M. Cousins, and J. Spengler, "Analytical Scale Ultrasonic Standing Waves Manipulation of Cells and Microparticles", *Ultrasonics*, vol. 38, 2000, pp. 638–641.

Dickinson, E., "Hydrocolloids at Interfaces and the Influence on Their Properties of Dispersion Systems", *Food Hydrocolloid.*, vol. 17, 2003, pp. 25–39.

Dickinson, E., M. Golding, and M.J.W. Povey, "Creaming and Flocculation of Oil-in-Water Emulsions Containing Sodium Caseinate", *J. Colloid Interf. Sci.*, vol. 185, 1998, pp. 515–529.

El-Aasser, M.S. and E.D. Sudol, "Miniemulsions: Overview of Research and Applications", *J. Coat. Technol. Res.*, vol. 1, 2004, pp. 21–31.

Gaikwad, S.G. and A.B. Pandit, "Ultrasound Emulsification: Effect of Ultrasonic and Physicochemical Properties on Dispersed Phase Volume and Droplet Size", *Ultrason. Sonochem.*, vol. 15, 2008, pp. 554–563.

Izquierdo, P., J. Esquena, T.F. Tadros, C. Dederen, M.J. Gracia, N. Azemer, and C. Solans, "Formation and Stability of Nanoemulsion Prepared Using the Phase Inversion Temperature Method", *Langmuir*, vol. 18, no. 1, 2002, pp. 6594–6598.

Jafari, S.M., Y. He, and B. Bhandari, "Nano-Emulsion Production by Sonication and Microfluidization: A Comparison", *Int. J. Food Prop.*, vol. 9, 2006, pp. 475–485.

Jena, S. and H. Das, "Modeling of Particle Size Distribution of Sonicated Coconut Milk Emulsion: Effect of Emulsifiers and Sonication Time", *Food Res. Int.*, vol. 39, no. 5, 2006, pp. 606–611.

Kentish, S., T.J. Wooster, M. Ashokkumar, S. Balachandran, R. Mawson, and L. Simmons, "The Use of Ultrasonics for Nanoemulsion Preparation", *Innov. Food Sci. Emerg.*, vol. 9, 2008, pp. 170–175.

Kralova, I. and J. Sjoblom, "Surfactants Used in Food Industry: A Review", *J. Disper. Sci. Technol.*, vol. 9, 2009, pp. 1363–1383.

Leong, T.S.H., T.J. Wooster, S.E. Kentish, and M. Ashokkumar, "Minimizing Oil Droplet Size Using Ultrasonic Emulsification", *Ultrason. Sonochem.*, vol. 16, 2009, pp. 721–727.

Li, P.H. and B.H. Chiang, "Process Optimization and Stability of D-Limonene-in-Water Nanoemulsions Prepared by Ultrasonic Emulsification Using Response Surface Methodology", *Ultrason. Sonochem.*, vol. 19, 2012, pp. 192–197.

Li, M.K. and H.S. Fogler, "Acoustic Emulsification. Part 1: The Instability of the Oil-Water Interface to Form Initial Droplets", *J. Fluid Mech.*, vol. 88, 1978a, pp. 499–511.

Li, M.K. and H.S. Fogler, "Acoustic Emulsification. Part 2: Breakup of the Large Primary Oil Droplets in a Water Medium", *J. Fluid Mech.*, vol. 88, 1978b, pp. 513–528.

Mason, T.J., J.N. Wilking, K. Meleson, C.B. Chang, and S.M. Graves, "Nanoemulsions: Formation, Structure and Physical Properties", *J. Phys. Condens. Mat.*, vol. 18, 2006, pp. 635–666.

McClements, D.J., *Food Emulsions: Principles, Practices and Techniques,* 2nd edition, CRC Press, Boca Raton, FL, 2004.

Ostertag, F., J. Weiss, and D.J. McClements, "Low-Energy Formation of Edible Nanoemulsions: Factors Influencing Droplet Size by Emulsion Phase Inversion", *J. Colloid Interf. Sci.*, vol. 388, 2012, pp. 95–102.

Rao, J. and D.J. McClements, "Lemon Oil Stabilization in Mixed Surfactant Solutions: Rationalizing Microemulsion & Nanoemulsion Formation", *Food Hydrocolloid.*, vol. 26, 2012, pp. 268–276.

Richards, W.T., "The Chemical Effects of High Frequency Sound Waves—A Study of Emulsifying Action", *J. Am. Chem. Soc.*, vol. 51, 1929, pp. 1724–1729.

Sajjadi, S., "Nanoemulsion Formation by Phase Inversion Emulsification: On Nature of Inversion", *Langmuir*, vol. 22, 2006, pp. 5597–5603.

Seekkuarachchi, I.N., K. Tanaka, and H. Kumazawa, "Formation and Characterization of Submicrometer Oil-in-Water (o/w) Emulsions Using High-Energy Emulsification", *Ind. Eng. Chem. Res.*, vol. 45, 2006, pp. 372–390.

Sharma, N., M. Bansal, S. Visht, P.K. Sharma, and G.T. Kulkarni, "Nanoemulsion: A New Concept of Delivery System", *Chronicl. Young Sci.*, vol. 1, 2010, pp. 2–6.

Solans, C., J. Esquena, and A.M. Forgiarini, "Nanoemulsion: Formation, Properties and Applications", *Surf. Sci. Ser.*, vol. 109, 2003, pp. 525–554.

Sonneville-Auburn, O., J.T. Simonnet, and F.L. Alloret, "Nanoemulsions: A New Vehicle for Skincare Products", *Adv. Colloid. Interfac.*, vol. 108–109, 2004, pp. 145–149.

Spengler, J. and M. Jekel, "Ultrasound Conditioning of Suspensions—Studies of Streaming Influence on Particle Aggregation on Lab and Pilot-Plant Scale", *Ultrasonics*, vol. 38, 2000, pp. 624–628.

Strawbridge, K.B., E. Ray, F.R. Hallett, S.M. Tosh, and D.G. Dalgleish, "Measurement of Particle Size Distribution in Milk Homogenized by a Microfluidizer: Estimation of Population of Particles With Radii Less than 100 Nm", *J. Colloid Interf. Sci.*, vol. 171, 1995, pp. 392–398.

Tadros, T.F., P. Izquierdo, J. Esquena, and C. Solans, "Formation and Stability of Nanoemulsion", *Adv. Colloid. Interfac.*, vol. 171, 2004, pp. 303–318.

Tal-Figiel, B., "Conditions for the Instability of Liquid-Liquid Interface in an Ultrasonic Field", *Int. Chem. Eng.*, vol. 30, 1990, pp. 526–534.

Tal-Figiel, B., "Emulsions in Cosmetics and Medicines", *Chem. Process Eng.*, vol. 27, 2005, pp. 403–409.

Tal-Figiel, B., "The Formation of Stable W/O, O/W, W/O/W Cosmetic Emulsions in an Ultrasonic Field", *Chem. Eng. Res. Des.*, vol. 85, no. 5, 2007, pp. 730–734.

Troncoso, E., J.M. Aguilera, and D.J. McClements, "Fabrication, Characterization and Lipase Digestibility of Food-Grade Nanoemulsions", *Food Hydrocolloid.*, vol. 27, 2012, pp. 355–363.

Uson, N., M.J. Garcia, and C. Solans, "Formation of Water-in-Oil (W/O) Nano-Emulsions in a Water/Mixed Non-Ionic Surfactant/Oil Systems Prepared by a Low Energy Emulsification Method", *Colloid. Surface. A*, vol. 250, 2004, pp. 415–421.

Wang, L., X. Li, G. Zhang, J. Dong, and J. Eastoe, "Oil-in-Water Nanoemulsions for Pesticide Formulations", *J. Colloid Interf. Sci.*, vol. 314, 2007, pp. 230–235.

Woods, R.W. and A.L. Loomis, "The Physical and Biological Effects of High-Frequency Sound-Waves of Great Intensity", *Phil. Mag.*, September, vol. 4, 1927, pp. 417–436.

Wulff-Perez, M., A. Torcello-Gomez, M.J. Galvez-Ruiz, and A. Martin-Rodriguez, "Stability of Emulsions for Parenteral Feeding: Preparation and Characterization of o/w Nanoemulsions with Natural Oils and Pluronic f68 as Surfactant", *Food Hydrocolloid.*, vol. 23, 2009, pp. 1096–1102.

13 Enhanced Oil Recovery
Case Study

To meet the global need of energy, crude oil, which is a limited source, is very essential. From time to time, the availability of crude oil has continued to reduce. This has resulted in serious oil crises accompanied by the rise in oil prices. This, in turn, has forced the industry to find new efficient ways to recover oil, using advanced techniques such as enhanced oil recovery processes. While these advanced techniques are being used today, they are also going through further advancement and development. Around two-thirds of crude oil remains trapped in the pore structure of conventional oil reservoirs after primary and secondary oil recovery processes (Rosen et al., 2005). Enhanced oil recovery (EOR) is required to recover the remaining oil. EOR processes increase the overall oil recovery to approximately 30%–60%.

EOR involves the application of methods or techniques in which extrinsic energy and materials are added to a reservoir to control:

- wettability
- interfacial tension
- fluid properties
- pressure gradients necessary to overcome retaining forces
- movement of remaining crude oil towards a production well

In order to assess the necessity for EOR, the determination of oil reserves must be made. Reserves refer to the amount of oil that can be obtained from a reservoir under existing economics and technology, which is given by the following material balance equation (Rashmi, 2016):

Present reserves = Past reserves + Additions to reserves − Production from reserves

$$(13.1)$$

In order to maintain oil reserves, large fields must be discovered, new wells drilled, or other techniques implemented to increase the percentage of recovery from known reservoirs. The probability of finding large fields is declining, making it necessary to increase the percentage of recovery from the presently known reserves using practical solutions; this can be possible only by applying EOR methods.

In general, the EOR processes involve the injection of fluids into the oil reservoir, displacing crude oil from the reservoir toward a production well. The injection processes supplement the natural energy present in the reservoir. The injected fluid also interacts

DOI: 10.1201/9781003283423-13

with rock and oil trapped in the reservoir creating favorable conditions for oil recovery. The classification of various EOR processes is given by Green and Willhite (1998) as:

- mobility-control
- chemical processes
- miscible processes
- thermal processes
- other (e.g., microbial EOR)

Mobility-control is a process based on maintaining favorable mobility ratios between crude oil and water by increasing water viscosity and decreasing its relative permeability. Sweep efficiency over water flooding can be improved during chemical processes.

Chemical processes involve the injection of a specific liquid chemical that effectively creates desirable phase behavior properties to improve oil displacement. Surfactant flooding is an example of chemical flooding. It is a complex process, where the displacement phase is immiscible, as water or brine does not mix with oil. However, this condition is changed by the addition of surfactants. This technique creates low interfacial tension (IFT), where an ultra-low IFT (0.001 mN/m) between the displacing fluid and the oil is a requirement in order to mobilize the residual oil. The liquid surfactant injected into the reservoir is often a complex chemical system, one which creates a micelle solution. During surfactant flooding, it is essential that the complex system forms micro-emulsions with the residual oil, as this supports the decrease of the IFT and increases mobility. There are also other chemical processes that have been developed, such as alkaline flooding and various processes where alcohols are introduced. In alkaline flooding, alkaline chemicals are injected into the reservoir, where they react with certain components in the oil to generate surfactants *in situ*. Alcohol processes have only been tested in laboratories so far and have not yet been applied in the field.

Miscible processes are based on the injection of a gas or fluid, which is miscible with the crude oil at reservoir conditions, in order to mobilize the crude oil in the reservoir.

Thermal processes are typically applied to heavy oils. These processes rely on the use of thermal energy. A hot phase of, e.g., steam, hot water, or combustible gas is injected into the reservoir in order to increase the temperature of the trapped oil and gas and thereby reduce the oil viscosity.

Surfactant flooding involves the injection of one or more liquid chemicals and surfactants. The injection effectively controls the phase behavior properties in the oil reservoir, thus mobilizing the trapped crude oil by lowering IFT between the injected liquid and the oil. Surfactants are added to decrease the IFT between oil and water. Certain physical characteristics of the reservoir, such as adsorption to the rock and trapping of the fluid in the pore structure, might lead to considerable losses of the surfactant. The stability of the surfactant system at reservoir conditions is also of great relevance. Surfactant systems are sensitive to high temperature and high salinity, and therefore, surfactants that can resist these conditions should be used (Green and Willhite, 1998). Polymers are also often added to the injected surfactant solution, to increase viscosity, thus maintaining mobility control. In general, there are three types of surfactant flooding for EOR as shown in Table 13.1 (Rosen et al., 2005).

In the work of Rashmi (2016), anionic surfactant sodium dodecyl sulfate (SDS) and cationic surfactant cetyltrimethylammonium bromide (CTAB) were taken

TABLE 13.1

Types of Surfactant Flooding

Type of Surfactant Flooding	Technique	Note
Micelle/polymer flooding	A micelle slug usually of surfactant, co-surfactant, alcohol, brine, and oil is injected into the reservoir.	Displacement efficiency close to 100% (measured in laboratory)
Micro-emulsion flooding	Surfactants, co-surfactants, alcohol, and brine are injected into the reservoir to form micro-emulsions to obtain ultra-low IFT.	Can be designed to perform well in high temperature or salinity or low permeable areas where polymer and/or alkali cannot work
Alkaline/surfactant/ polymer (ASP) flooding	The addition of alkaline chemicals reduces the IFT at significantly lower surfactant concentrations.	Lower concentration of surfactants involved in this process, which reduces the cost of chemicals

individually, and compared with their mixtures in the molar ratios of 0.3:0.7, 0.5:0.5, and 0.7:0.3. The combinations of surfactants were analyzed via IFT and EOR studies.

13.1   EXPERIMENTAL SETUP AND METHODOLOGY

The properties of the waxy crude used in the study are listed in Table 13.2.

Brine with a salinity of 3 wt.% NaCl was used for pre-flush (to clean the core samples), water flooding, and extended chase flooding. De-ionized water was used to prepare brine. NaCl (purity ≥ 99%) with a molecular weight of 58.44 g/mol was purchased from Merck Specialties Pvt. Ltd, Mumbai, India. Cationic surfactant, hexadecyltrimethylammonium bromide, with a purity of ≥99% and a molecular weight of 364.45 g/mol, was purchased from Sigma Aldrich Chemicals Pvt. Ltd, Bangalore, Karnataka, India. Anionic surfactant SDS with a purity ≥85% and molecular weight of 288.38 g/mol was purchased from Sisco Research Laboratories, Mumbai, Maharashtra, India. Polyacrylamide (PAM with a purity of ≥ 80%) with a molecular weight of 10^7 g/mol, used for polymer flooding, was purchased from Triveni Global Pvt. Ltd., Chennai, Tamil Nadu, India. The properties of PAM are listed in Table 13.3.

The sand particles were characterized using X-ray diffraction and Scanning Electron Microscopy techniques. Powder X-ray diffractometer (Bruker D8 Advance

TABLE 13.2

Bombay High Crude Properties

Specific gravity	0.8286
API gravity	39.3
Viscosity @ 40°C, cSt	2.70
Pour point, °C	30
Sulfur, wt.%	0.12
Ni, ppm	1.7
V, ppm	0.2
Acidity, mg KOH/g	0.12

TABLE 13.3

Properties of PAM

Product	Appearance	Bulk Density	pH of 1 % Solution at 25 °C	Anionic Charge	Apparent Viscosity (cP) at 25°C at Conc. Shown (%)		
					0.25	0.5	1.0
ZETAG 4100	Off-white, granular solid	Approx. 0.7 g/cm³	6.0–9.0	very low	0.50	200	1,100

instrument, with a reproducibility of ± 0.0001°C and an angular range (2θ) of −10° to 90°, NaI dynamic Scintillation detector with a maximum count rate of 2×10^6 s⁻¹) was used to estimate the minerals content of the sand used.

High-resolution Scanning Electron Microscopy (S4800 Type I, Hitachi, Japan, with an accelerating voltage of 0.5–30 kV (normal mode), with a magnification of 100–800,000×, and probe current of 1 pA-2 nA, finely adjustable, image processor up to 5120 × 3840 pixels) was used to identify the size and structure of the sand particles.

Surfactant stock solutions of desired concentration were prepared by dissolving weighed amounts of dried surfactant powder in distilled water. Samples were prepared by mixing stock solutions of SDS and CTAB in the desired ratio (0.5:0.5, 0.3:0.7, and 0.7: 0.3) for the IFT analysis (You et al., 2009). The used chemicals were weighed using an accurate RADWAG electronic balance (RADWAG Wagi Elektroniczne, Poland, with a repeatability of ±0.1 mg and readability of 0.1 mg). The surfactant solutions were prepared using a magnetic stirrer (IKA, Bangalore, Karnataka) with a speed range of 0–2500 rpm.

Schematic representation of the experimental setup used by Rashmi (2016) for the determination of crude oil-water interfacial tension is shown in Figure 13.1 (a–c).

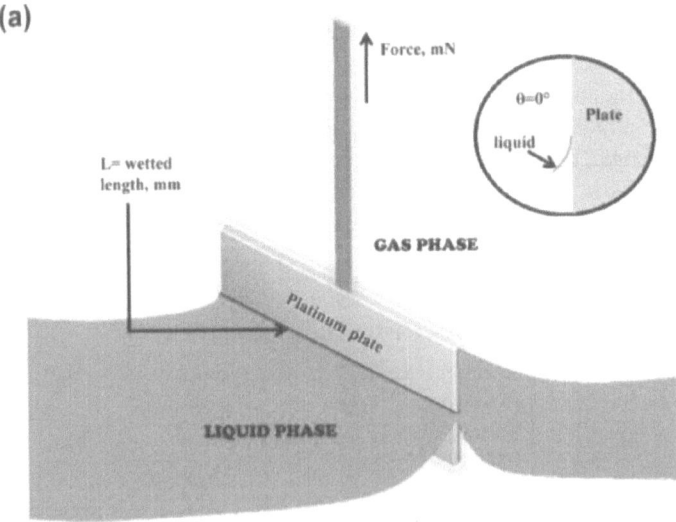

FIGURE 13.1 (a) Schematic diagram of the Wilhelmy plate at the liquid-vapor (gas) interface. (*Continued*)

(b)

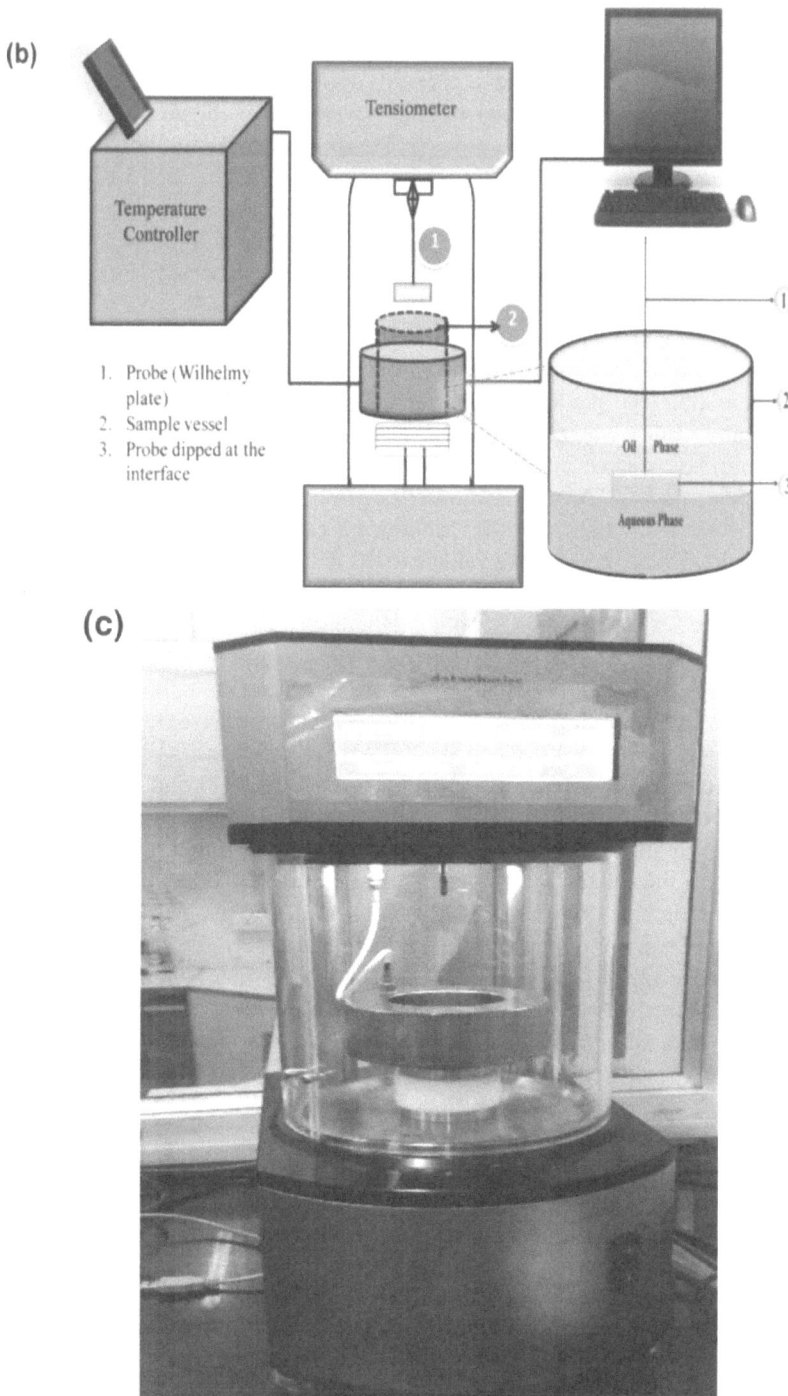

1. Probe (Wilhelmy plate)
2. Sample vessel
3. Probe dipped at the interface

Temperature Controller

Tensiometer

Oil Phase

Aqueous Phase

(c)

FIGURE 13.1 (*Continued*) (b) Schematic representation of the experimental setup for the determination of crude oil-water interfacial tension [27]. (c) Dataphysics tensiometer setup.

Dynamic contact angle tensiometer (Dataphysics DCAT 11EC, Germany) was used to measure the IFT between the crude oil-water-surfactant system wherein the Wilhelmy platinum-iridium plate of type PT- 11 having a thickness of 0.2 mm and area of 3.98 mm^2 with an accuracy of ±1.5% was used as the probe. IFT of water-crude oil, SDS aqueous solution- crude oil, CTAB aqueous solution-crude oil, a mixture of surfactants (SDS:CTAB)—crude oil was measured at 353.15 K maintained using Brookfield water bath (Model number: TC-650AP with operating temperature range of −20°C to +200°C). Before the measurements, the mixture of the individual systems mentioned above was stirred using a magnetic stirrer at 500 rpm for about 15 minutes at the temperature required for the experiment.

The sand pack reactor, also referred to as core flood reactor, had a bulk volume of 649 cm^3 with dimensions of 5.25 cm diameter and 30 cm length. Before column preparation, the reactor was cleaned thoroughly using solvent (hexane and acetone), and further with distilled water to get rid of any contaminants. Afterward, the column was packed compactly (rammed) using clean and dried silica sand (0.4–0.8 mm) with the gradual addition of distilled water. The water used to pack the sand pack column tight and 100% saturated is called the pore volume (PV) of the porous medium. The ratio of pore volume to the bulk volume is defined as the porosity of the particular sand pack column (in percentage). Then, the core flood reactor was maintained at the desired experimental temperature (353.15 K) for 1 hour and then used for the EOR studies.

Surfactant solutions were prepared with concentration three times of the critical micelle concentration value as obtained from IFT analysis. Five surfactant solutions were prepared, i.e., SDS, CTAB, and SDS:CTAB (0.5:0.5, 0.3:0.7, 0.7:0.3), using stock solutions in distilled water and by stirring vigorously at 1000 rpm for 1 hour. PAM solution of 5000 ppm concentration was prepared using powdered PAM in distilled water with vigorous and constant stirring at 1000 rpm using a mechanical stirrer for 24 hours. Brine solution was prepared using NaCl of 30,000 ppm concentration.

The experimental setup used by Rashmi (2016) for the EOR studies and its schematic are shown in Figures 13.2 and 13.3. The experimental setup consists of a syringe pump, a horizontal core-flood reactor, and a water bath and outlet collector flask. A syringe pump (D-series Syringe pump, Teledyne ISCO, of 500D model with a maximum capacity of 508.7 ml) was used for the injection of fluids such as brine, crude oil, polymer solution, surfactants, and aqueous solutions. The temperature required, i.e., 353.15 K for the EOR studies, was maintained using Brookfield water bath (Model number: TC-650AP with operating temperature range of −20°C to +200°C). The sand pack column was pre-flushed using brine of 30,000 ppm to ensure the uniformity of water saturation. During all the experiments, a constant flow rate of 5 ml/min was maintained. The absolute permeability (k_w) of the fluid (brine) was calculated using Darcy's law during the pre-flushing. Using a syringe pump, the pressure difference (ΔP) between the inlet and outlet of the core flood reactor was measured periodically for every 0.1 PV fluid injection and further used for permeability (k_w) calculation.

The core flood reactor was then subjected to waxy crude oil injection till there was no further production of water at the outlet. The amount of water collected (which is equal to the amount of crude oil saturated in the core flood reactor) during

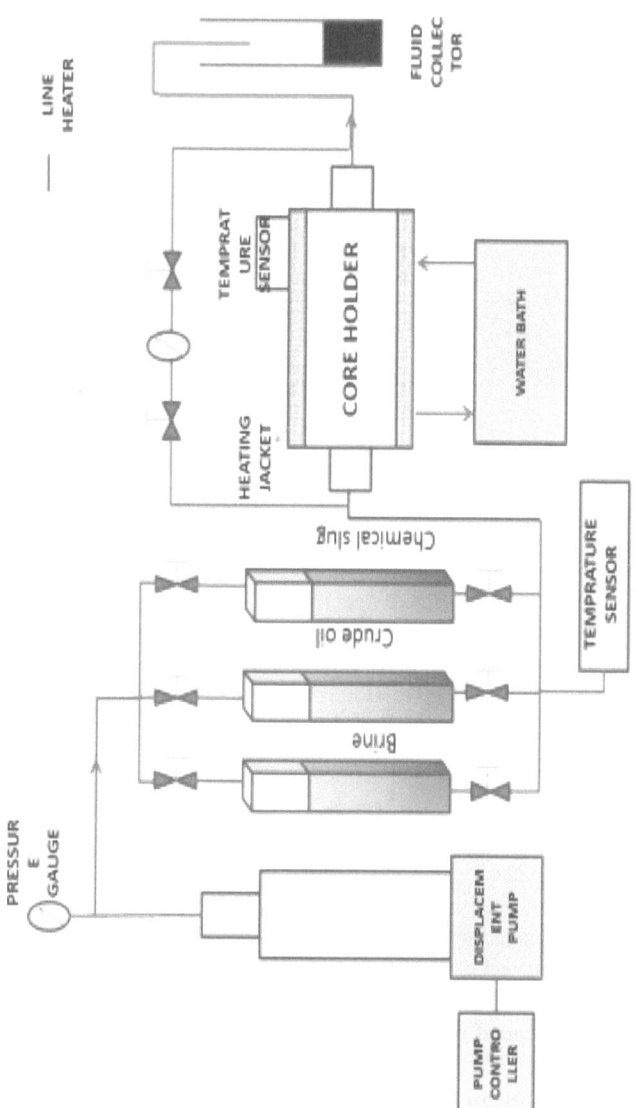

FIGURE 13.2 Schematic of the experimental setup used for EOR studies.

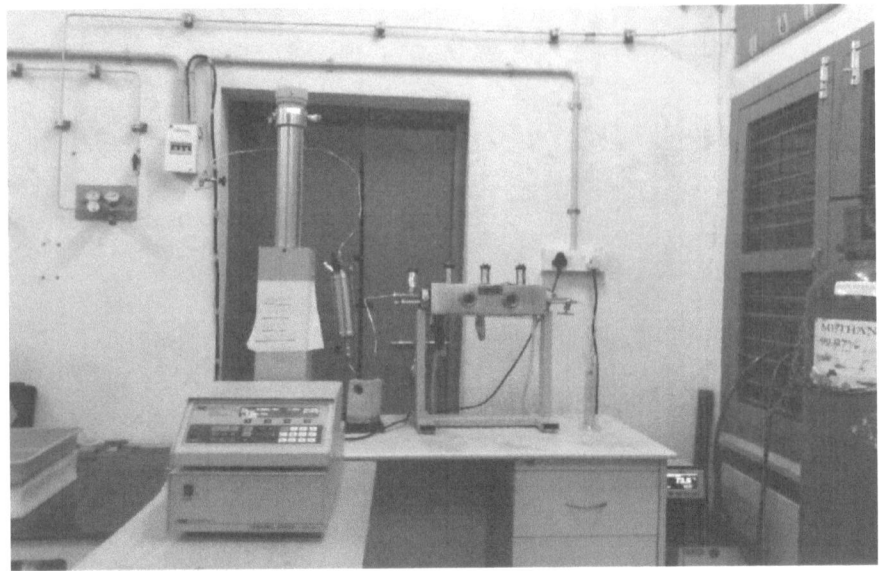

FIGURE 13.3   Experimental setup used for EOR studies.

the oil injection is termed "original oil in place (OOIP)". The initial water saturation and initial oil saturation were calculated using Equations (13.2) and (13.3):

$$S_{wi} = \frac{PV - OOIP}{PV} \times 100 \tag{13.2}$$

$$S_{oi} = \frac{OOIP}{PV} \times 100 \tag{13.3}$$

The effective permeability of the crude oil (k_o) during injection was measured using Darcy's law. The viscosity of crude oil and brine was measured using the Brookfield viscometer (model: DV2TLV using Spindle: SC4 - 18 with a measuring range: 1.5–30,000 cP) (Sakthivel et al., 2016).

The secondary oil recovery was mimicked using brine flood; 2 PV of brine was injected in the oil-saturated core flood reactor for the recovery of crude oil until there is no further production and water cut reaches 100%. The residual oil, or the amount of oil left in the core flood reactor after brine flood, is the objective function to be minimized for the enhanced oil recovery studies using various injection fluids, i.e., surfactant slug flooding (0.5 PV), subsequent polymer flooding (0.5 PV) followed by chase flood (2 PV) (Sakthivel et al., 2016).

13.2   RESULTS AND DISCUSSION

The enhanced oil recovery performance of the individual surfactants and combination systems was evaluated in sand pack flooding tests. Prior to surfactant flooding, water flooding was carried out as the secondary oil recovery to generate the residual oil-saturated porous media. The results show that after injecting 2 PV of brine, the water cut of the effluent was 100%. A surfactant solution (0.5 PV) was then injected in the enhanced oil recovery mode, followed by polymer flooding and then extended brine flooding (2 PV). The EOR experiments were carried out at 353.15 K to mimic reservoir conditions (details have been shown in Tables 13.4 and 13.5). Table 13.4 shows the various petro-physical properties of the sand-pack column used for the enhanced oil recovery experiment, such as initial water saturation (S_{wi}), initial oil saturation (S_{oi}), porosity (Φ) varying from 32.36% to 34.82%, permeability (k) varying from 167 to 197 mD, and OOIP varying from 115 to 130 ml. The S_{oi} (initial oil saturation) has been found to be in the range of 51%–50% for various experiments. Brine flood (2 PV) has further reduced S_{oi} and the subsequent residual oil saturation (S_{or}) has been found to be in the range of 22%–28% of the OOIP, thus resulting in the additional oil recovery of 50%–60% due to water flood. The high recovery of oil during brine floods is due to the high porosity and permeability of the sand pack. This S_{or} (residual oil saturation) represents the target for EOR studies for various experiments (Table 13.5).

Cumulative oil recovery results for the EOR experiments have been shown in Table 13.5. In the EOR experiments, cumulative oil recovery was found to be 65.59% and 69.46% for SDS and CTAB (neat), respectively, which is increased to 84.95%, 79.50%, and 90.09% for SDS (0.3): CTAB (0.7), SDS (0.5): CTAB (0.5), and SDS (0.7): CTAB (0.3), respectively (Bera et al., 2014; Ko et al., 2014). It has been observed that there is a considerable increment in the values of cumulative oil recovery using the combination of surfactants. But during the polymer recovery, the oil recovery increased from around 6% to 18%. This increment is shown during the chase brine flood, but again after some time, the recovery starts to decrease and finally reaches 0% and the water cut reaches 100%. During the chase water flood, it was found that 7%–10% of the oil was recovered in the case of SDS and CTAB individually but in the case of individual surfactants around 6%–26% oil was recovered. The surfactant effect can be assumed to be there till the extended brine flooding, and therefore the effect of the combination of surfactant systems can be assessed by taking into account total oil recovery.

Figure 13.4 (a–e) shows the cumulative oil recovery and water cut for the various surfactant systems. It may be observed from the graph that the oil recovery first increases with injected fluid during brine flood, but later reaches 0% and water cut reaches 100%. Hence, after 2 PV of brine flood, surfactant flood was started but did not lead to a significant change in the oil recovery; it remained in the range of 0.5%–2%.

TABLE 13.4

Initial Petro-Physical Properties of the Various Sand-Pack Columns Used in This Investigation

Expt. no./Sand pack no.	Surfactant System	OOIP (ml)	Porosity (%)	S_{wi} (%)	S_{oi} (%)	Permeability (mD)	
						k_w (S_w=1)	k_o (S_{wi})
1	SDS	121	34.51	43.30	56.70	544	184
2	CTAB	130	34.82	42.48	57.52	438	169
3	SDS(0.3):CTAB(0.7)	122	32.36	41.90	58.10	396	192
4	SDS(0.5):CTAB(0.5)	132	33.90	40	60	472	167
5	SDS(0.7):CTAB(0.3)	115	34.67	48.89	51.11	504	189

TABLE 13.5
Summary of the Results of Various EOR Experiments Performed

Expt. no.	Type of Flood	S_{oi} (%)	Residual Oil Saturation S_{or} (%)				Water Flood Recovery (% OOIP)	Additional Oil Recovery (% OOIP)				
			After Brine Flood (% OOIP)	After Surfactant Solution Flood (% OOIP)	After Polymer Flood (% OOIP)	After Chase Brine Flood (% OOIP)		After Surfactant Solution Flood (% OOIP)	After Polymer Flood (% OOIP)	After Chase Brine Flood (% OOIP)	Total Additional Oil Recovery (% OOIP)	Cumulative Oil Recovery (% OOIP)
1	SDS	56.7	27.81	27.14	23.62	19.51	50.94	1.18	6.22	7.24	14.65	65.59
2	CTAB	57.52	28.1	27.12	23.45	17.57	51.15	1.69	6.38	10.23	18.31	69.46
3	SDS:CTAB (3:7)	56.92	22.71	22.49	12.02	8.56	60.11	0.38	18.38	6.08	24.85	84.95
4	SDS:CTAB (5:5)	60	26.73	26.45	22.48	12.29	55.45	0.45	6.63	16.97	24.05	79.50
5	SDS:CTAB (7:3)	51.11	25.11	24.31	18.84	5.07	50.87	1.56	10.7	26.96	39.21	90.09

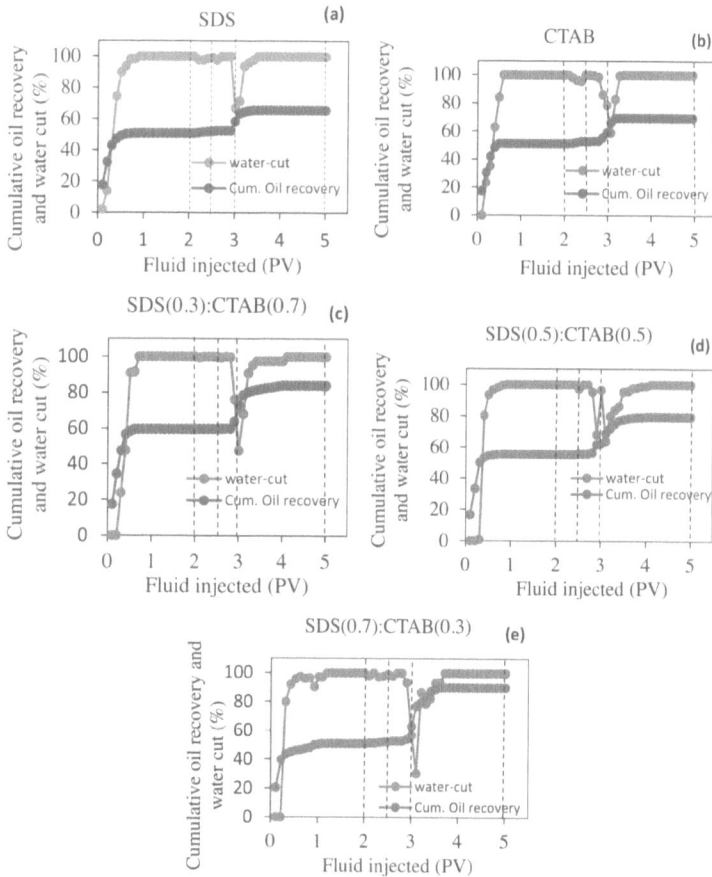

FIGURE 13.4 Variation of cumulative oil recovery and water cut (%) with fluid injected (PV) for various surfactant systems.

The total oil recoveries in the case of SDS and CTAB are 14.65% and 18.31%, respectively, but in the case of anionic-cationic (SDS:CTAB) surfactant systems, increase to 24.85%, 24.05%, and 39.21% for 3:7, 5:5, and 7:3, respectively. Cumulative oil recovery for SDS and CTAB was found to be 65.59% and 69.46%. It can be seen that CTAB performs better than SDS, but it is not widely used in the industry due to its cost. As compared to the cumulative oil recoveries of individual surfactants, anionic-cationic (SDS:CTAB) surfactant systems have much higher cumulative oil recovery, i.e., 84.95%, 79.50%, and 90.09% for 3:7, 5:5, and 7:3, respectively.

Figure 13.5 shows the pressure variation for various systems with fluid injected. The pressure drop did not change much during surfactant flooding (2–2.5 PV) also and was observed to be in the same range as water flooding. However, once the flooding of the polymer (2.3–3 PV) has been initiated, the pressure drop increased gradually in the range of 0.22–0.47 MPa, and kept increasing initially during

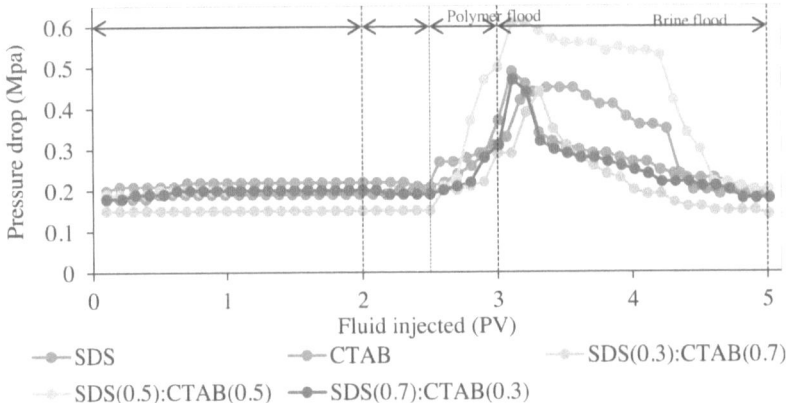

FIGURE 13.5 Graph showing pressure drop as a function of fluid injected (PV) in the core sample.

extended chase brine flood in the range of 0.47–0.61 MPa; after a certain point, it was observed to be gradually declining during the extended brine flood.

It has also been noted here that during the water flooding, the water cut increased rapidly (prompt breakthrough) and reached the maximum point (almost 100%), after which surfactant solution flooding was initiated followed by polymer injection. The water cut was reduced during the surfactant flooding, but after a certain point, it reached the maximum once again at the end of the surfactant flood and starts decreasing during polymer flood; this continues for some time during extended chase brine flood, and after a certain point, again starts increasing and reaches its maximum peak.

Figure 13.6 shows a comparison between the cumulative oil recoveries of the individual and the combination of surfactant systems. It can be observed from the

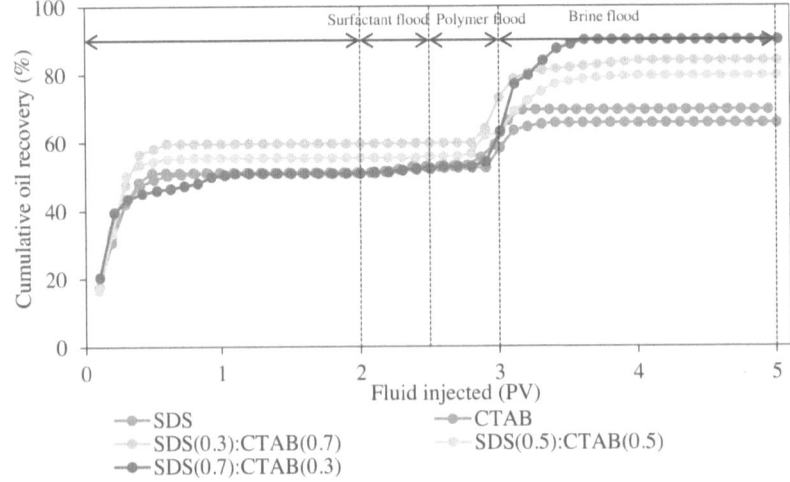

FIGURE 13.6 Comparison of cumulative oil recovery.

plot that the cumulative oil recoveries of the mixed surfactants systems are better right after the surfactant flooding starts, and this enhancement persists all the way through extended chase brine flooding.

PIF/CIF ANALYSIS

The common thread in the case studies presented in Chapters 11–13 is that process intensification is possible but comes at a cost. When several intensification options are available, the right one for the application at hand must be chosen through careful cost-benefit analysis, and a formal multi-objective optimization exercise must be undertaken. Benefit to society also needs to be a consideration. The cost of damaging Mother Earth's ecosystem may not be immediately apparent but will manifest itself over time. Running a profitable enterprise is a laudable activity, but the profits must not come at the expense of good citizenship. Corporate Social Responsibility is now a prime concern for many governments and watchdog organizations. Investors also place value on this aspect.

EXERCISE

There are several case studies reported in the literature for process intensification (e.g., Gogate, 2008; ; Keil, 2018; Lohokare et al., 2015; López-Guajardo et al., 2022). Review these and identify the Process Intensification Factors obtained in each study. Was there an attempt to estimate the Cost Impact Factor as well? If not, propose a methodology by which this can be done, and an optimization strategy for each Case based on both factors.

REFERENCES

Bera, A., A. Mandal, and B.B. Guha, "Synergistic Effect of Surfactant and Salt Mixture on Interfacial Tension Reduction between Crude Oil and Water in Enhanced Oil Recovery", *J. Chem. Eng. Data*, vol. 59, 2014, pp. 89–96.

Gogate, P.R., "Cavitational Reactors for Process Intensification of Chemical Processing Applications: A Critical Review", *Chem. Eng. Process.: Process Intensif.*, vol. 47, no. 4, 2008, pp. 515–527.

Green, D.W. and G.P. Willhite, "Enhanced Oil Recovery", *SPE Textbook Series*, vol. 6, 1998, pp. 1–10.

Keil, F.J., "Process Intensification", *Rev. Chem. Eng.*, vol. 34, no. 2, 2018, pp. 135–200; https://doi.org/10.1515/revce-2017-0085.

Ko, K.M., B.H. Chon, S.B. Jang, and H.Y. Jang, "Surfactant Flooding Characteristics of Dodecyl Alkyl Sulfate for Enhanced Oil Recovery", *J. Industrial and Engineering Chemistry*, vol. 20, 2014, pp. 739–734.

Lohokare, S.R., V.H. Bhusare, and J.B. Joshi, "Process Intensification: A Case Study", *Ind. Chem. Eng.*, vol. 57, no. 3–4, 2015, pp. 202–218.

López-Guajardoa, E.A., F. Delgado-Licona, A.J. Álvareza, K.D.P. Nigama, A. Montesinos-Castellanos, and R. Morales-Menendeza, "Process Intensification 4.0: A New Approach for Attaining New, Sustainable and Circular Processes Enabled by Machine Learning", *Chem. Eng. Process: Process Intensif.*, vol. 180, 2022, pp. 10867; https://doi.org/10.1016/j.cep.2021.108671.

Rashmi, "An Experimental Study of Mixed Surfactant System for Enhanced Oil Recovery", M.Tech. Project Report, Indian Institute of Technology Madras, 2016.

Rosen, M.J., H. Wang, P. Shen, and Y. Zhu, "Ultralow Interfacial Tension for Enhanced Oil Recovery at Very Low Surfactant Concentrations", *Langmuir*, vol. 21, 2005, pp. 3749–3756.

Sakthivel, S., S. Velusamy, R.L. Gardas, and J.S. Sangwai, "Effect of Alkyl Ammonium Ionic Liquids on the Interfacial Tension of the Crude Oil-Water System and Their Use for the Enhanced Oil Recovery Using Ionic Liquid-Polymer Flooding", *Energy Fuels*, vol. 30, 2016, pp. 2514–2523.

You, Y.L., L.S. Hao, and Y.Q. Nan, "Phase Behaviour and Viscous Properties of CTAB and SDS Aqueous Mixtures", *Colloids and Surfaces A: Physicochem. Eng. Aspects*, vol. 335, 2009, pp. 154–167.

14 In Conclusion...

Process intensification is a vast subject, with myriad facets investigated diligently by thousands of researchers. But at its core, it is a simple concept. A process has a baseline functional value that requires enhancement. Process intensification is a systematic attempt to do. The success of the effort may be defined on the basis of a "Process Intensification Factor" (PIF) which estimates the ratio of an enhanced process value to its baseline value. This may be simple, for example, a product quality or processing time parameter, or the monetized value of such a parameter—i.e., how much additional revenue would be generated at the enhanced quality value of a parameter versus at its baseline value.

The monetization approach lends itself to a direct comparison of the PIF to the Cost Impact Factor (CIF), though this may not always be feasible. Frequently, gains are intangible, whereas pains are very much tangible. In particular, when process improvements lead to societal benefits—such as pollution abatement—it is difficult to quantify the long-term revenue impact on the manufacturer. Process cost is a relatively simple and straightforward metric, one that senior management, and shareholders, can grasp the implications of rather quickly. Hence, the need to have clarity of thinking regarding process intensification vis-à-vis the associated cost impact. A multi-objective optimization is a useful approach in this regard (Gunantara, 2018; https://en.wikipedia.org/wiki/Multi-objective_optimization; https://www.sciencedirect.com/topics/engineering/multiobjective-optimization; https://www.youtube.com/watch?v=BEBsd0cK2xI; https://pymoo.org/).

This book has attempted to deal with both sides of the equation in an evenhanded manner. The contents presented in the preceding chapters may be summarized as follows:

Chapter 1 introduces the concept of Process Intensification as "any intervention that makes the process run better, faster, cheaper", provides definitions for Process Intensification Factor (PIF) and Cost Impact Factor (CIF), and highlights the need to consider them together to ensure techno-economic viability of any significant intervention.

Chapter 2 provides case studies—ranging from baking cookies to dimensional tolerancing to microcontamination control to BPOs to clean coal technology—to illustrate the point. In each, quantitative metrics were identified to estimate PIF and CIF unambiguously, and an approach to treat them together to make rational business decisions was laid out.

Chapter 3 presents the three most common intensifiers of chemical processes—temperature, pressure, and time—and contrasts their process intensification effects with their cost impact and complexity of implementation. The inflection point at which the effect deviates significantly from linearity is typically where real intensification begins; that must be the point at which the cost is evaluated.

DOI: 10.1201/9781003283423-14

Chapter 4 contains a short description of the augmentation effect of external fields—thermal, electrical, magnetic, acoustic, etc.—on transport phenomena. Equations governing convective mass flux are examined in particular detail to assess its enhancement under the influence of such fields. While this chapter contains primarily a theoretical treatment, it serves as a lead-in to a more detailed consideration of field effects in a later chapter.

Chapter 5 focuses on process intensification driven by high-frequency, high-intensity acoustic fields—the one intensifier that best fits the banner of "better, faster, cheaper" (based on the author's own experience). The two basic mechanisms present in these fields—cavitation and acoustic streaming—are introduced, and their consequences are outlined for physical and chemical phenomena. Illustrative applications are briefly touched on—surface cleaning, heat/mass/momentum transfer enhancement, acceleration of chemical reactions, leaching of sub-surface impurities, size reduction, and enlargement—and acoustic field parameter effects are tabulated. This chapter is again a prelude to a more detailed treatment of the topics in later chapters.

Chapter 6 presents more case studies depicting process intensification by applied fields—microwave, laser, thermophoretic, electrophoretic, magnetophoretic, and acoustophoretic. The common thread in these is the definition of process intensification and cost impact factors, and the need to consider both in decision-making regarding an optimum process. While the treatment here is largely descriptive, this chapter does reinforce the central theme.

Chapter 7 takes the reader on a deep dive into the realms of acoustic process intensification. Starting with a basic description of probe-type and tank-type ultrasonic/megasonic systems, the chapter presents detailed investigations into sono-driven applications in destratification, atomization, and nanoparticle synthesis by size reduction. The power of the technique is highlighted, while associated energy costs are also made visible.

Chapter 8 does likewise for applications in heat transfer, chemical reactions, and mass transfer with and without chemical reactions.

Chapter 9 presents heat transfer augmentation methods in depth, illustrating acoustic field effects via two case studies—one involving flow in boiler tubes, the other a tank fitted with side heaters. Nanofluids are also dealt with in the chapter, with a comparison between enhancement achieved by bottom-up and top-down synthesized nanoparticles.

Chapter 10 likewise deals with the intensification of mass transfer, presenting as case studies the dyeing of textiles, and oxygen dissolution in water.

Chapters 11–14 continue with the theme of describing process intensification and cost impact factors via case studies—wastewater treatment by ultrasonic irradiation, surface cleaning in the microelectronics industry, nanoemulsion formulation and stabilization, and enhanced oil recovery via surfactant flooding. In each case, PIF and CIF are enumerated.

Whether quality is king, or whether cost occupies that throne, is a matter decided by the market. In the business decision-making process, assigning appropriate weightage to PCF and CIF is the right way to go. In the spirit of "kaizen", an engineer's instinct is always to make processes run better and faster; "cheaper", however, can sometimes take an unwarranted backseat. When you see a process that is not running optimally, do intensify it, but keep in mind that for every PI, there is a CI. Evaluate both carefully, and make the right call to ensure that the business case is sound.

Process intensifiers, rejoice! The world is our oyster, the opportunities are limitless. We can build a sustainable world, and have fun doing it. This book, hopefully, facilitates the journey.

REFERENCES

Gunantara, N., "A Review of Multi-Objective Optimization: Methods and Its Applications", *Cogent Eng.*, vol. 5, no. 1, 2018; https://doi.org/10.1080/23311916.2018.1502242.

Index

A

Ablation, 56
Acoustic, 33
Acoustic capillary focusing, 71
Acoustic cavitation, 80
Acoustic contrast factor, 71
Acoustic energy density, 71
Acoustic radiation force, 71
Acoustic streaming, 29, 91
Acoustofluidics, 70
Active heat-transfer enhancement, 119, 147
Adhesion, 37
Advanced Oxidation Process, 39
Aerator, 95
Aerosol decomposition synthesis, 105
Agglomeration, 27
Aggregation, 27
Annualized Failure Rate, 14
Argon/nitrogen snow cleaning, 205
Arrhenius equation, 25
Ash, 18
Atomization, 94

B

Baking, 52
Ball milling, 107
Binary acoustophoresis, 72
Biomass waste, 54
Biomolecular separations, 64
Biosolids, 53
Blanching, 51
Blowing parameter, 34
Boiling point, 26
Boundary layer, 35
Breakthrough, 28
Brownian diffusivity, 26
Bubble implosion, 82
Bubble size distribution, 98
Bulb blackening, 58

C

Carbon dioxide snow, 205
Casting, 27
Cavitation, 19, 80
Cavitation intensity, 132
Channel curvature, 121
Chemical cleaning, 201
Chemical vapor synthesis, 105

Chemically frozen, 61
Cleanroom, 13, 197, 209
Coal beneficiation, 44
Collection efficiency, 66
Colloidal process route, 104
Comminution, 107
Compression, 37, 80
Contaminant, 13
Continuous flow concentration, 71
Cooking, 52
Corrosion, 26
Cost Impact Factor, 4
Cost of manufacturing, 7
Cost of quality, 7
Cryogenic liquid, 84
Crystallization, 27

D

Decolorization Activity, 192
Dehydration, 51
Destratification, 43, 84
Desulfurization, 44
Dew point shift, 61
Diffusiophoresis, 34
Disinfection, 53
Dispersion, 166
Drift, 33
Droplet size distribution, 223
Dry laser cleaning, 56
Drying, 51
Dual-frequency ultrasonics, 43
Dye exhaustion, 175
Dynamic viscosity, 25

E

Eckart streaming, 19, 83
Effervescent atomization, 43, 94
Electric perimittivity, 66
Electrical, 33
Electrophoretic, 34, 64
Electrostatic precipitator, 66
Emulsifier, 217
Emulsion, 213, 217
Emulsion stabilizer, 217
Enhanced cell sedimentation, 72
Enhancement ratio, 130
Entrance effect, 121
Erosion, 18
Evaporative drying, 27

F

Fenton's reagent, 42
Filtration, 71
Fine grinding, 108
Flame spray pyrolysis, 106
Flame synthesis, 106
Flow disruption, 119
Flow pulsation, 126
Fluid additives, 124
Fouling, 18
Fragmentation, 107
Free-flow acoustophoresis, 72

G

Gas-liquid mass transfer, 181
Gas-phase synthesis, 105
Global warming, 28
Grashof number, 59
Gravitational, 33
Grinding, 107

H

Hard error, 12
Heat transfer coefficient, 129, 154
Heat transfer enhancement, 42
Heat transfer enhancement ratio, 150, 156
High-energy ball milling, 107
High-energy emulsification, 215
Hot corrosion, 60
Hot spot, 191
Hydrodynamic cavitation, 80
Hydrodynamic focusing, 71

I

Ideal gas law, 26
Incineration, 54
Inert gas condensation route, 105
Initial oil saturation, 244
Initial water saturation, 244
Intensification, 2
Ion sputtering, 105

K

Killer defects, 41
Knudsen number, 63

L

Laminar flow, 42
Langmuir layer, 59
Laser, 55
Laser drilling, 57
Laser machining, 57

Laser pyrolysis synthesis, 105
Laser shock processing, 55
Liquid phase synthesis, 104
Low-energy emulsification, 215
Low-temperature reactive synthesis, 106

M

Magnetic, 33
Magnetic flux density, 67
Magnetic permeability, 67
Magnetic susceptibility, 67
Magnetophoresis, 67
Manual cleaning, 203
Mass transfer coefficient, 182
Mean free path, 63
Mean time between failures, 14
Mechanical agitation, 203
Medical waste, 53
Megasonic, 37, 79
Microchannel, 65
Microcontamination, 12
Microemulsion, 213
Microfluidic, 62
Micromixer, 69
Microstreaming, 83
Microwave, 51
Miscible processes, 238
Mixing, 69
Mixing effectiveness, 87
Mixing time, 87
Mobility control, 238
Molten salt, 60
Multi objective optimization, 251

N

Nanoemulsion, 213, 217
Nanoemulsion formulation, 215
Nanofluid, 163
Nanoparticle synthesis, 103
Nanoporous membrane, 64
Nebulization, 94
Neem oil nanoemulsion, 220
Nitriding, 55
Nucleation, 27
Nusselt number, 58

O

Oil reserves, 237
Oil in water emulsion, 213
Optic cavitation, 81
Original oil in place, 244
Osmophoresis, 34
Out-of-plane mixing, 124
Outgassing, 27
Oxide dissolution, 61

P

Particle breakage, 43
Particle cavitation, 81
Passive heat-transfer
          enhancement, 119, 147
Pasteurization, 51
Peclet number, 58, 69
Pelletization, 27
Photophoresis, 34
Photothermal synthesis, 105
Physical vapor deposition, 166
Plasma cleaning, 203
Power ultrasound, 79
Pressure, 26
Probe-type ultrasonic system, 80
Process, 1
Process capability index, 10
Process intensification, 3
Process Intensification Factor, 4
Protein binding, 62
Protein flux, 65
Pulsed laser ablation, 105
Pulsed laser propulsion, 57
Pyrolysis, 54

Q

Quality Loss Cost, 10
Quantum dot, 104
Quiescent conditions, 156

R

Rarefaction, 37, 80
Recirculation rate, 207
Re-entrant obstruction, 121
Reynolds number, 42

S

Schlichting streaming, 19, 83
Secondary flow, 122
Sedimentation, 34, 72
Sensitivity analysis, 26
Separation, 68
Settling, 34
Size reduction, 107
Slagging, 18
Soft error, 12
Sol-Gel, 104
Solubilization, 27
Solvent cleaning, 202
Sonochemical, 42
Sonochemical synthesis, 106
Sono-fragmentation, 42, 110
Soot, 63
Soret diffusion, 58

Sound wave, 79
Sparging, 203
Spark discharge generation, 105
Spin-rinse-dryer, 200
Spray cleaning, 199
Spray drying, 94
Spray pyrolysis, 105
Spraying, 27
Stability analysis, 229
Stabilization, 167
Steam laser cleaning, 56
Stefan flow, 34
Stefan blowing, 34
Sterilization, 51
Stokes-Einstein equation, 26
Stratification, 84
Stratification Index, 86
Supercritical fluid cleaning, 204
Surface acoustic wave, 68
Surface adsorption, 27
Surface cleaning, 40
Surface hardening, 55
Surface roughness, 119
Surfactant, 219
Surfactant flooding, 238
Susceptor, 54
Swabbing, 203

T

Tank-type ultrasonic system, 80
Temperature, 25
Tempering, 52
Thawing, 52
Thermal, 33
Thermal diffusion, 58
Thermal plasma synthesis, 106
Thermal stratification, 43
Thermogravitational column, 62
Thermophoresis, 34, 58
Time, 28
Total cost to business, 7
Transonic, 37
Transpiration, 62
Transport phenomena, 47
Trapping, 69
Tungsten-halogen cycle, 59

U

Ultrasonic, 37
Ultrasonic cleaning, 205
Ultrasonic generator, 80
Ultrasonic emulsification, 216
Ultrasonic transducer, 80
Ultrasound, 19, 73, 79
Undulation, 203
UV/ozone, 204

V

Vapor degreasing, 201
Vapor phase synthesis, 104
Variable roughness structure, 127
Vibration, 125

W

Waste treatment, 53
Water-in-oil emulsion, 213
Wave, 79

Weber number, 218
Weighting factor, 9
Wiping, 203

Y

Yield, 14

Z

Zero Element Flux, 59
Zeta potential, 66